D1529483

ASTRONOMY METHODS

Astronomy Methods is an introduction to the basic practical tools, methods and phenomena that underline quantitative astronomy. Taking a technical approach, the author covers a rich diversity of topics across all branches of astronomy, from radio to gamma-ray wavelengths. Topics include the quantitative aspects of the electromagnetic spectrum, atmospheric and interstellar absorption, telescopes in all wavebands, interferometry, adaptive optics, the transport of radiation through matter to form spectral lines, and neutrino and gravitational-wave astronomy. Clear, systematic presentations of the topics are accompanied by diagrams and problem sets. Written for undergraduates and graduate students, this book contains a wealth of information that is required for the practice and study of quantitative and analytical astronomy and astrophysics.

HALE BRADT is Professor Emeritus of Physics at the Massachusetts Institute of Technology. Over his forty years on the faculty, he carried out research in cosmic ray physics and x-ray astronomy, and taught courses in Physics and Astrophysics. Bradt founded the MIT sounding rocket program in x-ray astronomy, and was a senior or principal investigator on three NASA missions for x-ray astronomy. He was awarded the NASA Exceptional Science Medal for his contributions to HEAO-1 (High Energy Astronomical Observatory 1), the 1990 Buechner Teaching Prize of the MIT Physics Department, and shared the 1999 Bruno Rossi prize of the American Astronomical Society for his contributions to the RXTE (Rossi X-ray Timing Explorer) program.

Solutions manual available for instructors by emailing solutions@cambridge.org

Cover illustrations

Views of the entire sky at six wavelengths in galactic coordinates. The equator of the Milky Way system is the central horizontal axis and the galactic center direction is at the center. Except for the x-ray sky, the colors represent intensity with the greatest intensities lying along the equator. In all cases, the radiation shows an association with the galactic equator and/or the general direction of the galactic center. In some, extragalactic sources distributed more uniformly are evident. The captions below are listed in frequency order (low to high). The maps are also in frequency order as follows: top to bottom on the back cover followed on the front cover by top inset, background map, lower inset.

Radio sky at 408 Hz exhibiting a diffuse glow of synchrotron radiation from the entire sky. High energy electrons spiraling in the magnetic fields of the Galaxy emit this radiation. Note the *North Polar Spur* projecting above the equator to left of center. [From three observatories: Jodrell Bank, MPIfR, and Parkes. Glyn Haslam *et al.*, MPIfR, SkyView]

Radio emission at 1420 MHz, the spin-flip (hyperfine) transition in the ground state of hydrogen, which shows the locations of clouds of neutral hydrogen gas. The gas is heavily concentrated in the galactic plane and shows pronounced filamentary structure off the plane. [J. Dickey (UMn), F. Lockman (NRAO), SkyView; *ARAA* **28**, 235 (1990)]

Far-infrared (60–240 μm) sky from the COBE satellite showing primarily emission from small grains of graphite and silicates ("dust") in the interstellar medium of the Galaxy. The faint large S-shaped curve (on its side) is emission from dust and rocks in the solar system. Reflection of solar light from this material gives rise to the zodiacal light at optical wavelengths. [E. L. Wright (UCLA), COBE, DIRBE, NASA]

Optical sky from a mosaic of 51 wide angle photographs showing mostly stars in the (Milky Way) Galaxy with significant extinction by dust along the galactic plane. Galaxies are visible at higher galactic latitudes, the most prominent being the two nearby Magellanic Clouds (lower right). [©Axel Mellinger]

X-ray sky at 1–20 keV from the A1 experiment on the HEAO-1 satellite showing 842 discrete sources. The circle size represents intensity of the source and the color represents the type of object. The most intense sources shown (green, larger, circles) represent accreting binary systems containing a compact star, either a white dwarf, neutron star, or a black hole. Other objects are supernova remnants (blue), clusters of galaxies (pink), active galactic nuclei (orange), and stellar coronae (white) [Kent Wood, NRL; see *ApJ Suppl.* **56**, 507 (1984)]

Gamma-ray sky above 100 MeV from the EGRET experiment on the Compton Gamma Ray Observatory. The diffuse glow from the galactic equator is due to the collisions of cosmic ray protons with the atoms of gas clouds; the nuclear reactions produce the detected gamma rays. Discrete sources include pulsars and jets from distant active galaxies ("blazars"). [The EGRET team, NASA, CGRO]

ASTRONOMY METHODS

A Physical Approach to Astronomical Observations

HALE BRADT
Massachusetts Institute of Technology

CAMBRIDGE
UNIVERSITY PRESS

PUBLISHED BY THE PRESS SYNDICATE OF THE UNIVERSITY OF CAMBRIDGE
The Pitt Building, Trumpington Street, Cambridge, United Kingdom

CAMBRIDGE UNIVERSITY PRESS
The Edinburgh Building, Cambridge CB2 2RU, UK
40 West 20th Street, New York, NY 10011-4211, USA
477 Williamstown Road, Port Melbourne, VIC 3207, Australia
Ruiz de Alarcón 13, 28014 Madrid, Spain
Dock House, The Waterfront, Cape Town 8001, South Africa

http://www.cambridge.org

First published 2004

Printed in the United Kingdom at the University Press, Cambridge

Typefaces Times 11/14 pt. and Univers   *System* LaTeX 2_ε [TB]

A catalog record for this book is available from the British Library

Library of Congress Cataloging in Publication data
Bradt, Hale, 1930–
Astronomy methods: a physical approach to astronomical observations / Hale Bradt.
    p.   cm.
Includes bibliographical references and index.
ISBN 0 521 36440 X – ISBN 0 521 53551 4 (pbk.)
1. Astronomy.   I. Title.
QB45.2.B73 2003 520–dc21   2002041703

ISBN 0 521 36440 X hardback
ISBN 0 521 535514 paperback

To   Dottie, Elizabeth, Dorothy,
$(\text{Bart})^2$, Ben, and Rebecca

Contents

Figures

Tables

Preface

This volume is the first part of notes that evolved during my teaching of a small class for junior and senior physics students at MIT. The course focused on a physical, analytical approach to astronomy and astrophysics. The material in this volume presents methods, tools and phenomena of astronomy that the science undergraduate should incorporate into his or her knowledge prior to or during the practice and study of quantitative and analytical astronomy and astrophysics.

The content is a diverse set of topics ranging across all branches of astronomy, with an approach that is introductory and based upon physical considerations. It is addressed primarily to advanced undergraduate science students, especially those who are new to astronomy. It should also be a useful introduction for graduate students or postdoctoral researchers who are encountering the practice of astronomy for the first time. Algebra and trigonometry are freely used, and calculus appears frequently. Substantial portions should be accessible to those who remember well their advanced high school mathematics.

Here one learns quantitative aspects of the electromagnetic spectrum, atmospheric absorption, celestial coordinate systems, the motions of celestial objects, eclipses, calendar and time systems, telescopes in all wavebands, speckle interferometry and adaptive optics to overcome atmospheric jitter, astronomical detectors including CCDs, two space gamma-ray experiments, basic statistics, interferometry to improve angular resolution, radiation from point and extended sources, the determination of masses, temperatures, and distances of celestial objects, the processes that absorb and scatter photons in the interstellar medium together with the concept of cross section, broadband and line spectra, the transport of radiation through matter to form spectral lines, and finally the techniques used in neutrino, cosmic-ray and gravitational-wave astronomy.

I choose to use SI units throughout to be consistent with most standard undergraduate science texts. Professional astronomers use cgs units, probably because everyone else in the field does. Unfortunately, this precludes progress in bringing the various science communities together to one system of units. It is also a significant hindrance to the student exploring astronomy or astrophysics. In this work I vote for ease of student access and encourage my colleagues to do likewise in their publications. I do violate this in at least one respect. In avoiding the historical and highly specialized astronomical unit of distance, the "parsec", I use instead the better understood, but non-SI, unit, the "light year" (LY), the distance light travels in one year. This is a well defined quantity if one specifies the Julian year of exactly 365.25 days, each of exactly 86 400 SI seconds, or a total of 31 557 600 s per year.

Other features to note are. (*i*) problems are provided for each chapter; and approximate answers are given where appropriate; (*ii*) units are often given gratuitously (in parentheses) for algebraic variables to remind the reader of the meaning of the symbol; (*iii*) equation, table, figure, and section numbers in the text do not carry the chapter prefix if they refer to the current chapter, to improve readability; (*iv*) tables of useful units, symbols and constants are given in the Appendix, and (*v*) quantitative information is meant to be up to date and correct, but should not be relied upon for professional research. The goal here is to teach underlying principles.

In teaching this course from my notes, I adopted a seminar, or Socratic, style of teaching that turned out to be extremely successful and personally rewarding. I recommend it to teachers using this text. I sat with the students (up to about 20) around a table, or we would arrange classroom desks and chairs in a circular/rectangular pattern so we were all facing each other, more or less. I would then have the students explain the material to their fellow students ("Don't look at me," I often said). One student would do a bit, and I would move on to another. I tried very hard to make my prompts easy and straightforward, to not disparage incorrect or confusing answers, and to encourage discussion among students. I would synthesize arguments and describe the broader implications of the material interspersed with stories of real-life astronomy, personalities, discoveries, etc.

These sessions would often become quite active. During this discussion, the text is available to all and is freely referenced. To ease such referencing, all equations are numbered, labels are provided for many of them, and important equations are marked with a boldface arrow in the left margin. The students had to work hard to prepare for class, and thus got much out of the class discussion. And it was great fun for the teacher. In good weather, we would move outdoors and have our discussion on the lawn of MIT's Killian Court.

I hope to publish other portions of these notes in future volumes. The second should follow shortly; its working title and current chapter titles are:

Astrophysics Processes – Physical Processes that Underlie Astronomical Phenomena

Kepler's laws and the mass function
Special theory of relativity in astronomy
Kinetic theory and thermodynamics
Radiation from accelerating charges
Thermal bremsstrahlung radiation
Synchrotron radiation
Blackbody radiation
Compton scattering
Hydrogen spin-flip radiation
Propagation in phase space
Dispersion and Faraday rotation
Gravitational lensing

The author asks his readers forbearance with the inevitable errors in the current text and asks to be notified of them. Comments and suggestions are welcome.

Hale Bradt
Belmont MA
USA
bradt@mit.edu

Acknowledgments

I am indebted to many colleagues at MIT and elsewhere and to many students for their encouragement and assistance in hallway discussions, in class, and as readers of draft chapters, over the course of the two decades that this work has been evolving. It is impossible to fairly list all those who helped in these ways, but I will mention those who particularly come to mind. I apologize for omissions. I do not list those who helped specifically with chapters not included in this volume. Needless to say, those mentioned are not responsible for errors; I assume that role.

Colleagues: Marshall Bautz, Edward Bertschinger, Kenneth Brecher, Roberto Buonanno, Bernard Burke, Claude Canizares, Deepto Chakrabarty, George Clark, Charles Counselman, James Cronin, Alessandra Dicredico, Marco Feroci, Kathy Flanaghan, Peter Ford, Leon Golub, Mark Gorenstein, Marc Grisaru, Jackie Hewitt, Scott Hughes, Gianluca Israel, Garrett Jernigan, Erik Katsavounides, Alan Levine, Alan Lightman, Herman Marshall, Christopher Moore, James Moran, Edward Morgan, Philip Morrison, Stuart Mufson, Stan Olbert, Saul Rappaport, Ronald Remillard, Harvey Richer, Swapan Saha, Peter Saulson, Paul Schechter, Irwin Shapiro, David Shoemaker, Luigi Stella, Victor Teplitz, David Thompson, John Tonry, Jake Waddington, Joel Weisberg.

Graduate and undergraduate students (at the time): Stefan Ballmer, David Baran, James "Gerbs" Bauer, Eugene Chiang, Asantha Cooray, Yildiz Dalkir, Antonios Eleftheriou, James Gelb, Edgar Gonzalez, Karen Ho, Juliana Hsu, Rick Jenet, Jeffrey Jewell, Jasmine Jijina, Vishnja Katalinic, Edward Keyes, Janna Levin, Tito Pena, Jeremy Pitcock, Philipp Podsiadlowski, Antonia Savcheva, Robert Shirey, Donald A. Smith, Svetlin Tassev, Seth Trotz, Keith Vanderlinde, Emily Wang.

I am especially gratefull to colleagues Saul Rappaport and Stu Teplitz for their reading of the entire set of notes some years ago, and to graduate student Edward Keyes and undergraduate Keith Vanderlinde for their very recent reading of this volume in its current form. In the days before personal word precessors, secretaries

Trish Dobson, Ann Scales, Patricia Shultz, and Diana Valderrama did yeoman duty in typing revisions of the notes for my classes.

Much appreciated allowances have been made for my writing efforts by my family, by the Department of Physics at MIT, by my colleagues at the MIT Center for Space Research and by my associates in the Rossi X-ray Timing Explorer (RXTE) satellite program at MIT, the University of California at San Diego, and NASA's Goddard Space Flight Center. The hospitality of the Institute of Space and Astronautical Science (ISAS) in Japan and the Observatory of Rome (OAR) in Italy provided extended periods of quiet writing for which I am grateful. This volume began at ISAS and was completed at OAR.

Finally, it has been a pleasure to work with the staff and associates of Cambridge University Press, in particular, Miranda Fyfe, Jacqueline Garget, Carol Miller, Simon Mitton, Margaret Patterson and the folks at TechBooks in New Delhi. They have been encouraging, creative, patient and ever helpful.

1

Astronomy through the centuries

What we learn in this chapter

Celestial measurements reaching back 3000 years or more were carried out in many cultures worldwide. **Early astronomers** in Greece deduced important conclusions about the nature of the **earth** and the **solar system**. Modern astronomy began in the **renaissance** with the observations of **Tycho Brahe** and **Galileo** and the theoretical work of **Kepler** and **Newton**. The progress of our knowledge of the sky may be traced through a series of **major discoveries** which often follow the development of new **technologies** such as the **telescope, computers,** and **space observatories.** Astronomy is now carried out across the entire electromagnetic spectrum from the **radio** to the **gamma ray** (see cover illustrations) as well as with **cosmic rays, neutrinos,** and **gravitational waves.** The mutual dependence of **theory** and **observation** has led to major advances in the understanding of a wide diversity of celestial objects such as **stars, supernova remnants, galaxies,** and the **universe** itself. Current observations reveal important phenomena that are not understood. The promise of **new fundamental discoveries** remains high.

1.1 Introduction

This introductory chapter provides a brief sketch of the history of astronomy with emphasis upon some pivotal ideas and discoveries. The ideas presented here are covered more systematically in subsequent chapters of this or subsequent planned volumes.

1.2 Early development of astronomy

First astronomers

The rhythmic motions of the stars, the planets, and the sun in the sky have fascinated humankind from the earliest of times. The motions were given religious significance

Figure 1.1. Stonehenge, an early astronomical observatory used for tracking the sun and moon in their seasonal excursions. [© Crown copyright, NMR]

and were useful agricultural indicators. The sun's motion from south to north and back again marked the times of planting and harvesting. The annual motion of the sun against the background of the much more distant stars could also be followed and recorded, as could the motions of the moon and planets. This made possible predictions of the future motions of the sun and moon. Successful forecasters of the dramatic eclipses of the sun seemed to be in touch with the deities.

The periodic motions of the sun and moon were noted and described with calendars as early as the thirteenth century BCE in China. Surviving physical structures appear to be related to the motions of celestial bodies. Notable are an eighth century BCE sundial in Egypt and the assemblage of large stones at Stonehenge in England dating from about 2000 BCE (Fig. 1)[1]. The Babylonians and Assyrians in the Middle East are known to have been active astronomers in the several centuries BCE (The designations BCE "before common era" and CE "common era" are equivalent to BC and AD respectively.)

---

[1] In an attempt to minimize redundant numbers in the text, we omit the chapter designation in references to figures within the chapter in which the figure occurs, e.g. "Fig. 2". For references to figures in another chapter, say, Chapter 3, the reference is the conventional format "Fig. 3.2". Problem, equation, and section references are treated similarly. Equations are usually referenced in the text as a number within parentheses without the prefix "Eq.", for example: "as shown in (10)" for equations within the chapter, or "given in (5.10)" for Eq. 10 of Chapter 5.

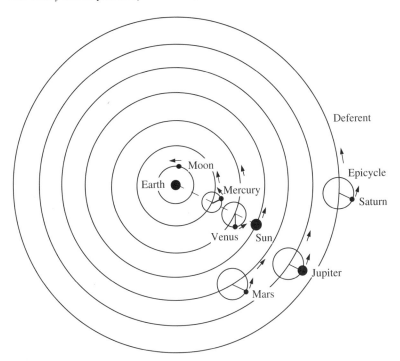

Figure 1.2. Ptolemaic system, not drawn to scale. The earth is at the center, the moon and sun follow circular paths and the planets follow (small) circular orbits (epicycles) the centers of which move regularly along large circular orbits known as deferents. Elliptical orbits are taken into account by offsetting slightly the centers of the deferents and also the earth itself from a geometrical "center".

Astronomy flourished under Greek culture (~600 BCE to ~400 CE) with important contributions by Aristotle, Aristarchus of Samos, Hipparchus, Ptolemy, and others. The Greek astronomers deduced important characteristics of the solar system. For example, Aristotle (384–322 BCE) argued from observations that the earth is spherical, and Aristarchus (310–230 BCE) made measurements to obtain the sizes and distances of the sun and moon relative to the size of the earth. Ptolemy (~140 CE) developed a complicated earth-centered model (Fig. 2) for the solar system which predicted fairly well the complicated motions of the planets as viewed from the earth.

The advance of astronomy in Europe faltered during the following 13 centuries. Nevertheless the sky continued to be observed in many cultures, e.g., the Hindu, Arabian, and Oriental. The sudden appearances of bright new and temporary "guest" stars in the sky were noted by the Chinese, Japanese, Koreans, Arabs, and Europeans. The most famous of all such objects, the Crab supernova, was recorded in 1054 by Chinese and Japanese astronomers. It is now a beautiful diffuse nebula in the sky (Fig. 3). The Mayan culture of Central America independently developed

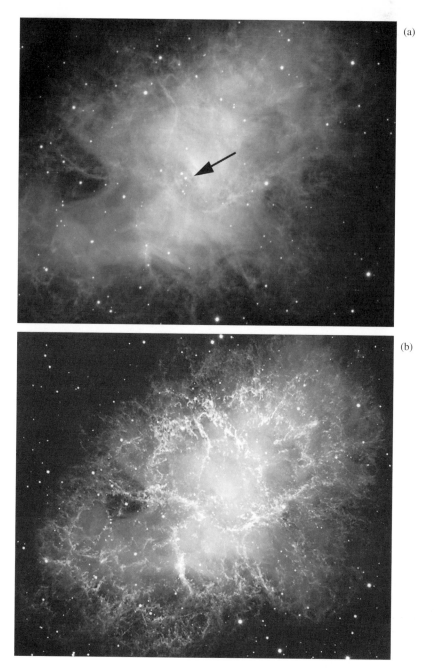

(a)

(b)

Figure 1.3. Crab nebula, the remnant of a supernova explosion observed in 1054 CE. The neutron-star pulsar is indicated (arrow). The filters used for the two photos stress (a) the blue diffuse synchrotron radiation and the pulsar and (b) the strong filamentary structure that glows red from hydrogen transitions. The scales of the two photos are slightly different. The Crab is about 4′ in extent and ~6000 LY distant. The orientation is north up and east to the left – as if you were looking at the sky while lying on your back with your head to the north; this is the standard astronomical convention. [(a) Jay Gallagher (U. Wisconsin)/WIYN/NOAO/NSF.; (b) FORS Team, VLT, ESO]

strong astronomical traditions, including the creation of a sophisticated calendar that could be used, for example, to predict the positions of Venus.

Renaissance

The Renaissance period in Europe brought about great advances in many intellectual fields including astronomy. The Polish monk Nicholaus Copernicus (1473–1543) proposed the solar-centered model of the planetary system. The Dane Tycho Brahe (1546–1601, Fig. 4) used elegant mechanical devices to measure planetary positions to a precision of $\sim 1'$ (1 arcminute)[1] and, over many years, recorded the daily positions of the sun, moon, and planets.

The German Johannes Kepler (1571–1630, Fig. 4), Brahe's assistant for a time, had substantial mathematical skills and attempted to find a mathematical model that would match Brahe's data. After much effort, he found that the apparent motions of Mars in the sky could be described simply if Mars' orbit about the sun is taken to be an ellipse. He summarized his work with the three laws now known as Kepler's laws of planetary motion. They are: (*i*) each planet moves along an elliptical path with the sun at one focus of the ellipse, (*ii*) the line joining the sun and a planet sweeps out equal areas in equal intervals of time, and (*iii*) the squares of the periods P (of rotation about the sun) of the several planets are proportional to the cubes of the semimajor axes a of their respective elliptical tracks,

$$P^2 \propto a^3 \qquad\qquad \text{(Kepler's third law)} \qquad (1.1)$$

This formulation laid the foundation for the gravitational interpretation of the motions by Newton in the next century.

Galileo Galilei (1564–1642, Fig. 4), a contemporary of Kepler and an Italian, carried out mechanical experiments and articulated the *law of inertia* which holds that the state of constant motion is as natural as that of a body at rest. He adopted the Copernican theory of the planets, and ran afoul of the church authorities who declared the theory to be "false and absurd". His book, *Dialog on the Two Great World Systems* published in 1632, played a significant role in the acceptance of the Copernican view of the solar system. Galileo was the first to make extensive use of the *telescope*, beginning in 1609.

The telescope was an epic technical advance in astronomy because, in effect, it enlarged the eye; it could collect all the light impinging on the objective lens and direct it into the observer's eye. Since the objective lens was much larger than the lens of the eye, more light could be collected in a given time and fainter objects could be seen. The associated magnification allowed fine details to be resolved. Galileo was the first to detect the satellites (moons) of Jupiter and to determine

---

[1] The measures of angle are degree ($^\circ$), arcmin ($'$), and arcsec ($''$) where $60'' = 1'$ and $60' = 1^\circ$.

(a) Tycho Brahe (1546–1601)

(b) Galileo Galilei (1564–1642)

(c) Johannes Kepler (1571–1630)

Figure 1.4. The three astronomers who pioneered modern astronomy. [(a) Tycho Brahe's Glada Vänner; (b) portrait by Justus Sustermans; (c) Johannes Kepler Gesammelte Werke, C. H. Beck, 1937. All are on internet: "Astronomy Picture of the Day", NASA/GSFC and Michigan Tech U.]

their orbital periods. He showed that the heavens were not perfect and immutable; the earth's moon was found to have a very irregular surface and the sun was found to have dark "imperfections", now known as *sunspots*.

The Englishman Isaac Newton (1643–1727; Gregorian calendar) was born 13 years after the death of Kepler and almost exactly one year after Galileo died.

His study of mechanics led to three laws, *Newton's laws*, which are stated here in contemporary terms: (*i*) the vector momentum $\boldsymbol{p} = m\boldsymbol{v}$ of a body of mass m moving with velocity $\boldsymbol{v}$ is conserved in the absence of an applied force[1] (this is a restatement of Galileo's law of inertia), (*ii*) a force applied to a body brings about a change of momentum,

$$\boldsymbol{F} = \frac{\mathrm{d}\boldsymbol{p}}{\mathrm{d}t} \qquad \text{(Newton's second law)} \qquad (1.2)$$

and (*iii*) the force $\boldsymbol{F}_{1,2}$ on one body due to second body is matched by an opposing force of equal magnitude $\boldsymbol{F}_{2,1}$ on the second body due to the first,

$$\boldsymbol{F}_{1,2} = -\boldsymbol{F}_{2,1} \qquad \text{(Newton's third law)} \qquad (1.3)$$

These laws are the bases of *Newtonian mechanics* which remains the essence of much modern mechanical theory and practice. It fails when the speeds of the bodies approach the speed of light and on atomic scales.

Newton was able to show that a *gravitational force* of a particular kind described perfectly the planetary motions described by Kepler. This force is an attractive gravitational force $\boldsymbol{F}$ between two bodies that depends proportionally upon the masses (m_1, m_2) and inversely with the squared distance r^2 between them,

$$\boldsymbol{F} \propto -\frac{m_1 m_2}{r^2}\hat{\boldsymbol{r}} \qquad \text{(Newton's law of gravitation)} \qquad (1.4)$$

where $\hat{\boldsymbol{r}}$ is the unit vector along the line connecting the two bodies. Such a force leads directly to the elliptical orbits, to the speeds of motion in the planetary orbit, and to the variation of period with semimajor axis described by Kepler's three laws. Thus, all the celestial motions of the earth, moon, and sun could be explained with a single underlying force. This understanding of the role of gravity together with the invention of the telescope set astronomy solidly on a path of quantitative measurements and physical interpretation, i.e., *astrophysics*.

1.3 Technology revolution

Telescopes, photography, electronics, and computers

The study of the sky continued with the development of larger and larger telescopes. Generally these were refractive instruments wherein the light passes through the lenses, as in a pair of binoculars. The glass refracts the different colors of light slightly differently (*chromatic aberration*) so that perfect focusing is difficult to attain. This led to reflecting telescopes that make use of curved mirrors. In this case, all wavelengths impinging at a position of the mirror from a given angle are

---

[1] We use italic boldface characters to signify vector quantities and the hat symbol to indicate unit vectors.

reflected in the same direction. The 5-m diameter mirror of the large telescope on Palomar Mountain in California (long known as the "200 inch") was completed in 1949. It was the world's largest telescope for many years. In the 1960s and 1970s, several 4-m diameter telescopes were built as was a 6-m instrument in the Soviet Union. At this writing, the two Keck 10-m telescopes in Hawaii are the largest, but other large telescopes are not far behind. New technologies which allow telescopes to compensate for the blurring of starlight by the earth's atmosphere are now coming on line.

Photography was an epochal development for astronomy in the nineteenth century. Before this, the faintest object detectable was limited by the number of *photons* (the quanta of light) that could be collected in the integration time of the eye, ~30 ms (millisecond) to ~250 ms if dark adapted. If a piece of film is placed at the focus of a telescope, the photons can be collected for periods up to and exceeding 1 hour. This allowed the detection of objects many orders of magnitude fainter than could be seen by eye. A photograph could record not only an image of the sky, but also the *spectrum* of a celestial object. The latter shows the distribution of energy as a function of wavelength or frequency. The light from the object is dispersed into its constituent colors with a prism or grating before being imaged onto the film. Large telescopes together with photography and spectroscopy greatly enlarge the domains of quantitative measurements available to astronomers.

Since the mid-twentieth century, more sensitive electronic detection devices have come into use. Examples are the *photomultiplier tube*, the *image intensifier*, and more recently, the *charge-coupled detector* (CCD). Computers have come into wide use for the control of the telescope pointing and for analysis of the data during and after the observation. The greatly increased efficiencies of data collection and of analysis capability go hand in hand in increasing the effectiveness of the astronomer and his or her ability to study fainter and more distant objects, to obtain spectra of many objects simultaneously, or to measure bright sources with extremely high time resolution. In the latter case, changes of x-ray intensity on sub-millisecond time scales probe the swirling of ionized matter around neutron stars and black holes.

Non-optical astronomy

Electromagnetic radiation at radio frequencies was discovered by Heinrich Hertz in 1888. This eventually led to the discovery of radio emission from the sky by Carl Jansky in 1931. This opened up the field of *radio astronomy*, an entirely new domain of astronomy that has turned out to be as rich as conventional optical astronomy. Entirely new phenomena have been discovered and studied. Examples are the distant *quasars* (described below) and the neutral hydrogen gas that permeates interstellar space. The invention of the maser and the use of supercooled detectors have greatly increased the sensitivity and frequency resolution of radio telescopes. Multiple radio telescopes spread over large distances (1 km to 5 000 km or more) are now used in concert to mimic a single large telescope with angular resolutions down to better than $0.001''$ (arcseconds).

The Very Large Array (VLA) of 27 large radio telescopes extending over about 40 km of New Mexico desert operates on this principle. With its large area it has excellent sensitivity. It has produced many beautiful images of radio objects in the sky with angular resolution comparable to that of ground-based optical astronomy ($\sim 1''$).

The atmosphere of the earth is a great impediment to many kinds of astronomy. Photons over large bands of frequencies can not penetrate it. The advent of space vehicles from which observations could be made opened up the field of *x-ray astronomy*. Like radio astronomy, this field led to the discovery of a variety of new phenomena, such as neutron stars in orbit with ordinary nuclear-burning stars, high-temperature shock waves in supernova remnants, black holes (described below), and high-energy phenomena in distant quasars.

Space vehicles have also made possible the study of the *ultraviolet radiation* from nearby stars and distant galaxies and *gamma-ray emission* from pulsars and from the nuclei of active galaxies. *Infrared astronomy* can be carried out at only a few frequencies from the ground, but in space a wide band of frequencies are accessible. Infrared astronomers can peer into dust and gas clouds to detect newly formed stars. Space vehicles also carry optical/ultraviolet telescopes above the atmosphere to provide very high angular resolutions, $\lesssim 0.05''$ compared to the $\sim 1''$ normally attained below the atmosphere. This is a major feature of the Hubble Space Telescope.

The space program also has provided a platform for *in situ* observations of the planets and their satellites (moons); the spacecraft carries instruments to the near vicinity of the planet. These missions carry out a diversity of studies in a number of wavebands (radio through the ultraviolet) as well as magnetic, cosmic-ray, and plasma studies. The Voyager missions visited Jupiter, Saturn, Uranus and Neptune. One of them will soon leave the solar-system *heliosphere* and thus be able to carry out direct measurements of the *interstellar medium.*

A given celestial object can often be studied in several of the frequency domains from the radio to gamma rays. Each provides complementary information about the object. For example, the x rays provide information about very hot regions (~ 10 million kelvin) while infrared radiation reflects temperatures of a few thousand degrees or less. The use of all this information together is a powerful way to determine the underlying nature of a class of celestial objects. This type of research has come to be known as *multi-frequency astronomy*. Sky maps at various frequencies (cover illustrations) illustrate the variation of the character of the sky with frequency.

Signals other than the electromagnetic waves also provide information about the cosmos. Direct studies of *cosmic rays* (energetic protons, helium nuclei, etc.) circulating in the vast spaces between the stars are carried out at sea level and also from space. These high-energy particles were probably accelerated to such energies, at least in part, by the shock waves of supernova explosions.

Neutrinos, neutral quanta that interact very weakly with other matter, have been detected from the nuclear reactions in the center of the sun and from the spectacular implosion of a star in the Large Magellanic Cloud, an easily visible stellar system in the southern sky. The outburst is known as *supernova 1987A*. Neutrino detectors are placed underground to minimize background.

The detection of gravitational waves predicted by the theory of general relativity is still a challenge. Observatories to search for them with high sensitivity are now beginning operations, such as the US Laser Interferometer Gravitational-wave Observatory (LIGO) with interferometer "antennas" in Washington State and Louisiana or the German–UK GEO-600. A likely candidate source of gravitational waves is a binary system of two neutron stars in the last stages of spiraling into each other to form a black hole.

1.4 Interplay of observation and theory

The objective of astronomical studies is to learn about the nature of the celestial objects, including their sizes, masses, constituents, and the basic physical processes that take place within or near them. Progress is made through an interplay of observational data and theoretical insight. Observations guide the theorist and theories suggest observations. The pace of this interplay greatly accelerated in the late nineteenth and twentieth centuries due to the rapid increase in technical capability described above. The recent history of astronomy is replete with examples of this symbiosis of observation and theory.

Stars and nebulae

Dark absorption lines were discovered in the solar spectrum in 1802, and Joseph Fraunhofer (1787–1826) recorded the locations of about 600 of them. Comparison to spectra emitted by gases and solids in earth laboratories showed that the gaseous outer layer of the sun contains elements well known on earth. The quantum theory developed in the 1920s yields the frequencies of radiation emitted by atoms as well as the probabilities of emission under various conditions of density and temperature. This allowed astronomers to diagnose the conditions in the solar atmosphere and in the atmospheres of other more distant "suns", the stars in the sky.

The light from the surface of our sun does not show us directly what is happening inside it. However, we now know that the energy source of the sun is nuclear fusion, a concept developed and demonstrated by nuclear physicists. Hydrogen nuclei under high pressures and temperatures fuse to become helium nuclei and other heavier elements. Theoretical models of stars that incorporate a nuclear energy source closely match the observed characteristics of real stars. This understanding has been verified by the measurement of neutrino fluxes from the sun as mentioned above; also see below.

In 1862, a faint companion star to Sirius was first seen with the aid of a new, large (0.46-m) telescope. Observations much later showed it to be very compact, about the size of the earth, but about 350 000 times more massive than the earth (about as massive as the sun). It was called a *white dwarf*. It was thought that the huge inward pull of gravity of such a compact object would prevent it from remaining stable at its observed size. (In a normal star like the sun, the pressure due to the hot gases prevents such a collapse.) In the mid 1920s, the newly developed quantum theory showed that *electron degeneracy pressure* would support such a star from further collapse, thus providing a physical basis for the existence of white dwarfs. Degeneracy pressure is strictly a quantum mechanical effect for which there is no classical analog.

Nuclear and quantum physics also led to the demonstration in 1939 that an extremely compact *neutron star* could be a stable state of matter. It would be about as massive as the sun, but $\sim$1000 times smaller than the white dwarf! It would be as dense as the nucleus of an atom and would consist almost solely of neutrons. (The high pressures within the star force the electrons to combine with the protons.) These stars were finally discovered in 1967, first as *radio pulsars*, and a few years later as *x-ray pulsars*. The neutron stars are typically spinning and shine in our direction once each rotation, as does a lighthouse beam. Their periods range from about 1 ms to about 1 ks (1000 s). This pulsing of the radiation gave them their name ("pulsars"). Note that, in this case, the underlying theory of neutron stars existed before their discovery, unlike the case of the white dwarf companion of Sirius where the observations spent decades in search of a theory.

These and other developments brought forward a general picture of the lives of stars. They form from the condensation of large gas and dust clouds in the interstellar medium that appear as beautiful colorful *nebula* such as the Trifid and Orion nebulae (Figs. 5, 6). A newly formed star stops shrinking and stabilizes when it becomes sufficiently dense and massive so that nuclear burning starts in its center.

Figure 1.5. Trifid nebula, a star-formation region of gas and dust. Newly created hot massive stars radiate ultraviolet light that ionizes the surrounding gas which then emits radiation as it recombines. It is at distance 5500 LY and is 15′ in angular extent. The image is about 22′ square. North is up and east is left. [T. Boroson/NOAO/AURA/NSF]

Stars often form in groups of 10 or more; these groups are seen as *open clusters* of stars such as the Pleiades or "Seven Sisters" (Fig. 7). The more massive stars burn out quickly, in a few million years. Intermediate-mass stars like our sun live for ~10 billion years, and lower mass stars would live longer than the age of the universe (10–20 billion years).

As the nuclear fuel in the star is expended, the star goes through several phases of size and color changes. It may expel a cloud of gas to become a beautiful *planetary nebula*. Radiation from the star excites the atoms in the expanding cloud so they fluoresce, as in the Ring nebula (Fig. 8). The star eventually becomes a compact object, a white dwarf or neutron star. The latter may occur by means of a spectacular supernova implosion/explosion to produce the Crab pulsar and nebula as noted above. If the original star is sufficiently massive, it could instead collapse

Figure 1.6. Orion nebula, a star formation region in the sword of Orion. The nebula is 1300 LY distant and of full optical extent 1.5°. Only the northern half is shown here. The famous trapezium stars are in the southern half. North is up and east is left. [Gary Bernstein, Regents U. Michigan, Lucent Technologies]

to become a *black hole*, an object so dense that a light beam can not escape from its gravitational pull. Observations and theory together point strongly toward the existence of black holes, but there is still room for more definitive evidence.

Galaxies and the universe

The fuzzy, irregular band of diffuse light that extends across the night sky is known as the Milky Way. Astronomers determined that this light consists of a dense collection of many isolated stars. The Milky Way was thus found to be a large "universe" of stars of which the sun is a member. It is called the *Galaxy* (with capital *G*)[1] after the

---

[1] We generally follow astronomical practice and use "the Galaxy" to describe the Milky Way system of stars. On occasion, we use "(MW) Galaxy" as a reminder where there could be confusion with other "galaxies".

(a) Pleiades (optical)

(b) Pleiades (x ray)

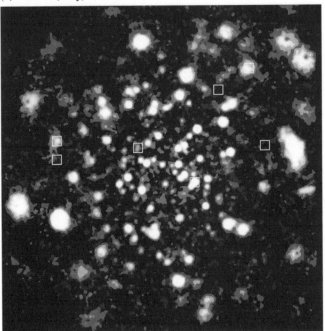

Figure 1.7. The Pleiades, or "Seven Sisters", an open cluster of ~100 young stars in (a) optical light and (b) x rays. The haze in (a) is blue light scattered by dust in the cluster. The boxes in (b) indicate the positions of the brightest optical stars. The Pleiades are about 400 LY distant and about 2° in angular extent. North is up and east is left. [(a): © Anglo-Australian Obs./ Royal Obs., Edinburgh; photo from UK Schmidt plates by David Malin. (b) T. Preibisch (MPIfR), ROSAT Project, MPE, NASA]

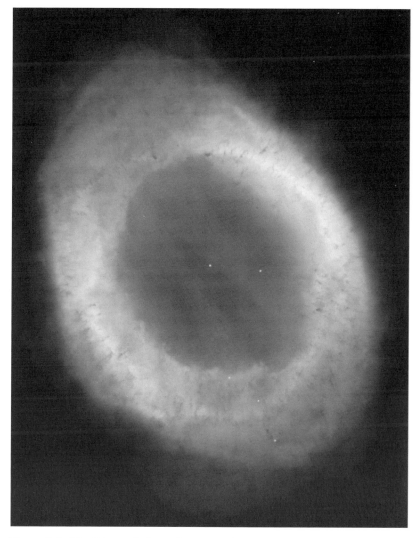

Figure 1.8. The Ring nebula, a planetary nebula. The central star in a late stage of its evolution has ejected gas which it fluoresces. It is about 2300 LY distant and 1.3′ in extent. [H. Bond *et al.*, Hubble Heritage Team (STScI/AURA)]

Greek word *gála* for "milk". It was in 1917 that Harlow Shapley determined the distance to the center of the Galaxy to be ~25 000 LY[1] (current value). He did this by measuring the locations and distances of *globular clusters* (tightly clustered groups of 10^5 or 10^6 old stars; Fig. 9), which he realized must surround the center

---

[1] One light year (LY) is the distance light travels in one year in a vacuum. It is not an SI unit, but we choose to use it because of its natural physical meaning. There are several definitions of the year (i.e. Tropical, Julian and Sidereal) which differ slightly in duration, but each is consistent with the conversion, 1.0 LY = 0.946 × 10^{16} m ≈ 1 × 10^{16} m. We use the symbol "yr" for the generic year of ~365.25 d where "d" is the non-SI unit for the mean solar day which is about equal to 86 400 SI (atomic) seconds. See Chapter 4.

Figure 1.9. The globular cluster, M10. Globular clusters are remnants from the formation of galaxies. There are about 160 associated with the (MW) Galaxy. Each contains 10^5 to 10^6 stars and orbits the center of the Galaxy. M10 is about 65 000 LY from the center of the Galaxy. This photo is 26′ full width; the cluster is about 69′ in diameter. North is up and east is left. [T. Credner & S. Kohle, Hoher List Observatory]

of the Galaxy. The discus-shaped Galaxy has a diameter of roughly 100 000 LY. It contains about 10^{11} stars.

The nature of certain diffuse nebulae of small angular extent in the sky was hotly debated: were they diffuse clouds of gas within the Galaxy or were they very distant giant *galaxies* (with lower case *g*) similar to the Galaxy? Edwin Hubble obtained a distance to the *Andromeda nebula* (Fig. 10) in 1924 which turned out to be very large – the current value is 2.5 million light years – which placed the nebula well outside the Galaxy. This distance and its apparent angular size on the sky ($\sim$3.4°) demonstrated that Andromeda is another huge galaxy like the Milky Way, of size $\sim$100 000 LY.

Figure 1.10. Andromeda nebula M31, our sister galaxy, is about 2.0° in angular extent and is distant 2.5 × 10⁶ light years. North is 26° counterclockwise (left) of "up", and east is 90° further ccw, toward the lower left. [Jason Ware]

Galaxies are found out to great distances; more than 10^{11} galaxies (or precursor galaxy fragments) are in principle detectable with the Hubble Space Telescope. Some have cores that emit intense radiation at many wavebands. These *active galactic nuclei* (AGN) may be powered by a massive black hole of mass $\sim 10^8$ solar masses. The most luminous of these cores are known as quasars; they can now be detected to great distances, up to $\sim 90\%$ the distance to the "edge of the observable universe", about 12 billion light years distant.

A great theoretical advance was Albert Einstein's *general theory of relativity* (1916). This provided a dynamical description of motions in space-time that allowed for accelerating (non-inertial) frames of reference. In this context, gravity can be viewed as a distortion of space. One consequence of this theory is that light rays from a distant star should bend as they pass through a gravitational field, i.e., near a star or galaxy. This effect and its magnitude were first measured in 1919 during a solar eclipse; it made Einstein famous. With more powerful telescopes, *gravitational lensing* is found to be a prevalent phenomenon in the sky. Distant quasars are sometimes seen as double images or as narrow crescents because an intervening galaxy or group of galaxies serves as a gravitational lens. General relativity also predicts the existence of black holes.

Another consequence of Einstein's theory is that the universe as a whole should evolve. It is expected to be expanding and slowing, or instead, it could be contracting

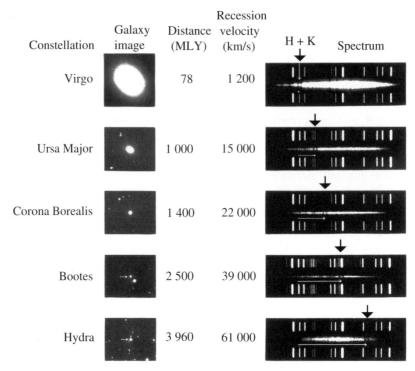

Constellation	Galaxy image	Distance (MLY)	Recession velocity (km/s)	Spectrum
Virgo		78	1 200	
Ursa Major		1 000	15 000	
Corona Borealis		1 400	22 000	
Bootes		2 500	39 000	
Hydra		3 960	61 000	

Figure 1.11. Images and spectra of distant galaxies (nebulae). The fainter (and hence more distant) objects show the double absorption (dark) H and K lines of ionized calcium shifted to longer and longer wavelengths, i.e., to the red. If the redshift is interpreted as a Doppler shift, this would indicate a correlation between recession velocity and distance (See Chapter 9). More properly stated, the redshifts are due to the effect of an expanding universe on photons from the distant galaxy as they travel to the earth. *Redshift* is defined as the fractional wavelength shift, $z \equiv (\lambda_{obs} - \lambda_0)/\lambda_0$ where λ_{obs} and λ_0 are the observed and rest wavelengths respectively. The redshifts are given here in terms of the equivalent velocities. For speeds much less than the speed of light, these are the Doppler velocities, cz. The distances to the galaxies are derived from the redshift assuming a velocity/distance ratio (Hubble constant) of 15.3 km/s per mega light year. [Palomar Observatory/ Caltech]

and speeding up. The analog is a ball in the earth's gravitational field: it rises and slows, or it falls and speeds up; it does not remain motionless. This aspect of the theory was not appreciated until *after* Edwin Hubble's observations of distant galaxies showed in 1929 that, indeed, galaxies are receding from one another (Fig. 11); the universe is expanding! This expansion is similar to that of a raisin bread baking in an oven; every raisin moves away from its neighbors as the bread rises. In the universe, the galaxies are like the raisins. This discovery opened up the entire field of cosmology: the story of the birth, life, and death of our universe.

There is strong evidence that the universe was once hot and dense, some ~10 billion years ago. Such a beginning is often called the *big bang*. The proportions of light elements (Li, Be, He) in the solar system and Galaxy relative to hydrogen are nicely consistent with those expected from nuclear interactions in the hot early universe. Also, a diffuse radiation at microwave frequencies is seen to arrive from all directions of the sky. It is known as the cosmic microwave background radiation, CMB, and has a blackbody spectrum with temperature 2.7 K. This radiation is expected theoretically to be the remnant of the early hot phases, and its characteristics match the theory extraordinarily well. Also geological information yields ages of rocks and meteorites that are comparable to 10^{10} years, an age inferred independently from the expansion of the galaxies.

The final fate of the universe is not known. According to viable theories, it could expand forever, it could slow to a stop and start contracting, or it could be in between, slowing "critically" forever. Continuing interplay of observation and theory are providing further progress on this issue; see below.

New horizons

Another consequence of the general theory of relativity is that oscillating masses should radiate gravitational waves. (Recall that oscillating electric charges radiate electromagnetic waves.) A *binary radio pulsar* can consist of two neutron stars in orbit about their common center of mass. If they radiate away enough energy through gravitational radiation, they will move closer together as they orbit around their common center of mass. In so doing they lose potential energy and gain a lesser amount of kinetic energy. At this writing, the rates of decay of several such systems, including PSR 1913 + 16 (the *Hulse–Taylor pulsar*) and PSR 1534 + 12, confirm to very high precision the predictions of Einstein's general relativity. There are efforts underway to detect gravitational waves directly as mentioned above. Very sensitive and large detection systems are required due to the small amplitude of the expected signals.

The physics of the interior of the sun may not be completely resolved. The neutrino experiments mentioned above do not show the expected numbers of neutrinos; the observed rate is about half that expected. The neutrinos are a measure of the nuclear reactions taking place within the sun. Are the conditions of temperature, density, and composition in the nuclear-burning regions of the sun not well understood? Are the neutrinos changing form (*neutrino oscillations*) as they pass through the solar material so some of them become undetectable in existing instruments? Neutrino oscillations of a different type of neutrino (the *muon neutrino*) have been detected in the flux of neutrinos created by cosmic ray interactions in the earth's atmosphere. This is a major advance in neutrino physics. The resolution of the solar neutrino puzzle has important ramifications in astrophysics and particle physics, and ongoing experiments are addressing it.

There is a well-received theory of the early universe (the *inflationary universe*) wherein it expanded by many orders of magnitude very early, 10^{-33} s, after the "big bang". This suggests the universe is now expanding at just the *critical* rate where it asymptotically approaches zero expansion speed; it neither completely "escapes", nor does it start falling inward. A major difficulty with this view is that the visible matter in the universe falls far short of the mass required to yield the critical condition.

If the theory is correct, there must be some new kind of *dark matter* of an unknown type. Such speculation gains credence from the motions of stars in galaxies and of galaxies in *clusters of galaxies*. In both cases, the objects move so rapidly in a confined volume that unseen matter would seem to be holding them in their orbits. The nature of this dark matter is one of the great questions now facing astrophysicists. The picture is further complicated by indications from observations of supernovae that the expansion of the universe is increasing, due to some (unknown) type of *dark energy*.

The distribution of galaxies in space is found to be very clumpy with huge voids and "walls" of galaxies. This can be compared to the very smooth distribution of the CMB. It is generally believed that the galaxy clustering arises from small density fluctuations in the early universe, and these should be visible as tiny fluctuations in the brightness (temperature) of the CMB as a function of angle on the sky. In fact, such fluctuations have been detected by the Cosmic Background Explorer (COBE) and WMAP satellites, and with experiments carried out in Antarctica (high-altitude balloon and ground based). These warm and cool spots and the postulated existence of large amounts of *cold dark matter* could well lead to the formation of galaxies with the observed clustering. The dominant angular scale of the CMB fluctuations, $\sim 1°$, indicates that the universe is expanding at the critical rate, or equivalently that it has a "flat" geometry. This gives support to the existence of an episode of inflation in the early universe. Confidence in these ideas is growing at this writing.

These are only some of the challenges facing astronomers today. The nature of gamma-ray bursts, the most energetic explosions known to man, and the mechanisms that give rise to jets of material from many different types of celestial objects are among them. If the past has been any indication, some of the answers will prove to be quite surprising. Also serendipitous discoveries will surely provide new surprises and new questions. It is likely that many of the questions asked today will be shown to be off the mark. The truth will lie in other unexpected directions.

Problems

There are no problems assigned for this chapter. However, two problems in Chapter 9 (9.51 and 9.52) illustrate the methods the ancients used as described here. The student may wish to attempt these now.

2

Electromagnetic radiation

What we learn in this chapter

Astronomers learn about the cosmos through the study of signals arriving at the earth in the form of **electromagnetic radiation** or as **neutrinos, cosmic rays, meteorites**, and, hopefully in the near future, **gravitational waves**. Electromagnetic radiation travels at speed c and can behave either as a **wave** or as a flux of **photons** each of energy $E = h\nu$. One can **convert** between wavelength, frequency and photon energy through **algebraic** or **numerical relations**. The **bands** of electromagnetic radiation extend from **radio waves** at the lowest frequencies to **gamma rays** at the highest. The average photon energy, or frequency, of radiation from an object is an indicator of the **temperature** of the emitting source if the radiation is thermal. **Absorption** of photons in the **earth's atmosphere** is frequency dependent so observations of some bands must be carried out from **high altitude balloons** or **space vehicles**. Similarly, absorption in the **interstellar medium** by dust and atoms renders the cosmos more or less transparent, depending upon the frequency band (see also Chapter 10).

2.1 Introduction

Electromagnetic radiation is the primary source of our knowledge of the cosmos. Its characteristics (e.g., speed and frequency) are briefly summarized in this chapter. At some frequencies, the radiation can penetrate the atmosphere and ground-based observations are feasible; at other frequencies the atmosphere is opaque and observations must be carried out from space. Particulate matter (e.g., cosmic ray protons and meteorites) also brings us information about the solar system and the Galaxy. In addition, neutrinos and gravitational waves carry information about the cosmos to astronomers on earth, though gravitational waves have yet to be detected directly.

2.2 **Photon and non-photon astronomy**

The several types of signals received from space are outlined here.

Photons (electromagnetic waves)

The astronomical "light" that arrives at the earth from a distant source is known as electromagnetic radiation. This radiation can be described in terms of waves or in terms of *photons*. Electromagnetic waves are propagating electric and magnetic fields whereas photons are discrete bundles of energy. These two descriptions of light are difficult to reconcile with one another intuitively. Both are correct; the radiation can behave as one or the other under different circumstances.

The refraction of light in a prism is usually described in terms of waves, while the ejection of electrons from an illuminated surface (*photoelectric effect*) may be understood in terms of momentum transfers from individual photons. The detection of radio sources with high angular resolution by means of a technique called *interferometry* arises from the interference of waves detected with two or more separate telescopes. In gamma-ray astronomy, individual photons are detected one-by-one with scintillation crystals that emit a detectable pulse of light when a single gamma ray interacts with the atoms of the crystal.

The electromagnetic waves described by Maxwell's equations are encountered as radio waves, infrared radiation, optical light, ultraviolet radiation, x rays and gamma rays. These different names simply specify ranges of wavelengths or frequencies. The lowest frequencies (or longest wavelengths) are radio waves, and the highest frequencies (or shortest wavelengths) are gamma rays. In the discrete picture, the photons, or *quanta*, carry energy and momentum much as a mass-bearing particle does. A radio photon has very low energy while a gamma-ray photon has very high energy. In both pictures, the energy is propagated at the *speed of light* (3.0×10^8 m/s in a vacuum), and the signal has the following properties: intensity (number of photons), frequency ("color"), polarization, and direction of travel.

Astronomers usually refer to radio and optical radiation in terms of waves, characterizing them with a wavelength λ or a frequency ν, while they refer to x rays and gamma rays as photons, characterizing them with an energy E. The terminologies are completely interchangeable; presented below are the relations between frequency of a wave and the energy of a photon.

Photons are invaluable for astronomy because they travel in straight lines, for the most part. Photons thus appear to come directly from the spot on the sky whence they originated. Not all of these photons arrive at the earth undisturbed however. Low-frequency radio waves undergo refraction in ionized interstellar plasma; radiation from distant quasars is bent by intervening galaxies which act as *gravitational lenses*; optical photons can be scattered by tiny grains of graphite, silicates and ice (called *dust*); ultraviolet photons and x rays can be absorbed by neutral atoms. Nevertheless, under many circumstances, photons do travel along a straight path from the source to the earth and are not strongly absorbed in the interstellar medium. Such photons provide astronomers with a relatively clear view of the cosmos.

Cosmic rays and meteorites

Information about the cosmos can also be gleaned from the detection and measurement of particulate matter. *Cosmic rays* are bits of matter (protons and heavier atomic nuclei) that travel with high energies, arriving at the earth from distant celestial regions (e.g., the sun, a supernova, an active galactic nucleus). Since most cosmic ray particles are charged, the weak and irregular magnetic fields that lie between the stars will change their directions of travel through the action of the magnetic $F = q(v \times B)$ force. The particles from a given source will spiral around the magnetic fields in the Galaxy for millions of years, circulating in the company of particles from many other sources.

Most cosmic rays thus arrive at the earth from a direction that bears little if any relation to the direction of their point of origin. That is, cosmic rays from many different sources can arrive at the earth from the same apparent direction. It is as if you were very near-sighted, took off your glasses, and looked at the sky. The light from many stars would be mixed together on your retina, and you would not be able to study individual stars. However, even with this mixing, cosmic rays provide important information about the details of supernova explosions and the interstellar spaces through which they travel. (See Section 12.3.)

Macroscopic chunks of matter, known as meteoroids, arrive at the earth from the solar system. As they penetrate the earth's atmosphere, they heat up due to atmospheric friction and lose material by vaporization; in the night sky, they are easily visible as bright rapidly moving objects. In this stage, they are called *meteors* ("shooting stars"). Some burn up completely before reaching the earth's surface. The remains of others reach the earth's surface (*meteorites*). Analyses of their chemical composition provide valuable data on the quantities of the chemical elements in the solar system and on the ages of those elements. Meteorites are reminiscent of the moon rocks brought to earth by Apollo astronauts except that they are a lot less trouble to obtain; they arrive on their own and from much greater distances!

Neutrino and gravitational-wave astronomy

As noted in Chapter 1, the universe can be probed through observations of neutrinos and gravitational waves. These are relatively new branches of astronomy, and substantial effort is now being expended on their development. Neutrino observations of the sun have been in progress for a number of years and new more sensitive experiments are now in progress. A burst of neutrinos was detected from the implosion of Supernova 1987A in the Large Magellanic Cloud, giving us a surprising and dramatic interior view of the core of a star at the instant of its collapse to a neutron star. (See Section 12.2.)

Gravitational waves have not been detected directly. Large (several kilometer) sensitive detectors are now operating and attempting detections. One candidate emitter that should be detectable is the gravitational "chirp" expected in the last moments in the life of a binary star system consisting of two neutron stars that are gravitationally bound to one another. The large accelerations of the two stars lead to the emission of *gravitational radiation*. This causes the system to lose energy. The stars thereby decrease their separation and orbit each other at increasing speeds. In turn, this increases the rate of energy loss to gravitational radiation. The inward spiraling of the orbits thus increases at a faster and faster pace. In the last few seconds before they finally merge to become a black hole, they should emit a strong chirp of about 1-kHz gravitational radiation. (See Section 12.4.)

2.3 Electromagnetic frequency bands

The "color" of light can be described in three equivalent ways: the energy of a single photon E, the frequency of the electromagnetic wave ν, or the wavelength λ. The several *bands* of the electromagnetic spectrum, e.g., radio, optical and x-ray, may be described in terms of each of these parameters as shown in the logarithmic display (Fig. 1)[1]. The relationships among these quantities will now be presented.

Wavelength and frequency

The frequency ν of an electromagnetic wave is described in units of cycles per second, or hertz (Hz). In a vacuum, the frequency is related to its wavelength λ as

$$\lambda \nu = c \tag{2.1}$$

where c is the speed of light in a vacuum,

$$c = 2.9979 \times 10^8 \text{ m/s} \approx 3.00 \times 10^8 \text{ m/s} \tag{2.2}$$

---

[1] It will be our convention on logarithmic plots to label the axis (or axes) as "log y" and to indicate the logarithm of the value rather than the value itself; that is 12 instead of 10^{12}. We also use the conventions log $\equiv \log_{10}$ (base 10) and ln $\equiv \log_e$ (base e).

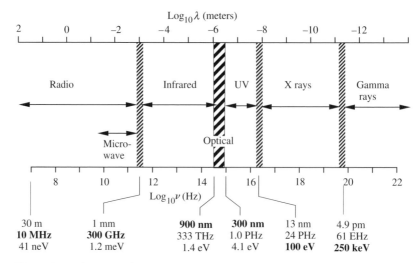

Figure 2.1. The approximate "bands" of the electromagnetic spectrum. The bound-aries are not well defined. Radio waves extend further to the left, but the earth's ionosphere is opaque for the most part below about $\nu = 30$ MHz. Gamma rays extend to the right by many more factors of 10. Values of the boundaries are given in several units below the figure. Among these, commonly used values are shown in boldface. Note that 1 nm = 10 angstroms.

Wavelength in the visible (optical) range has traditionally been expressed in angstroms (1 Å $= 10^{-10}$ m), but the preferred SI unit is the nanometer (1 nm $= 10^{-9}$ m) or micrometer (1.0 μm $= 10^{-6}$ m) traditionally called "microns". Thus,

$$1 \text{ μm} = 10\,000 \text{ Å} \qquad\qquad \text{(infrared/optical)} \qquad (2.3)$$

$$1 \text{ nm} = 10 \text{ Å} \qquad\qquad\qquad\qquad \text{(x ray)} \qquad (2.4)$$

The human eye is sensitive from about 420 nm (blue) to 680 nm (red). Optical ground-based astronomers work from about 320 to 900 nm.

Frequency–wavelength equivalences over 12 decades of frequency from (1) are given in Table 1. The values given in Table 1 are placed in the context of the observational bands through comparison to Fig. 1. In this text, frequency units will be used extensively.

Photon energy

The energy E (units of joules, J, in the SI system) of a photon is[1]:

$$E = h\nu \qquad\qquad\qquad\qquad\qquad\qquad \text{(J)} \qquad (2.5)$$

$$h = 6.626\,069 \times 10^{-34} \text{ J s} \qquad\qquad \text{(Planck constant)} \qquad (2.6)$$

---

[1] We often choose to give SI units (in parentheses) when they are not required, as in (5), to remind the reader of the dimensions. The dimensions often clarify the meaning of the quantity in question.

Table 2.1. *Frequency–wavelength correspondence*

Frequency[a]	Wavelength[a]	Band
3×10^8 Hz = 300 MHz	1 m	(radio)
3×10^{11} Hz = 300 GHz	1 mm	(microwave)
3×10^{14} Hz = 300 THz	1 μm	(IR/optical)
3×10^{17} Hz = 300 PHz	1 nm	(x ray)
3×10^{20} Hz = 300 EHz	1 pm	(gamma ray)

[a] The SI prefixes used here are: m = milli = 10^{-3}, μ = micro = 10^{-6}, n = nano = 10^{-9}, p = pico = 10^{-12}, M = mega = 10^6, G = giga = 10^9, T = tera = 10^{12}, P = peta = 10^{15}, E = exa = 10^{18}. See complete list in the Appendix, Table A3.

where h is the Planck constant (J s = joule second) and ν is the frequency (Hz). The energy of a photon can be given in units of electron volts (eV), where 1 eV is the energy gained by an electron that is accelerated through a 1 volt potential change,

$$1 \text{ eV} = \text{charge of electron} \times 1 \text{ volt} \tag{2.7}$$
$$= 1.602\,176 \times 10^{-19} \text{ (coulombs)} \times 1.0 \text{ (volt)}$$

➡ $$1 \text{ eV} = 1.602\,176 \times 10^{-19} \text{ J} \qquad \text{(electron volt)} \quad (2.8)$$

Astrophysicists often use the energy units of electron volts when dealing with photons and fundamental particles. From the relations (5) and (8), the frequency (Hz) is related to the photon energy (eV) as follows:

$$\nu(\text{Hz}) = \frac{E(\text{J})}{h(\text{J s})} = \frac{E(\text{eV}) \times 1.602\,176 \times 10^{-19} \text{ (J/eV)}}{6.626\,069 \times 10^{-34} \text{ (J s)}} \left(= \frac{E(\text{eV})\,e}{h} \right)$$
$$\tag{2.9}$$

➡ $$\nu(\text{Hz}) = E_{\text{photon}}(\text{eV}) \times 2.418\,0 \times 10^{14} \tag{2.10}$$

Thus a photon of energy 1 eV has frequency of 2.4×10^{14} Hz. The rightmost term of (9) leads directly to the relation between the wavelength $\lambda(\text{m})$ and the photon energy $E(\text{eV})$,

$$\lambda(\text{m}) = \frac{c}{\nu} = \frac{1}{E(\text{eV})} \frac{hc}{e} \tag{2.11}$$

$$\lambda(\text{m}) \times E_{\text{photon}}(\text{eV}) = hc/e = 1.239\,842 \times 10^{-6}$$

➡ $$\lambda(\text{nm}) \times E_{\text{photon}}(\text{eV}) = 1\,239.842 \tag{2.12}$$

which tells us that 1 eV corresponds to $\lambda = 1.24 \times 10^{-6}$ m and that radiation in the visible range at $\lambda = 620$ nm corresponds to photon energies of 2.0 eV. According to (10) or (1), the frequency associated with these 2.0-eV photons is $\nu = 4.8 \times 10^{14}$ Hz. The reader should remember that optical radiation is characterized by the values ~ 2 eV and $\sim 5 \times 10^{14}$ Hz. The x-ray wavelength of $\lambda = 1$ nm corresponds to 1240 eV or 1.24 keV. From (10), this corresponds to a frequency of 3×10^{17} Hz.

Figure 1 shows frequency increasing to the right whereas wavelength increases to the left, according to (1). The photon energy also increases to the right according to (5). The boundary between the infrared and optical bands is roughly at $\lambda = 900$ nm or $\nu = 333$ THz and the radio/infrared boundary is roughly $\lambda = 1$ mm or $\nu = 300$ GHz. The high frequency end of the radio band is called the microwave band, from about 5 GHz ($\lambda = 60$ mm).

These names and boundaries of the wavelength bands are historical and not precisely defined. They reflect in part the detection apparatus first used, how the photons are created, or the context of their original discovery. For example, medical radiologists probably call 2-MeV photons "x rays" rather than gamma rays because, I gather, they are generated by energetic electrons crashing into a metal target, but they would call them "gamma rays" if they were emitted by an atomic nucleus (radioactivity). In contrast, astrophysicists would call 2-MeV photons "gamma rays" simply because they are in the gamma-ray frequency range.

The very narrow range of the visible or optical portion of the spectrum is notable. It is only about a factor of three in bandwidth, compared to the overall range of observed photon energies which is more than a factor of 10^{15}! It is a happy circumstance that the peak of the solar radiation falls at the same frequency as a transparent band in the atmosphere (see below). Our eyes evolved, naturally, to operate in this same narrow range. The explosion of astronomical knowledge in this century has been possible, in large part, because we have learned to "see" at wavelengths outside the visible band. The dynamic ranges encountered are so large that many of the figures in this text will show axes in logarithmic units, as in Fig. 1.

Temperature

Photons are often emitted from a body that can be characterized by a temperature T. The SI unit of thermodynamic temperature is the *kelvin* which refers to the scale known as *absolute temperature*. In this system, the freezing temperature of water is at $T = 273.15$ K (0 °C). Temperature can be defined for a gas in thermal equilibrium. In such a gas, the particles have kinetic energies of order kT where k is the *Boltzmann constant* (units J/K = joules/kelvin), according to the kinetic theory of gases. If these particles emit photons and if the photons are in thermal equilibrium with the particles (i.e., the object is a "blackbody"), the average photon energy $h\nu_{av}$ will be roughly of order kT, or more precisely,

$$E_{photons,av} = h\nu_{av} = 2.70kT \qquad \text{(thermal equilibrium)} \qquad (2.13)$$

$$k = 1.380\,650 \times 10^{-23} \text{ J/K} \qquad \text{(Boltzmann constant)} \qquad (2.14)$$

It is thus sometimes convenient, as a reference, to know the relation between frequency and temperature according to the approximation $h\nu \approx kT$,

$$\nu = \frac{kT}{h} \qquad (2.15)$$

➡ $$\nu(\text{Hz}) = 2.084 \times 10^{10} \, T(\text{K}) \qquad \text{(for } h\nu = kT) \qquad (2.16)$$

From (16) one finds that a temperature of 10 000 K is associated with a frequency of 2×10^{14} Hz. This is in the near-infrared band (meaning near to the optical band). However, since the average photon frequency of a blackbody is actually about a factor of three higher than kT (13), much of the radiant energy from a star of $T = 10^4$ K will fall in the optical band where the human eye can sense it.

The relation between energy kT (eV) and temperature T (K) is also useful.

$$kT(\text{eV}) = \frac{kT(\text{J})}{e(\text{J/eV})} = 8.617\,34 \times 10^{-5} \, T(\text{K}) \qquad (2.17)$$

➡ $$T(\text{K}) = 11\,605 \times kT(\text{eV}) \approx 12\,000 \times kT(\text{eV}) \qquad (2.18)$$

Thus our 10 000-K stellar surface contains particles of energy $E \approx kT = 0.86$ eV. X rays of energy $kT \approx 10^4$ eV correspond to temperatures of $T \approx 10^8$ K. Do remember that 1 eV corresponds to about 12 000 K.

2.4 Photons and the atmosphere

Atmospheric absorption

Astronomy can be performed only if the photons can reach our instruments. If they are absorbed or scattered en route from their source, they may never arrive. Whether they do or not depends upon the frequency of the radiation and the contents of the space they traverse. In a pure vacuum, the photons will travel unimpeded. Unfortunately (or fortunately), the regions between the stars contain dilute gases that can absorb photons of certain frequencies (See Chapter 10). Closer to home, the earth's atmosphere is absolutely opaque to many wavelengths and highly absorbing to others. For this reason, astronomers often clamor to obtain access to space vehicles for their telescopes.

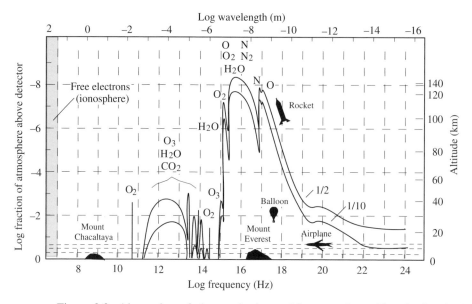

Figure 2.2. Absorption of photons in the earth's atmosphere. The abscissa is the frequency (lower scale) or wavelength (upper scale) of the radiation, and the ordinate is the height above the earth's surface in kilometers (right axis) or fraction of the atmosphere *above* the corresponding height (left axis). The plotted curves represent the approximate altitude (or fractional atmospheric depth) at which either 1/2 or 1/10 of the radiation will survive the passage through the atmosphere above this point. The specified elements and molecules provide the absorption in the regions so labeled. The ionospheric electrons at altitudes extending well beyond those in the figure are highly opaque to radio waves below about 10 MHz. [Adapted from M. Oda, *Proc. Int. Cosmic-Ray Conference London*, 1965, v. 1, p. 68]

The transparency of the earth's atmosphere to photons is illustrated as a function of frequency in Fig. 2. The ordinate represents height in the atmosphere, or equivalently, the fraction of the atmosphere above the indicated height. The abscissa indicates the frequency (Hz) or the wavelength (m) of the incoming radiation. The two solid lines indicate the positions (height) in the atmosphere to which 50% and 10% of the photons from a celestial source will penetrate; the other 50% or 90% are absorbed at higher elevations. To use the curves of Fig. 2, pick a frequency on the lower horizontal axis and read vertically until you intercept the lower curve. The ordinate for this point indicates the altitude where one would detect 10% of the incoming photons. If one moves further vertically (to the second curve), one finds the altitude that would yield 50% of the flux. One can then hire a skyhook (a mountain top, balloon, rocket, or satellite) to raise the telescope to the appropriate altitude to carry out the desired observations.

Most radio astronomy is ground based because of the broad region of atmospheric and ionospheric transparency from ~10 MHz to ~100 GHz. Below about 10 MHz, the ionosphere is quite opaque: incoming radio waves are reflected back out to space and man-made AM radio waves reflect from it to yield extended ranges. The ionosphere is variable in its absorption so that substantially lower frequencies can occasionally be observed from the ground. Space missions allow studies to frequencies <10 MHz. They also yield large separations between radio telescopes for measurements with extremely high angular resolution (*interferometry*).

At the highest radio frequencies, at about 300 GHz ($\lambda \approx 1$ mm; in the *microwave band*) the atmosphere is becoming opaque. It is in this region that the blackbody radiation from the early universe is detected. Space missions have been used in this region with great success.

Infrared (IR) astronomy (10^{12} to 3×10^{14} Hz) must contend with water vapor, carbon dioxide, and ozone in the atmosphere. These permit ground-based observations at only a few narrow bands of frequency. In spite of this, ground-based observations have been carried out with great productivity for many years. Space missions have greatly extended this field.

Radiation in the optical band passes relatively unimpeded through the atmosphere; see the gap just below 10^{15} Hz in Fig. 2. Humans on the earth's surface thus can see the stars and carry out astronomical studies of them. The thermal turbulence of the atmosphere does cause light to be irregularly refracted, thus limiting the resolution normally to about $1''$. Space missions can (and have) overcome this limitation as will ground-based *adaptive optics* systems that are now coming into operation.

Ultraviolet (UV) astronomy (10^{15} to $\sim 2 \times 10^{16}$ Hz) can not be carried out from the ground because of the atmospheric opacity due to ozone (O_3) and other molecules such as H_2O (water vapor). Progress in this field has been through rockets and satellites.

X-ray astronomy ($\sim 2 \times 10^{16}$ to $\sim 6 \times 10^{19}$ Hz), similarly, can not be carried out on the ground. The nitrogen and oxygen in the earth's atmosphere absorb the x rays by means of the *photoelectric effect*; the x ray knocks an electron out of the atom and is absorbed in the process. As a consequence, this branch of astronomy has been carried out solely from rockets, balloons and orbiting satellites.

At gamma-ray frequencies ($\gtrsim 6 \times 10^{19}$ Hz), the atmosphere remains opaque due to interactions with electrons and atomic nuclei. Thus this field has been exploited with high altitude balloons and satellites with great success. At extremely high frequencies, $\sim 10^{27}$ Hz (or energies ~ 4 TeV), the secondary products of the atmospheric interactions release sufficient energy in the atmosphere (*extensive air showers*) to permit their detection from the ground. Telescopes detect the optical light (*Cerenkov light*) released by the energetic secondary electrons as they traverse the atmosphere at speeds greater than the speed of light in air.

Interstellar absorption

A transparent atmosphere is not sufficient for effective astronomy. The interstellar medium (matter between the stars) must also be sufficiently transparent along the line of sight so the photons can reach the earth (See Chapter 10). In some cases it is not. For example, *interstellar grains* or "*dust*" absorb optical light making the interiors of star formation regions impenetrable to optical astronomers, but not to radio and infrared astronomers. Above the ionization energy of hydrogen (13.6 eV), in the "far" ultraviolet, the interstellar medium is highly opaque because the probability that a photon will ionize a hydrogen atom is very high. Atoms of higher atomic numbers absorb low-energy x rays. Dense gases near some star systems can block radio waves and x rays.

However, even in bands where our view is quite obscured, studies may be carried out of objects, or in directions, that have less of the blocking material, or simply of objects that are close enough so that the absorption is not total. Telescopes in different frequency bands complement each other because they can see into different regions of the cosmos.

Problems

2.3 Electromagnetic frequency bands

Problem 2.31. (a) Verify that the several values given in Fig. 1 for each band boundary are in accord with the expressions given in the text for unit conversions. Check the values on the upper and lower axes as well as those expressed numerically below the figure. (b) Make a table which indicates the values of frequency (Hz), wavelength (m) and energy (eV) for each of the following: 100-GHz radio waves, 2-μm IR radiation, 450-nm light, 300-nm UV radiation, 4-keV x rays, and 100-MeV gamma rays. Express the entries in exponential notation, e.g., 3×10^{12}, rather than logarithms.

Problem 2.32. The ground state of atomic hydrogen is split into two energy levels separated by an energy $\Delta E = 5.87 \times 10^{-6}$ eV, the so-called hyperfine splitting. A hydrogen atom undergoing a transition from the upper to the lower state emits a photon of energy ΔE. Calculate the wavelength of this photon. To what part of the electromagnetic spectrum does this photon belong? Speculate on the astrophysical significance of this radiation.

2.4 *Photons and the atmosphere*

Problem 2.41. (a) Use the plots of Fig. 2 to verify the statements in Sect. 2.4 regarding the altitudes from which the various branches of astronomy can be carried out. (b) Where would you set up a gamma-ray observatory operating in the region of 100 MeV if you needed to detect at least 50% of the photons? (c) How high should a sounding rocket go if it is to effectively detect 2-keV x rays from an astronomical object?

Problem 2.42. Find the approximate altitude where the transmission in the atmosphere is only 1% at 100 MeV. Use the 50% curve of Fig. 2 as a reference. Consider that it takes a certain amount of material to absorb 1/2 of the radiation, that the same amount of material will absorb 1/2 of the remaining, etc. until only 1% is left. Ignore the 1/10 curve which is plotted too low at this energy for this model.

3

Coordinate systems and charts

What we learn in this chapter

Stars are located on the sky with **two angular coordinates**. Distances to them may be ignored by visualizing all of them as being on a **celestial sphere** at "infinite" distance. The angular coordinates define the star's location on the sphere. Any number of **coordinate systems** can be defined on this sphere. Astronomers use the **equatorial**, **galactic**, **ecliptic**, and **horizon** systems. The coordinates of a star differ from coordinate system to coordinate system so **transformations** between them are needed. In the equatorial system, the coordinates of a given star vary steadily and slowly due to **precession** of the earth, so one must define the **epoch**, e.g., J2000.0, of any quoted coordinates. "Areas" on the sky are defined as **solid angles**. Cataloging of stars is accomplished through **photographic surveys**, **printed sky charts**, and printed lists ("**catalogs**"). Unnamed stars can be specified unambiguously by marking the star on a **finding chart**, a sky photograph of the local region. The **name of a star or galaxy** may depend on its location and brightness within a constellation, its equatorial coordinates ("**telephone number**"), or simply its sequential number in a published catalog of objects together with the catalog name, e.g., Messier 42 is the Orion nebula.

3.1 Introduction

A casual look at the sky confronts one with uncountable pinpoints of light. It is imperative that we be able to refer to particular celestial objects without waiting for a cloudless and moonless night. Sky charts and coordinate systems are intrinsic to this task. It turns out, fortunately, that the motions of most celestial bodies relative to one another on the sky are quite slow; the Big Dipper will still look nearly the same in 50 years as it does now. It is therefore quite feasible to define coordinate systems on the sky and to make charts and maps of the brighter objects. In this chapter, we discuss some of the more common coordinate systems used by astrophysicists, the concept of solid angle and some of the charts and maps in use today.

3.2 Coordinates on a celestial sphere

Mathematical sphere at "infinity"

The position of a star or galaxy should properly be specified in three dimensions, i.e., the two angular coordinates of the object from the perspective of the observer and a distance from the observer. Knowledge of distance to a given star is not as readily obtained as its angular position which can be measured directly. Thus star charts usually show only the two angular coordinates of each star. The x, y coordinates on a given chart correspond directly to the angles, say θ, ϕ, which define the direction of the star. A photograph of the sky is such a chart.

It is convenient to imagine a sphere at great distance ("infinity") upon which all stars lie. This is called the *celestial sphere* (Fig. 1). The positions of stars on this sphere may be specified with two angles, analogous to the way latitude and longitude specify a position on the earth's surface. This celestial sphere is an artificial construction; stars are *not* all at the same distance. Stars in the (MW) Galaxy range in distance from 4 LY to more than 50 000 LY from the earth while the farthest quasars may be as distant as 10^{10} LY. Nevertheless, the concept of celestial sphere is useful for charting the sky as one sees it.

Motions of stars on the celestial sphere arise from their intrinsic motion relative to each other in three-dimensional space. The motion on the celestial sphere (i.e., normal to the line of sight) is called *proper motion*, which is typically less than $1''$/year. Catalogs quote the position of the star at a fixed time (*epoch*), e.g., 1950 or 2000, and also give the rate and direction of motion of the star. For many purposes, the motions are so small that it is appropriate to think of the stars as fixed (almost) on the celestial sphere. See Section 4.3 for more on proper motions.

There are also small *apparent* motions of the stars due to the earth's annual motion about the sun. The distance to α Cen, the bright companion of the closest star Proxima Cen (in the constellation Centaurus), is 4.4 LY or ∼140 000 times the diameter of the earth's orbit about the sun. As the earth moves from one side of the sun to the other, the direction of the line of sight to α Cen will change by ∼1/(140 000) radians, or ∼$1.5''$ relative to distant background stars. The star α Cen will thus appear to move back and forth against the more distant background stars once a year, a phenomenon called *parallax*. (Parallax motion is used to determine distances to nearby stars; see Sections 4.3 and 9.5.) For catalogs of sufficient precision, the quoted celestial position (actually *direction*) is that which would be viewed from the barycenter (center of mass) of the sun.

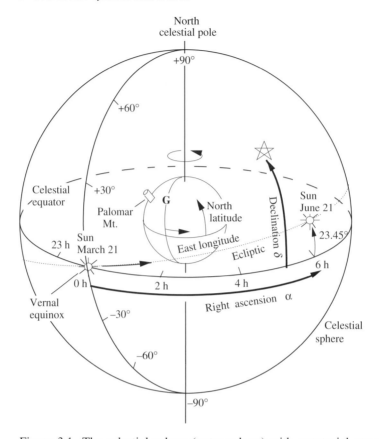

Figure 3.1. The celestial sphere (outer sphere) with equatorial coordinates and the spinning earth within it. The letter "G" indicates Greenwich England. The mathematical celestial sphere is "at infinity"; stars and galaxies are considered to be fixed upon it with equatorial coordinates right ascension α and declination δ as shown by the heavy arrows. The sun follows a great-circle path (ecliptic; dotted line) inclined 23.4 degrees to the celestial equator. The zero of right ascension is the intersection of these two planes where the sun is proceeding from south to north on about March 21. The motions of the stars relative to one another (*proper motion*) and the slow *precession* of the earth's spin axis both lead to changes in the equatorial coordinates of a given object as time progresses.

In contrast to these slow motions, the planets of the solar system move along the celestial sphere, sometimes at very high rates. This motion is discussed briefly in Section 4.4.

Celestial coordinate systems

The two angular coordinates indicating position on the celestial sphere are two of the three coordinates of a spherical coordinate system, the radial (distance) component being ignored. An equator must be defined; it can be any *great circle* about the sphere. (A great circle is any circle whose plane passes through the center of the sphere; it encompasses the full diameter of the sphere.) The two angles in such a coordinate system can be chosen to be latitude and longitude where the zero of latitude is at the equator. In addition, the zero of longitude must be defined at some point on the equator. The lines of constant longitude are called *meridians*. For any sphere, there are an infinite number of possible great circles that could serve as the equator. Thus there are an infinite number of possible coordinate systems. Any such system is known as a *celestial coordinate system*.

The system most used by astronomers and navigators makes use of *equatorial coordinates* which are illustrated in Fig. 1 and discussed below. When astronomers use the more general term *celestial coordinates*, they often have in mind equatorial coordinates. Three other useful systems are *horizon coordinates*, *galactic coordinates* and *ecliptic coordinates*. The latter system is of particular use in solar-system studies and in space-astronomy missions. We now discuss briefly these several systems, beginning with the system most natural for a person viewing the stars from some position on the earth.

Horizon coordinates

The horizon coordinate system locates a star relative to the observer's horizon which becomes the equator of this coordinate system. Formally, the horizon is the great circle defined by the intersection of the plane tangent to the earth's surface at the observer's location with the celestial sphere. The two angles are the *altitude*, measured from the horizon up to the star, and the *azimuthal* measured along the horizon from north in the easterly direction (Fig. 2). These angles define the local *horizon coordinates*. Instead of altitude, one could equally well use the *polar angle* from the zenith to the star. The origin of these angles is the observer.

Horizon coordinates are convenient for specifying the position where a person would find a star at a given time on a clear night. As the earth rotates under the stars, the stars appear to move rapidly across the sky from east to west, much as the sun and moon do. (See Chapter 4 for a description of these motions.) Thus the altitude and azimuthal angles depend upon the time of night as well as upon the observer's location on the earth. For most uses, however, astronomers prefer a coordinate system that is tied to the stars so that the coordinate of a given star does not change solely because the earth rotates.

Equatorial coordinates

In equatorial celestial coordinates, the *celestial equator* is the projection of the earth's equator onto the celestial sphere (Fig. 1); it lies in the plane defined by the earth's equator. The celestial latitude, indicated by the Greek character delta δ, is

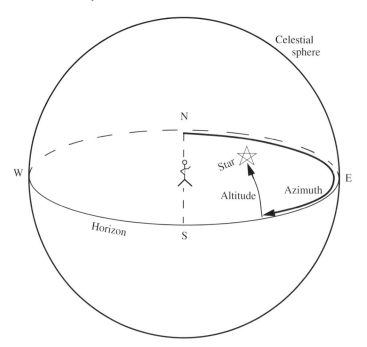

Figure 3.2. The horizon coordinate system on the celestial sphere. A celestial object is located by the two angles, altitude and azimuth, in a coordinate system with the equator defined by the plane of the earth's surface at the observer's position.

called *declination* and is measured in degrees from $-90°$ to $+90°$ with zero at the equator. The celestial longitude (Greek alpha, α) is called *right ascension* and is measured from west to east in units of time, hr-min-sec from 0 h to 24 h (equivalent to 0° to 360°), with zero at the sun's position when it crosses the celestial equator on March 21. This position ($\alpha = 0$, $\delta = 0$) on the celestial sphere is called the *vernal equinox*. The directions of increasing α and δ define a "right handed" coordinate system.

These two coordinates, α and δ, are angles subtended from the center of the earth; they are spherical coordinates. Most stars are so far from the solar system that it would not matter if the origin of the system were the center of the earth, the top of Palomar Mountain, or the barycenter (center of mass) of the solar system. However, for closer stars and planets, the motion of the earth about the sun leads to varying apparent positions on the celestial sphere, e.g., the parallax discussed above. Thus, as noted, current precise catalogs of stellar positions report the positions in a *barycentric coordinate system* wherein the motion of the earth is removed.

The equatorial coordinates of a star change slowly year by year due to the slow *precession* (period 25 770 years) of the earth. The earth's axis precesses slowly about an average direction perpendicular to the orbital plane of the sun–earth system, much

Table 3.1. *Examples of celestial equatorial coordinates*

Object	Epoch	Right asc. (α) h m	Declination (δ) ° ′
Sirius (α CMa)	B1950	06 42.9	−16 39
(bright star)	J2000	06 45.1	−16 43
Andromeda nebula[a]	B1950	00 40.0	+41 00
(M31; close galaxy)	J2000	00 42.7	+41 16
Crab nebula[a]	B1950	05 31.5	+21 59
(supernova remnant)	J2000	05 34.5	+22 01
Proxima Cen	B1950	14 26.3	−62 28
(closest star; 4.3 LY)	J2000	14 29.7	−62 41
Tarantula nebula[a]	B1950	05 39	−69 09
(in Large Magellanic Cloud)	J2000	05 39	−69 06
Vernal equinox	B1950	00 00	00 00 (by definition)
	J2000	00 00	00 00 (by definition)

[a] Diffuse objects ranging in size from ∼5′ (Crab) to ∼2° (Andromeda).

as a spinning top precesses about the vertical. Because the equatorial coordinate system is, by definition, locked to the earth's equator and the vernal equinox, the coordinate system slowly slides around the sky as the earth precesses. The coordinates of any given star will therefore change slowly ($\lesssim 1'$ per year), and one must assign an *epoch* or date to any such coordinates. (See Section 4.3 for more on precession.)

Epoch B1950.0 was the common system for catalogs and charts in the latter part of the past century. It is based on the *Besselian year* and refers to the earth's orientation at (approximately) midnight of New Year's eve in Greenwich, or more precisely 22 h 09 m UT (Universal Time, essentially the time in Greenwich England) of 1949 Dec. 31. Epoch J2000 (based on the *Julian year*) is now the standard for current astronomy. It refers to the orientation of the earth at about noon in Greenwich on 2000 Jan. 1. (See Chapter 4 for more on the keeping of time and the definition of "epochs".)

The celestial positions of several important objects at the two standard epochs are shown in Table 1.

Why equatorial coordinates?

Why do astronomers use the continually changing equatorial coordinates? In early telescopes, the mechanical rulings that indicate the two angular scales, right ascension and declination, are fixed to the mount which is fixed to the earth; thus the mechanical pointers (attached to the movable telescope) show the pointing direction relative to the mount and hence relative to the earth.

Traditionally the telescope has been mounted so one of its rotation axes is aligned parallel to the earth's rotation axis and the other perpendicular to it. This permits a star to be tracked throughout the night with rotations about only one axis, the one parallel to the earth's spin axis. The indicator for telescope rotation angle about this axis gives the angle of the pointing east or west relative to the meridian that passes through the zenith. This angle is called the *hour angle* and is typically indicated in units of "hours". With knowledge of the time of day, this angle yields the right ascension of the pointing direction.

The indicator for the other axis reads the pointing angle relative to the earth's spin axis, or equivalently relative to the earth's equator. This is the declination of the pointing direction for the epoch of the observation, e.g., epoch 2002.50 for the ~2002 July 1. Conversely, if one wishes to use the mechanical indicators to point a telescope toward a given (faint) star on this date, one must know or calculate its epoch 2002.5 celestial coordinates and take into account the time of day.

This rationale for the use of equatorial coordinates is becoming irrelevant for the many computer-driven telescopes. Nevertheless the system shows no signs of being superseded. Conversions of coordinates from epoch to epoch are becoming trivial; they are routinely carried out on the computers that control telescope pointing at most observatories. Thus, if you know the position of the desired star in any epoch, e.g., B1900, the computer will point the telescope there on your observation date. Computers remove the pain associated with this cumbersome system, so why change it?

Galactic coordinates

Another coordinate system in common use is the *galactic coordinate system*. In this system, the equator on the celestial sphere is defined to be a great circle that runs along the Milky Way.

A schematic of the Galaxy is shown in Fig. 3. It is a disk-shaped cluster of some 10^{11} stars. The visible stars tend to cluster in *spiral arms*, and the sun is one unimportant star well removed (~25 000 LY) from the center. The central plane of the Galaxy contains a high concentration of stars. It thus appears to us as a band in the sky that (more or less) follows a great circle on the celestial sphere. It is this great circle that is chosen to be the equator for the galactic coordinate system as illustrated in Fig. 4.

The two coordinates used to define a celestial position are *galactic longitude* (ℓ) and *galactic latitude* (b). These are analogous to latitude and longitude on the earth. The angles are measured in degrees with latitude increasing toward the north and longitude running from $0°$ to $360°$ in the direction shown (counter-clockwise when viewing from the north galactic pole), with zero in the approximate direction of the center of the Galaxy. This is also a right-handed system. One often sees the

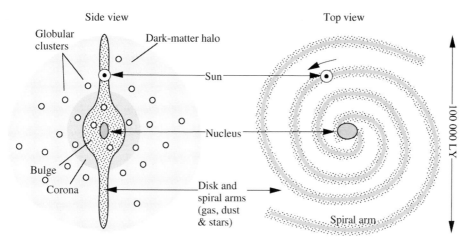

Figure 3.3. Simplified sketches of our Milky Way system of stars showing spiral arms of increased matter density, the bulge, the corona, a dark matter halo, globular clusters, and the position of the sun. The stars, gas, and dust rotate in the counter-clockwise direction about the center. As the dust and gas pass through the slowly rotating spiral arms, shocks cause new star formation to take place. [Adapted from Abell, *Exploration of the Universe,* 3rd Ed., Holt Rinehart Winston, 1975, p. 484]

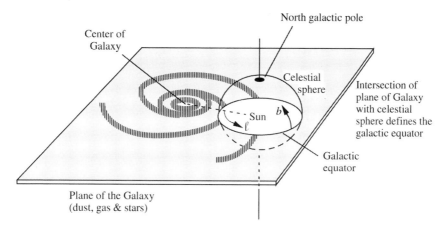

Figure 3.4. Galactic coordinates. The plane of the Galaxy defines the galactic equator on the celestial sphere. The angles that specify the location of a celestial body are measured from the sun. Galactic longitude, ℓ, is measured approximately eastward from the direction of the galactic center in units of degrees ($0°$ to $359.9°$) as shown. Galactic latitude, b, is measured in degrees ($0°$ to $\pm 90°$) from the galactic equator, similar to latitude on the earth's surface. The north galactic pole (dark circle) is shown. The celestial sphere is quite small in this figure; in fact, its radius is infinite. The earth observer is located close to the sun.

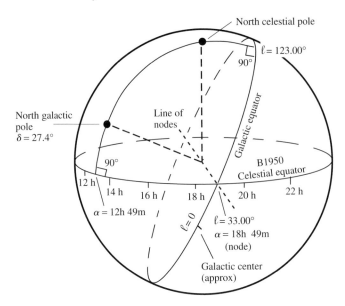

Figure 3.5. Celestial sphere showing the celestial (B1950) and galactic equators, their north poles, and the line of nodes.

superscript "II" applied to the ℓ and b to designate the modern second definition of the galactic coordinates. We omit the "II" in this text as there is no longer any ambiguity; the "II" system has been in use for several decades.

The galactic system was defined by the International Astronomical Union (IAU) in terms of B1950.0 equatorial coordinates as follows (Fig. 5). The north galactic pole (NGP) is defined to be precisely at

$$\alpha_{\mathrm{NGP}}(\mathrm{B1950}) = 12\,\mathrm{h}\,49\,\mathrm{m} = 192.25° \quad \text{(North galactic pole)} \qquad (3.1)$$

$$\delta_{\mathrm{NGP}}(\mathrm{B1950}) = +27°24' = +27.4° \qquad (3.2)$$

The values given are not rounded off; they are exact values which implies that the celestial and galactic equators are tilted relative to each other by $90-27.4 = 62.6°$.

The usual convention in such a coordinate transformation is to assign zero of longitude to an intersection of the two planes (the *ascending node*), but this is not at the galactic center. Accordingly, the zero is set such that the galactic longitude of the north *celestial* pole (NCP) is precisely at

$$\ell_{\mathrm{NCP}} = 123° \qquad (3.3)$$

which places the zero of galactic longitude near, but not exactly at, the galactic center. The two great circles intersect at two *nodes*. The *line of nodes* (Fig. 5) is

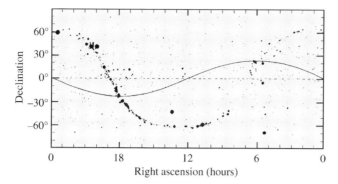

Figure 3.6. Distribution of discrete radio sources plotted in equatorial coordinates, in continuum radiation near $\nu = 1420$ MHz. Larger diameters are brighter sources. The brightest sources lie in the galactic plane and dramatically delineate the galactic equator. [Courtesy G. Verschuur]

the axis of rotation that transforms one plane to the other. It then becomes apparent that the two equators cross (ascending node) at

$$\ell_{node} = 123 - 90 = 33° \qquad\qquad\qquad\qquad \text{(Node)} \qquad (3.4)$$

$$\alpha_{node} = 12\,\text{h } 49.0\,\text{m} + 6\,\text{h} = 18\,\text{h } 49.0\,\text{m} = 282.25° \quad \text{(Node)} \qquad (3.5)$$

This specification of the relation between the two coordinate systems leads to formulae for conversion of B1950 equatorial coordinates α, δ to galactic coordinates ℓ, b and vice versa.

The significance of the galactic coordinate system can be seen from a map of discrete radio sources in *equatorial coordinates* (Fig. 6). This plot represents a sphere projected onto a flat surface. The great circle of the *galactic* equator is clearly seen as a sinusoidal curve of bright radio sources. Clearly these sources are associated with the Galaxy. It therefore is often convenient to use galactic coordinates. In such a plot they would lie along the equator as in the cover illustrations. The general background of discrete sources in Fig. 6 represents an isotropic distribution of sources, mostly (distant) extragalactic sources.

Ecliptic coordinate system

A third coordinate system on the celestial sphere is the *ecliptic coordinate system*. It is used less often but is very convenient for planning observations from an earth-orbiting satellite that must keep its solar panels pointed toward the sun. In this system, the equator is simply the path on the sky that the sun follows, i.e., the ecliptic. The plane of the ecliptic lies $23.4°$ from the celestial equator (Fig. 1). The ecliptic latitude of a star directly indicates the closest that the sun can come to the source; the ecliptic longitude indicates directly the date (month and day) that the sun is at this closest point.

Reference frames

An important complement to a celestial coordinate system is a catalog of reference stars with precisely measured positions. These are equivalent to surveyors' permanently placed markers (bounds) in your neighborhood which are used for later detailed surveys of particular lots, etc. Without them every survey would have to work all the way from some basic reference possibly miles away. These permanent bounds would not be necessary if there were lines of longitude and latitude painted all over the earth, say every 100 meters or so.

Similarly, the sky is not marked with J2000.0 lines of constant right ascension and declination. Thus a set of well measured star positions and proper motions is invaluable for the determination of precise coordinates of stars in a local region. The proper motions (rate of change of α and δ in units of milliarcseconds per year (mas/yr) allow one to calculate precise positions for a number of years after the measurements. The systems of well measured stars have become of higher and higher quality as technology improves. The FK4 catalog adopted in 1976 contained the precise equatorial coordinates and proper motions of 3522 stars. The newer FK5 catalog with fainter stars and more accuracy gives its positions and motions in equatorial celestial coordinates J2000.0.

The current (since 1995) reference frame is the *International Celestial Reference System* (ICRS). It is based on the extremely accurate positions, $\sim \pm 0.5$ mas, of about 250 extragalactic radio sources. Radio observations with widely separated telescopes yield these great precisions. The great distance of extragalactic sources ensures that their angular motions (proper motions) on the sky will be so small as to be undetectable. In contrast the optical stars in the FK4 and FK5 catalogs are stars in our part of the Galaxy; they do show proper motions.

More recently, the *Hipparcos* earth-orbiting satellite (1989–1993) measured the positions and proper motions of 118 000 optical stars with high precisions, ~ 0.7 mas and ~ 0.8 mas/yr respectively. The positions were determined in the Hipparcos Reference Frame. This frame was aligned to the radio ICRS frame with high precision through ground-based optical position measurements of the extragalactic radio sources. Since the entire sky contains $\sim 40\,000$ sq. deg (see below), the Hipparcos catalog contains, on average, ~ 3 very well-located stars per square degree on the sky. The positions are in epoch J1991.25 but these can easily be precessed to J2000.0 or any other epoch. If you discover an important optical object in the sky, you can measure its location relative to several nearby Hipparcos stars in order to get its precise celestial coordinates.

Transformations

The several coordinate systems are simply redefinitions of the coordinates on the same celestial sphere. In all cases the stars are defined on a two-dimensional surface with two angles, a latitude and a longitude. One can convert the coordinates of a given star from one coordinate system to another with standard spherical coordinate transformation formulae. As noted the galactic coordinate system was defined in terms of the B1950 equatorial coordinates, α (B1950.0) and δ (B1950.0). According to this definition, the celestial (B1950.0) to galactic transformations (and vice versa) are

$$\cos b \cos(\ell - 33°) = \cos \delta \cos(\alpha - 282.25°) \qquad (3.6)$$

$$\cos b \sin(\ell - 33°) = \cos \delta \sin(\alpha - 282.25°) \cos 62.6° + \sin \delta \sin 62.6°$$
$$(3.7)$$

$$\sin b = \sin \delta \cos 62.6° - \cos \delta \sin(\alpha - 282.25°) \sin 62.6° \qquad (3.8)$$

$$\cos \delta \sin(\alpha - 282.25°) = \cos b \sin(\ell - 33°) \cos 62.6° - \sin b \sin 62.6°$$
$$(3.9)$$

$$\sin \delta = \cos b \sin(\ell - 33°) \sin 62.6° + \sin b \cos 62.6° \qquad (3.10)$$

where the angles in the formulae are exact (not rounded off).

The first two of these equations yield values for b and ℓ, given values of α and δ. However, there remains an ambiguity as to the quadrant of angle b. The third equation resolves this. Recall that b can reside in only two quadrants since it ranges only from $-90°$ to $+90°$. Similarly, the final two equations yield values for α and δ, given values of b and ℓ. In this case, quadrant ambiguity remains for α which can be resolved with (6). Similar transformations allow one to convert between equatorial coordinate systems for different epochs. For precise telescope pointing, one must further correct the coordinates for parallax and aberration (Chapter 4).

3.3 Solid angle on the celestial sphere

The concept of *solid angle* Ω is fundamental to all of astronomy. It is simply an "angular area" on the sky, or equivalently, on the celestial sphere. This area can be expressed as "square degrees" or "square radians"; the latter unit is called the *steradian*.

The solid angle is expressed in terms of two angular displacements, e.g., $d\theta$ and $d\phi$ in Fig. 7. The beam of an antenna, e.g., $1° \times 1°$, or $1'' \times 2''$, or a "fan beam" of $1° \times 100°$ can be characterized by its solid angle, approximately 1 deg^2, 2 arcsec^2, and 100 deg^2 for these cases respectively.

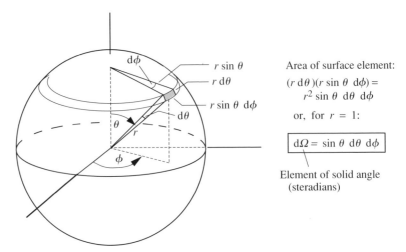

Figure 3.7. The concept of *solid angle*, Ω, which is the "angular" area on a sphere. The element of solid angle dΩ (hatched area) can be expressed in terms of steradians (or "square radians") which is numerically equal to an element of surface area on a sphere of unit radius. Integration over the entire sphere yields a total solid angle of 4π steradians.

The unit of solid angle, the steradian (sr), is a dimensionless quantity of magnitude 1 rad $\times$ 1 rad where 1 radian $= 360/(2\pi) = 57.3°$. The equivalent number of square degrees is

$$1.0 \text{ sr} \equiv \frac{360}{2\pi} \times \frac{360}{2\pi} = (57.296)^2 = 3282.8 \text{ deg}^2 \quad \text{(Unit of solid} \atop \text{angle)} \qquad (3.11)$$

We refrain from saying that a region of 1 rad $\times$ 1 rad on the celestial sphere has a solid angle of 1.0 sr. Such a statement is incorrect because the celestial sphere is not a flat surface.

The correct statement is that the solid angle Ω is the area of a region on an imaginary sphere divided by the square of the radius r of the sphere, or equivalently, it is the area A_1 on a sphere of unit radius,

$$\Omega = \frac{A(\text{m}^2)}{r^2} \underset{r=1}{\to} A_1 \qquad \text{(sr)} \qquad (3.12)$$

In general, one must integrate over the region on the imaginary sphere to obtain the correct solid angle. For a complete sphere, the result is simply, from (12),

$$\Omega = \frac{4\pi r^2}{r^2} \underset{\substack{\text{entire} \\ \text{sphere}}}{\to} 4\pi \text{ sr} \qquad (3.13)$$

This provides a more basic definition of the steradian. It is $1/4\pi$ the area of a sphere of unit radius.

For small solid angles ($\lesssim 100$ deg^2 = 0.03 sr), the approximate solid angle in steradians is obtained simply by calculating the area on the sky, in square radians, as if the sky were a flat piece of paper. Thus a square $1° \times 1°$ region on the sky would have solid angle close to $(1/57.3)^2 = 3 \times 10^{-4}$ sr. For solid angles that are this small, the approximation is excellent.

Integration to find a solid angle that is a substantial portion of the celestial sphere makes use of the differential increment of solid angle $d\Omega$ shown in Fig. 7 as a small shaded rectangular area. The solid angle is simply the product of two angular displacements each of which is a linear displacement on a sphere of unit radius. From Fig. 7,

➡ $\quad d\Omega = \sin\theta\, d\theta\, d\phi \quad$ (sr; differential solid angle element) $\qquad$ (3.14)

This can be integrated over any desired portion of the sphere to obtain the solid angle, keeping in mind that the limits on θ, ϕ must be given in radians. The integration is not difficult for areas with certain symmetries so that one integration variable can be held constant while the other varies. An example is the calculation of the solid angle of a polar cap of $30°$ radius ($\theta_{max} = 30°$). For arbitrary shapes, piecewise summation may be required.

The solid angle of the entire celestial sphere can be calculated as follows. Apply the definition of the element of solid angle (14),

$$\Omega_{\text{sphere}} = \int_{\text{sphere}} d\Omega = \iint_{\text{sphere}} \sin\theta\, d\theta\, d\phi \qquad (3.15)$$

$$= \int_0^{2\pi} d\phi \int_0^{\pi} \sin\theta\, d\theta \qquad (3.16)$$

➡ $\quad \Omega_{\text{sphere}} = 2\pi(-\cos\theta)|_0^{\pi} = 4\pi \text{ sr} \qquad (3.17)$

Thus, the solid angle of the entire celestial sphere is 4π sr in agreement with (13). In units of square degrees, from (11),

$$\Omega_{\text{sphere}} = 4\pi \text{ sr} = 41\,253 \text{ deg}^2 \qquad (3.18)$$

It is important to have a sense of solid angle. Star and galaxy numbers are often quoted as the number per steradian. The number of a certain object one might find in a piece of the sky depends on the size of the solid angle searched. Solid angle also is the basis of the formal description of the *brightness* of diffuse objects described in Chapter 8.

3.4 Surveys, charts, and catalogs

Astronomers must be able to refer to a given star or other celestial object after
studying it. This can be done with the star coordinates or with a name. Our knowledge of the existence of the fainter stars in the optical sky derives from surveys of
the sky, such as a series of large-area photographs or the counting rate data from an
x-ray detector that scans the entire sky. Such surveys typically yield the celestial
coordinate of each located object. Interesting objects of a given type that are found
in such surveys can be plotted on maps of the sky (*charts*) or listed with positions
and other information in printed *catalogs*.

Photographs and charts

Palomar, SRC, and ESO sky surveys

The true comprehensive charts of the faint stars in the sky at optical wavelengths
are actually deep (i.e., sensitive) photographs of the sky. The first and most famous
of these is the Palomar Observatory Sky Survey (POSS-I) carried out in the early
1950s. It consists of 936 pairs of large glass plates (350 mm × 350 mm) taken with
the large Schmidt telescope at Palomar Mountain in California. The plates cover
declinations +90 to −30. At each position, a red-sensitive and a blue-sensitive
plate were exposed. The glass plate provides stability against bending and thermal
expansion. Each plate covers a large portion of the sky, about 6.5° × 6.5°, and shows
stars as faint as ∼21st magnitude. (The brightest stars have magnitude $m \approx 0$, and
the faintest stars visible to the naked eye on a dark night have magnitude $m \approx 6$.
See Section 8.3 for more.)

The red and blue plates were taken with specific types of emulsion and filters.
The developed plate is a black and white photograph, but the relative "brightnesses"
of the stellar images on a given plate give the relative star-to-star intensities in
the color band of the plate. Comparison of the intensities of a given star on two
different plates, i.e., in two different color bands, provides a good indication of the
temperature of the object. A hot object will be relatively more intense in the blue
than in the red, as compared to a cool object which will be more intense in the red
than in the blue. The shape of images on such photographs also conveys a great
deal of information. Point-like objects (stars) and diffuse nebulosities (supernova
remnants, globular clusters, galaxies, etc.) are seen in abundance on such plates.

In the recent decades, similar surveys with somewhat better sensitivity have been
carried out in the southern hemisphere, at observatories in Chile and Australia.
(ESO-R and SRC-J surveys). A second, more sensitive Palomar survey (POSS-II)
of the northern sky is now being completed at Palomar and digitized at the Space
Telescope Institute.

Astronomical libraries have collections of paper prints of these plates, and digitized versions are now available on compact disks. These are produced with machines that either scan or photograph digitally. A reproduction of the original photograph can then be produced in a computer from these data. Computer analysis of the digitized reproduction can yield celestial positions (α, δ) and star colors of each object. From this information one can generate a catalog. The POSS-I/SRC/ESO survey has yielded a catalog of 526 million stars, the US Naval Observatory (USNO) catalog. The USNO also provides the digitized images.

Often a star or other object will reside anonymously on one of these sky photographs or charts for many years until it suddenly becomes of interest to astronomers. This interest arises because it is found to have new or unusual characteristics. This new-found fame can be due to careful surveys of the optical sky with different or more sensitive instrumentation. Often, observations in *other* wavelength bands, e.g., radio, infrared, ultraviolet or x-ray, will call attention to a previously anonymous optical object.

Quasars, for example, are point-like objects that appear to be ordinary stars on all but the highest quality images. They first attracted attention because of their radio emission. Quasars are now also discovered in other frequency bands such as the optical and x-ray and are known to be the very active nuclei at the centers of galaxies. Neutron-star binary systems were first discovered in the x-ray band. In many such cases, when the celestial position of the radio, infrared, or x-ray object is known to sufficient precision, an examination of the optical survey plates will reveal the optical counterpart to be a star, a galaxy, or other interesting object that can then be studied at many wavelengths.

The Schmidt plates are very sensitive to nebulosities because the telescopes have a short focal length and large aperture (see Section 5.3 for definitions). A large solid angle of sky is compressed onto a single plate; hence a single grain of film receives a lot of energy from a diffuse object. The prints are quite beautiful and fascinating when examined by eye and with a small magnifier. They reveal dramatically the complexity and richness of the sky, e.g., the Pleiades in Fig. 1.7a.

Once an interesting astronomical object is identified on a survey plate, astronomers usually wish to study it in detail with powerful astronomical instruments mounted on a telescope, such as a spectrometer that disperses the light to reveal its spectrum. To do this, they need the equatorial coordinates of the object so the telescope can be properly pointed. The celestial coordinates of a faint object may be uncataloged. One can measure its physical (x, y) position on a survey photograph along with the x, y positions of nearby reference stars that have precise cataloged celestial coordinates. One can then solve to obtain the celestial position (α, δ) of the unknown object.

A less precise, but simple, measurement technique is to make use of the plastic transparent overlays that are provided (by Ohio State Univ.) for the Palomar Schmidt plates. The overlay for a given plate contains a coordinate grid (black lines) and also crosses at the locations of the brightest stars. If one places the overlay precisely over a paper print of the plate, using the bright-star positions for alignment, the celestial position of a previously uncataloged object may be obtained to about $20''$. Today, this can be done on the computer with the scanned (digitized) images.

Finding charts

Astronomers traditionally went to telescopes with less precise, $\sim 1'$, celestial positions for a couple of reasons. It was quite an onerous task to obtain a very precise $1''$ position of a previously unmeasured star. Also, many telescopes do not point to a position on the sky more accurately than about $1'$. In these cases, astronomers point the telescope in the approximate direction of the star and then use a *finding chart* to locate the exact star they wish to study.

A finding chart is simply an enlarged photographic print of the sky (e.g., a plate scale of $4''$/mm) with the star of interest indicated by a pen mark. When the telescope is pointed in the approximate direction of the desired star, the astronomer looks through the eyepiece (or nowadays, at the TV monitor) to examine the stars in the field of view of the telescope. The pattern of stars should look similar to those on the finding chart, and the star of interest can then be identified. The telescope is then pointed more precisely so the light from this star falls precisely on the correct position for measurement, for example, the entrance slit of a spectroscope. Data accumulation can then begin.

The astronomical literature (and now the internet) abounds with finding charts; each is simply a photo of a piece of the sky with the star or galaxy of interest indicated. With such a chart, other astronomers can tell which star the author has studied. For example, the photo of the Crab nebula with the Crab pulsar marked (Fig. 1.3) would serve as a finding chart for the pulsar. These photographs made the astronomical literature more attractive and fun to peruse. One can see the beauty of the sky while sitting in the library. Nowadays, of course one does all this on the internet, and in color too!

Printed charts

Charts, or maps, of the sky historically complement photographs of the sky. These are printed maps that are produced from the measured positions and brightnesses of the objects of interest with the brighter stellar objects shown as larger dark circles. In times past, it could be a large project to make such plots. Now with computers and a file listing of star coordinates and intensities, one can quickly generate charts of any desired portion of the sky or of the whole sky. A chart of the POSS/SRC/ESO survey could in principle be created, but it would be no more valuable than the reproduced digitized images one can now call up at will.

Historically, charts of the sky have been a staple of astronomy; perusal of old charts is intriguing indeed. A modern useful set of charts for the amateur astronomer is *Norton's 2000.0 Star Atlas*. It includes stars down to 6.5 magnitude on 14 charts, a total of ~8700 objects. This convenient atlas contains all the stars visible to the naked eye and includes much useful information. It serves as finding charts for the brighter stars also.

Charts can also be made from catalogs of other types of astronomical objects, e.g., radio sources, infrared sources, x-ray sources, etc. An example is the x-ray sky shown as the large background illustration on the front cover of this text.

Catalogs of celestial objects

The brightest of objects or those that exhibit characteristics of particular interest are often cataloged in a book, journal, or in computer-readable form. There are numerous catalogs or lists of different kinds of stars (e.g., white dwarfs, emission-line stars, variable stars, etc.) and of interesting extragalactic objects (Seyfert galaxies, BL Lacertae objects, quasars, etc.). Other catalogs contain a mix of different types of objects that were detected in a survey of the sky by a single instrument. Catalogs typically list the intensity (brightness), the equatorial coordinates, the type of object (if known), special characteristics (e.g., the proper motion, the spectral type and/or broad band colors, or period of a pulsar), and references to other work on the object.

The goal of understanding the cosmos requires the classification of the types of objects in it. Catalogs of objects of a given type allow astronomers to study the class of objects as a whole to ferret out the extent of their characteristics and behavior.

A standard catalog of bright stars is the *Yale Bright Star Catalog* and its Supplement, now in its 5th Edition. Together, they contain about 11 700 stars down to 7.1 magnitude. It is the basis of the Norton star charts described above. The *Smithsonian Star Catalog 2000* contains 260 000 stars down to about 9th magnitude. The contemporary *Hipparcos Catalog* lists coordinates to ~1 mas and proper motions for 118 000 stars mostly brighter than ~10 mag. The *US Naval Observatory–A2 Megastar Catalog* was obtained from the digitizing scans of the POSS/SRC/ESO Schmidt plates. It reaches down to 20th magnitude and lists the positions of 526 million stars with ~0.2″ precision. In the infrared, a ground-based survey at wavelength 2 μm has yielded the *Two-Micron All Sky Survey (*2MASS) catalog of 20 million objects. Stars known to have variable brightness are listed in the *General Catalog of Variable Stars* with 31 918 entries in 1998 (Edition 4.1).

In the past, catalogs were created from hand measurements of individual objects, and this limited the number of objects one could practically list. The huge numbers of objects in current catalogs has become possible only recently with automated computer data acquisition, measurements, and listing. Rather than books, catalogs are now distributed on compact disks, or simply made available on the internet.

Objects that appear to have finite extent on a photographic plate are called nebulae. Some nebulae cataloged in early times are now known to be galaxies external to the Galaxy. Nebulae were first listed by Messier (1781). His list has 110 items and includes both galactic and extragalactic objects according to our present knowledge. (See *Norton's 2000 Star Atlas*, p. 156.) Messier made his list to keep track of objects that could be confused with comets and thereby would slow him in his competitive searches for new comets. Today the Andromeda nebula (our sister galaxy) is known as Messier 31, the 31st entry on Messier's list. It is also known as NGC 224 because it is listed as #224 in a much more complete *New General Catalogue of Nebulae and Clusters of Stars* published by J. Dreyer in 1888 or its two supplements, *Index Catalogue* (IC; 1895 and 1908). Currently one can refer to the *Third Reference Catalog of Bright Galaxies* (*RC3*) with 23 022 galaxies and the *APM Bright Galaxy Catalog* of the Southern Sky which lists 14 681 galaxies. The *First Byurakan Survey* lists 1469 galaxies with strong ultraviolet continuum which is a characteristic of active galactic nuclei (AGN).

Catalogs of the radio sky include the *Fourth Cambridge Radio Survey* (4C) which lists 4844 northern sources at 178 MHz, two California Institute of Technology lists at 960 MHz (CTA and CTB), and the Australian Parkes Observatory survey of the southern sky with 8264 sources at several frequencies between 178 MHz and 22 GHz. A survey of the x-ray sky by the HEAO-1 satellite in 1977–8 yielded a catalog of the brightest x-ray sources at 1–20 keV; it contains 842 sources and is the origin of the x-ray cover chart. A soft (<2 keV) x-ray survey of the sky has been carried out by the German ROSAT satellite; it yielded $\gtrsim 50\,000$ sources.

Many catalogs are maintained on the internet for use by astronomers together with tools for using them. The Strasbourg Astronomical Data Center (CDS) is a repository for about 4000 of them. There are so many that there are actually catalogs of catalogs! Try http://cdsweb.u-strasbg.fr for the VizieR catalog service and the Simbad service which provides coordinates and references for your favorite celestial object.

Names of astronomical objects

The catalogs must give a name or number to each entry. In many cases this number becomes the name of the object. If the object was evident and conspicuous to early civilizations it probably has a historical name given in classical times. Subsequent catalogs can result in new names for the old objects, e.g., M 31 = NGC 224 as noted above. Thus the names of astronomical objects are rich and diverse; they derive from many researchers working in a variety of frequency domains (radio, optical, etc.). Many names derive from the celestial constellation in which the object resides.

Constellations

The constellations are 88 regions on the sky noted for particular arrangements of the stars. The origins of the names of the constellations in large part are lost in antiquity. Over the centuries, new constellations have been defined and other large ones broken up into smaller ones. The most recent activity of this sort applied mostly to the southern sky which until recently had been less familiar to northern cultures. The boundaries of the constellations were not specified with care until 1930 when the International Astronomical Union (IAU) defined constellation boundaries that are fixed among the stars and are in accord with the traditional constellations. The boundaries were chosen to follow the lines of constant right ascension and declination for epoch 1875. Due to the earth's precession, these now deviate somewhat from the current lines of constant right ascension and declination, but they do remain fixed relative to the stars and continue to specify the constellation boundaries.

The most well known constellation names are the 12 signs of the zodiac, well known to horoscope readers. These are the constellations that lie along the ecliptic through which the sun passes on its annual cycle, but not at the times suggested by horoscopes (See Section 4.3).

Stars

The brightest stars have classical or medieval names like Sirius, Castor, Diphda, etc. A system by Mayer in 1603 named stars according to the constellation within which they reside, the brightest being given the first letter of the Greek alphabet, α, the second β, etc., more or less in order of brightness. The stars Sirius, Castor, and Diphda are thus also known as α CMa (alpha Canis Majoris), α Gem (alpha Geminorum), and β Cet (β Ceti). These names are the genitive forms for the constellations Canis Major (big dog), Gemini (twins), and Cetus (whale).

When the Greek alphabet was exhausted in a given constellation, small roman letters were used (b Cygni) and then capital letters, (A Cygni) up through the letter Q. A general star catalog by Flamsteed published in 1725 named faint stars that had no Greek-Roman name with numbers, e.g., 61 Cyg. For the most part, these were ordered by increasing right ascension within each constellation. Precession of the earth since 1725 causes this ordering to not be strictly preserved in today's coordinates.

Variable stars have been cataloged by constellation with a similar naming convention, but with the unused capital letters R, S, T, etc. (R Cyg). When the single capital letters were exhausted, double capital letters were invoked. For example, the object AM Her (Herculis) is the first-discovered (prototype) example of a highly magnetic binary stellar system. It consists of a very compact star (a white dwarf) that is accreting gas from a companion star. Subsequent star systems of the same type are given their own names, but, as a class called *AM Her-type* objects. Variable star objects are added to the General Catalog of Variable Stars as they are discovered and reported in the literature.

Modern names ("telephone numbers")

The earliest radio detections of the brightest discrete sources in a given constellation were named after the constellation with the suffix A, B, etc. in order of descending intensity. Thus Taurus A, and Sagittarius A, Cyg A are, respectively, the radio sources associated with the Crab nebula, the center of the Galaxy, and a distant active galaxy with luminous radio jets emerging from it.

The sources in modern catalogs are often listed in order of right ascension (with the epoch specified) with no heed paid to constellations. The name given the source is very likely to be something exciting like 1956 + 350, another name for Cygnus X-1, a bright x-ray source that was the first to provide evidence for the existence of black holes. The digits often refer to the epoch B1950 celestial position, α (B1950) = 19 h 56 m, δ (B1950) = +35.0°. Many important objects are known by their seven-digit name, jokingly called a "telephone number".

The original Twin quasar, an example of gravitational lensing, is often called QSO 0957 + 561, where the label stands for Quasi Stellar Object, an early name for quasars. The famous binary radio pulsar that is losing energy to gravitational radiation is known as PSR 1913 + 16 where the label stands for "PulSaR". In other cases, the number of the source in a catalog will become the commonly used name. The first discovered quasar is always called 3C273 because it is the 273rd source listed in the *Third Cambridge Catalogue of Radio Sources*. The 85th cluster of galaxies listed in the catalog prepared by Abell is commonly called Abell 85.

The telephone numbering system was initiated when the epoch B1950 was in use. Unfortunately, in epoch J2000 coordinates, the familiar telephone numbers of our favorite sources would change. It would be like changing the street number in your house address just because someone changed the coordinate system in your town.

The solution is that the sources that were labeled in the former epoch keep their old names, possibly with a "B" precursor, e.g., QSO B0957 + 561, to indicate epoch B1950.0 while sources newly discovered or recently found to be interesting are given names based on their epoch J2000 coordinates. An example of this is the x-ray and radio source XTE J1748-288, a binary system probably containing a black hole. X-ray astronomers still assign prefixes that indicate the satellite that made the discovery of the object, e.g., "XTE" for the Rossi X-ray Timing Explorer.

It helps x-ray astronomers remember the object, but prefixes that identify the type of object would probably be more helpful to other astrophysicists.

Problems

3.2 *Coordinates on a celestial sphere*

Problem 3.21. (a) What are the approximate equatorial coordinates of the north celestial pole, of the sun on March 21, the star shown in Fig. 1, and the nominal galactic center at $\ell = b = 0$? In each case, specify the epoch of your answer. What are the J2000 coordinates of the Crab nebula? (b) Locate *Norton's 2000.0 Star Atlas* in the library and find the J2000.0 coordinates (to the nearest arcminute) of the variable and bright star γ Cassiopeiae, the Seyfert galaxy NGC 1068, the variable star η Carinae, and the globular cluster M 30? You can measure from the charts or use the "2000" coordinates listed in the several tables of the Atlas; the Atlas suppresses the "J" prefix. [Ans. (a) galactic center $\alpha \approx 18$ h, $\delta \approx -30°$]

Problem 3.22. (a) Plot the following objects on a copy of Fig. 5 and determine by inspection the galactic coordinates of the objects as best you can (e.g., to about 20° precision, though in some instances one can do better): the Crab nebula at α (B1950) = 5 h 32 m, δ (B1950) = $+21°$ 59′, the radio galaxy M87 (12 h 28 m, $+12°$ 40′), the nearby Andromeda galaxy M31 ($+0$ h 40 m, $+41°$ 00′), and the central star in Cassiopeia called Gamma Cas = γ Cas (00 h 54 m, $+60°$ 27′). (b) Program your calculator (or the equivalent) with Eqs. (6)–(10) to obtain accurate values (to about 0.1° precision) of the galactic coordinates of the above objects. Pay attention to the signs of your trig functions. (c) If an astronomical plastic globe with celestial equatorial coordinates indicated on it is available to you, determine the above answers by making measurements (with a piece of string) along the surface of the globe. Your answers should be good to about 2 or 3 degrees.

Problem 3.23. (a) Make a sketch of a celestial sphere with the ecliptic as the equator. Indicate the path followed by the sun. Show the track scanned by the view direction of an x-ray detector on a satellite in space if its view direction is normal to the sun direction and it rotates once per hour about the axis directed toward the sun. The spin axis remains pointed toward the sun as the sun moves along the ecliptic. The field of view (FOV, full width) is circular with diameter 2°. (b) What happens to the scan track as the year progresses? Approximately how many times a year is a celestial source transited by at least some part of the FOV if the source lies on the ecliptic equator, or if it is at ecliptic latitude $+80°$? For some transits, the view will be blocked by the earth; assume this never happens. [Ans. (b) ~100, ~600]

3.3 Solid angle on the celestial sphere

Problem 3.31. (a) Derive an expression for the solid angle of one polar cap on the celestial sphere, of angular radius θ. (On the earth, the polar cap could be encompassed by the arctic circle.) What is the *fraction* of the entire sky that is contained in this polar cap of radius θ? What is this fraction if $\theta = 30°$? (b) In the limit of small θ, i.e., $\theta \ll 1$, show that the solid angle is simply the geometric flat-space "area" $\pi\theta^2$ measured in square radians. (c) Use this result to recalculate the solid angle of the $30°$ polar cap. By what fraction is it in error? (d) Repeat (a) and (c) for $\theta = 1°$ and for $\theta = 60°$. [Ans. $\sim 7\%$; —; $\sim 2\%$; $\sim 10^{-3}$ sr and 0.003%]

Problem 3.32. (a) What is the approximate solid angle of the sun as viewed from the earth, in square arcmin?, square degrees?, steradians? The sun's angular diameter is $32'$. (You may use the geometric flat-space approximation.) (b) Estimate the solid angle (in steradians and sq. arcsec) of the sun-like star α Centauri of radius $\sim 7 \times 10^8$ m, which is at a distance of 4.4 LY. [Ans. $\sim 10^{-4}$ sr; $\sim 10^{-15}$ sr]

3.4 Surveys, charts and catalogs

Problem 3.41. Compare the problems you would encounter in trying to keep track of the positions of all the trucks in the United States as time progresses with the problems facing astronomers in keeping track of all stars as time progresses.

Problem 3.42. Use *Norton's 2000.0 Star Atlas* as a finding chart for the bright naked-eye objects. Locate a constellation of interest to you, e.g., Orion, on the star chart and pick out a faint 5th or 6th magnitude star near one of the bright stars on the sky map. Try to find it with your naked eye or with binoculars. Do this on a dark night (moon set) away from city lights if possible, and be sure to let your eyes become dark adapted. (Use a faint red-light flashlight when examining the catalog.) Find the Orion nebula, labeled 42^M for Messier 42, in the sword of Orion. Can you see its nebulosity with binoculars? Mount your 35-mm camera on a tripod and take several time exposures with the focus at infinity and the aperture wide open. Try exposures: 1 s, 10 s, 60 s, 90 s, 2 min, 3 min, 5 min, 10 min. Can you see more or fewer stars in the photos than with binoculars? Do the lengths of the tracks in the prints match the rotation rate of the earth?

4

Gravity, celestial motions, and time

<table>
<tr><td>

What we learn in this chapter

Gravity is the underlying reason for the **spin of the earth**, the motions of stars within galaxies, and the evolution of stars and the universe. The **apparent motions** of stars in the equatorial coordinate system arise from **precession** and **nutation** of the coordinate system, from **parallax** and **stellar aberration** due to the orbital motion of the earth about the sun, and from **proper motion**, the projection onto the celestial sphere of the **peculiar motion** of a star relative to the **local standard of rest**. **Precession** and **nutation** of the earth arise from gravitational torques on its equatorial bulge applied by the sun, moon and planets.

The **calendar** is tied to the seasons such that the first day of spring occurs when the sun moving north crosses the (precessing) **vernal equinox**. The non-integral number of days in the **tropical year** (equinox to equinox) was accommodated with the addition of a leap day every 4 yr in Caesar's **Julian calendar** (46 BCE). The **Gregorian calendar** (1542) of Pope Gregory XIII removes some of these leap years to obtain a more precise agreement.

Eclipses of the sun and **moon** are a consequence of the motions of the earth and moon in their respective orbits about the sun and earth. The 18-yr **saros cycles** of lunar eclipses allowed the ancients and early astronomers to predict when they would occur. **Total solar eclipses** are wonderful to behold. They have revealed to scientists the nature of the **solar chromosphere** and **corona**.

Systems of **time** depend upon transits of the sun (**mean solar time**) or upon transits of the stars (**sidereal time**). Today's standard is **Universal Coordinated Time (UTC)** which at midnight at zero longitude (Greenwich England) reads zero hours. It counts seconds of **atomic time** with occasional leap seconds inserted or removed to keep it synchronized with the solar day. Astronomers keep track of time intervals over long periods by making use of **Terrestrial Time (TT)** and **Julian date** which flow continuously in units of seconds and days, respectively, without leap second or leap day insertions. **Precise timing of astronomical events** requires corrections for the relativistic effects due to the location and speed of the earth in its orbit about the sun. **Barycentric Coordinate Time (TCB)** was introduced in 1991 for this purpose.

</td></tr>
</table>

4.1 Introduction

The apparent motions of the celestial bodies on the sky are a consequence of their motion in a three-dimensional inertial space combined with that of the observer. The earth is a platform that spins with one rotation per day, orbits the sun once per year, and precesses with a period of 25 770 yr. The motions of the celestial bodies include the orbiting of the planets about the sun, random motions of stars relative to the sun, and the rotations of stars about the center of the (MW) Galaxy with rotation speed dependent on distance from the center.

This results in considerable apparent motions of the stars and planets from our viewpoint. The driving force for all these motions is gravity; it can truly be said that gravity drives the universe. We therefore begin with a brief discussion of the role of gravity in astrophysics. We will describe how the motions of the moon and earth relative to the sun lead to solar and lunar eclipses. Finally, we conclude with a discussion of time standards which play an essential part in studies of celestial motions.

4.2 Gravity – Newton to Einstein

The motions of the celestial bodies may be understood in terms of two laws articulated by Newton. His *second law* states that a vector force $\boldsymbol{F}_1$ applied to a mass m_1 brings about a vector acceleration $\boldsymbol{a}$ of m_1 according to

$$\boldsymbol{F}_1 = m_1\boldsymbol{a} \qquad \text{(Newton's second law)} \qquad (4.1)$$

if the observer is in an inertial (non-accelerating) frame of reference. His *law of gravitation* gives the force due to gravity,

$$\boldsymbol{F}_{\text{gravity}} = -\frac{Gm_1m_2}{r_{1,2}^2}\hat{\boldsymbol{r}} \qquad \text{(Newton's law of gravitation)} \qquad (4.2)$$

The force on point mass m_1 in the presence of m_2 is proportional to the product of the masses and the gravitational constant $G = 6.67 \times 10^{-11}$ N m^2 kg^{-2}, and is inversely proportional to the square of the distance between the masses. The direction of the force is along the line joining the two masses and the force is attractive. This is indicated with the unit radial vector $\hat{\boldsymbol{r}}$ and the minus sign.

The acceleration of m_1 can be obtained by substituting the force (2) into (1). If m_2 is much more massive than m_1, and if we ignore the effects of other bodies such as planets, the resultant differential equation represents the classical planetary problem which applies quite well to the earth–sun system. The sun is about 330 000 times heavier than the earth. The solution of the differential equation for the earth's motion indicates that the earth should move in an elliptical orbit about the sun, as indeed it does. The orbit is nearly circular, but not quite; it is slightly elliptical. This results in a changing angular size of the sun as viewed from the earth. It is about 3% smaller on July 6 than it is six months later in January.

Gravity and Newton's second law are also responsible for the spinning of the earth, the orbital motion of the planets about the sun, and the formation of the planets from clumps of gas and dust left over from the formation of the sun. This material coalesced under the influence of the *self gravity*; each part of a clump finds itself attracted to all other parts. In this process, gravitational potential energy is converted into heat (infrared radiation), kinetic energy of outflowing material, and rotational kinetic energy of the resultant planet.

The spin of the earth has not been constant since its formation; it is gradually slowing due to gravitational *tidal forces*. The moon and the oceans are mutually attracted to one another by gravity giving rise to oceanic tides. Friction between the oceans and the earth results in a long-term average increase in the length of the day of $\sim$0.7 ms per century. Other irregular effects (ocean currents, atmospheric motions, etc.) contribute comparable and unpredictable changes to this slowdown rate, for a long-term average of $\sim$1.4 ms per century.

Gravity similarly drives the formation of stars like the sun. It also is the force that causes stars to burn atomic nuclei to provide the radiation pressure against further collapse. Without the inward pull of gravity and the conversion of gravitational potential energy to kinetic energy, the centers of stars would not have the pressure and temperature necessary to sustain nuclear burning. When the nuclear fuel is expended at the center of a star, it will collapse to a white dwarf, neutron star, or black hole, again under the influence of gravity. The resultant supernova explosion (in the collapse of a stellar core to a neutron star) thus is powered by gravity.

On larger scales, the motions of stars and gaseous clouds in galaxies are clearly dominated by gravity, as are the motions of individual galaxies in clusters of galaxies. In fact, there appears to be an excess of gravitational force on these large scales; the current view is that an unknown type of *dark matter* is the source of this excess force.

On the largest scale of all, gravity runs the universe. The universe itself is expanding, but as it does, it is subject to self gravity which would tend to slow the expansion. In contrast to this expectation, recent work suggests an opposing repulsive force that accelerates the expansion, due to some sort of *dark energy*.

Much current research is directed toward determining the rate of expansion of the universe. Will it expand forever, or will it slow down eventually coming to rest (much like a rising ball), and then begin to fall inward toward a Big Crunch? The presence of gravitational matter on such large scales can not be described with Newtonian geometry nor with Einstein's special theory of relativity. Einstein's general theory of relativity, *general relativity* or GR, is required to properly describe the universe insofar as we know it; it pulls together the concepts of gravity and space-time into a unified whole. In fact, it contains a *cosmological constant* that results in a long-range repulsive force such as that attributed just above to dark energy.

Thus we find that gravity does indeed underlie the entire field of astrophysics.

4.3 Apparent motions of stars

Horizon coordinate system

An observer on the earth's surface at low latitudes notes that the sun rises in the east and sets in the west. During the night, the same observer would note that the stars and planets also rise in the east, move across the sky, and set in the west. These motions are simply an effect of the earth's daily rotation about its axis. For an observer at the earth's north pole, the north celestial pole (NCP) of the celestial sphere is directly overhead (at the *zenith*). For this same observer, the stars just above the horizon move all around the horizon at a fixed elevation once per day, never dipping below it. At other northern latitudes, the pole is not directly overhead, but stars sufficiently near the NCP will still appear to rotate around the NCP without setting. Stars further south do rise and set (e.g., the sun), and stars even farther south are never seen by the northern observer.

The motion of stars in an earth-based *horizon system* is illustrated in Fig. 1. The view is from the west looking east. The figure is drawn for an observer at an intermediate northern latitude (e.g., Kitt Peak, Arizona, USA at Latitude 32 N; $\theta_{\mathrm{lat}} = +32°$). The horizontal base line is tangent to the earth's surface at the observer's position, and the half circle is the celestial sphere. The observer's latitude θ_{lat} represents the angle from the earth's equator to the observer's position. On the celestial sphere, this is the angle from the celestial equator to the observer's *zenith*, or equivalently the angle between the horizon and the polar axis.

As the earth rotates about its axis, the earth-based observer sees the celestial sphere rotating about the polar axis (Fig. 1). The tracks of two stars in the horizon system are shown as heavy lines A and B. Consider star B. As the earth rotates, it follows a path parallel to the celestial equator since, on the celestial sphere, it lies at a fixed angle δ (declination) from the celestial equator. On the figure, the star rises somewhat north of east, moves south as it rises further, crosses directly south of the observer's zenith (i.e., crosses the observer's *meridian*), then descends as it moves westward, and finally sets north of west. This path is actually a circle around the NCP, but part of the circle is below the horizon; the star sets for part of each 24-hour day, as does the sun. If the star is very near the NCP (star A in Fig. 1), it never sets, and the complete circle could be observed if one could see the star during daylight.

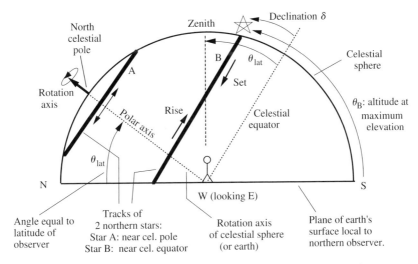

Figure 4.1. Sketches of stars rising and setting in the horizon coordinate system; the view is from west to east. The horizontal line is the horizon, the plane tangent to the earth at the observer's position. The half circle is the longitudinal meridian (on the celestial sphere) that passes through the observer's zenith at the time shown. The tracks of two stars A and B, both with northern declinations, are shown.

The altitude θ_B of star B shown in Fig. 1 is the maximum altitude reached by the star, which occurs when it transits the observer's meridian. At this point, θ_B is directly related to the star's declination δ and the observer's latitude θ_{lat},

$$\theta_B = 90° - \theta_{lat} + \delta \qquad (4.3)$$

Thus, given the declination of the star and the latitude of the observer, one can deduce the elevation angle at the time it reaches its maximum elevation.

Navigators at sea can observe the sun to determine their latitude with the aid of (3). The altitude of the sun is measured repeatedly with a sextant at about the time of *local apparent noon* when the sun is near its highest point. The highest elevation reached by the sun yields, through (3), the latitude of the ship. The declination of the sun changes negligibly due to its annual motion during the observations. Thus it is well known at the time of the observation, even in the absence of a precise clock. A navigator can obtain latitude accurate to $\sim 1'$ or, by definition, 1 *nautical mile* in this manner (1 nautical mile = $1'$ along a meridian, or 1.852 km). Unfortunately, the longitude is completely undetermined. Star sightings at sunset and sunrise may be used for a precise ($\sim 1'$) fix of both longitude and latitude if the times of the sightings (Greenwich time or UTC time, see below) are known to within a few seconds.

Annual motion

The annual motion of the earth about the sun has important consequences for astronomers who view the sky from the earth.

Sun and the ecliptic

The earth's orbital motion yields an apparent motion of the sun relative to the celestial sphere (Fig. 3.1). At some date, e.g., March 21, an observer on the earth would find the sun to be in front of a given constellation, if the sun weren't so bright as to obscure our view of the constituent stars.

At some later date, say June 21, the earth will have moved 90° in its orbit about the sun, and according to an earth-observer, the sun would appear to have moved 90° around the celestial sphere as shown in Fig. 3.1. The observer would then see a different group of stars near (and beyond) the sun, such as the large star in Fig. 3.1. Throughout the year, the sun would thus *appear* to move steadily through the stars at the rate of 360°/365 days ≈ 1 deg/day, passing sequentially through the constellations that are well known as the signs of the *zodiac* and returning to its original position relative to the stars (not to the slowly moving vernal equinox) after 1.00 *sidereal year*,

$$1.00 \text{ sidereal year} = 365.2564 \text{ days} \tag{4.4}$$

The track of the sun on the sky (or the celestial sphere) is called the *ecliptic*. It is a great circle tilted 23.45° relative to the celestial equator. A portion of it is shown in Fig. 3.1. See Table A7 in the Appendix for the several types of "years".

Sun and dark-sky observations

More relevant to astronomers than the signs of the zodiac are the months when certain regions of the sky are accessible for study. Ground-based optical astronomers can not make observations when the sun is above the horizon because the sun is very bright at optical wavelengths, and the atmosphere scatters the light severely. Because of this, the regions that can be observed are those more or less opposite to the sun on the celestial sphere.

A star directly opposed to the sun on the celestial sphere, 180° away, can typically be observed all night (7–11 hours at mid to low latitudes), depending on the season of the year and the latitude of the observatory. When the sun is below the horizon (night time), the star will be above the horizon, and vice versa. Three months later, the sun will be roughly 90° from the star, and it can be observed for only about 1/2 the night or 3 to 5 hours. The star is above the horizon for the usual ~12 hours but for half this time the sun is too. Observers can look up the times of sunset and the end of *astronomical twilight*. The latter is the time when the sky is quite dark (sun 18° below the horizon); observations usually can begin somewhat before this.

The accessibility of stars for observation also depends upon a number of other factors, including an absence of clouds. Stars usually can not be observed if they are more than ~60° from the zenith. At this angle there are about two atmospheric thicknesses (*air masses*) of atmosphere along the line of sight, with its refractive properties, its absorbing and scattering dust, etc. By definition, there is 1.0 air mass along the line of sight from sea level toward the zenith.

Since the celestial sphere rotates past the observer at $360°/24\,\text{h} = 15°/\text{h}$, a star that passes more or less overhead would be within the 60° angle of the zenith for a full 8 hours, from *hour angle* –4 h to +4 h. (The hour angle of a star at a given time is the right ascension of the star less the right ascension of the observatory zenith at that time.) For a northern telescope, stars that pass south of the zenith may be observed for lesser times than those to the north. Stars far toward the north do not set at all and thus may be observed for even longer times, if not all night; see Fig. 1.

Parallax of star positions

As noted earlier (Section 3.2), parallax is due to the changing position of the earth as it moves about the sun. A star relatively close to the solar system will appear to move relative to the background stars (Fig. 2). Parallax is the same phenomenon that can be observed while riding in an automobile; the foreground objects (trees and telephone poles) appear to be moving relative to the distant mountains.

The earth's orbit about the sun is only slightly elliptical, so we will approximate it as a circle here. If a foreground star lies normal to the ecliptic plane (plane of earth's orbit), its apparent motion on the celestial sphere throughout the year is a circle as shown with star A in Fig. 2. Define the parallax angle θ_{par} to be the radius of the circular track on the sky. If the distance to the source is r and the radius of the earth's orbit is a, the parallax angle θ_{par} is, from Fig. 2,

$$\tan\theta_{\text{par}} = a/r \tag{4.5}$$

For small angles ($\theta \ll 1$ rad), $\tan\theta \approx \theta$,

$$\theta_{\text{par}} \approx a/r \qquad\qquad \text{(rad; parallax angle)} \tag{4.6}$$

The radius a can be set to the *semimajor axis* of the earth's orbit, known as the *astronomical unit* (AU),

$$a \equiv 1.00\ \text{AU} = 1.496 \times 10^{11}\ \text{m} \qquad \text{(Astronomical unit)} \tag{4.7}$$

One can use (6) to solve for the distance r to the star if θ_{par} is measured.

If the foreground star lies *in* the ecliptic plane, the path of the star on the sky is a "straight line" (actually a portion of a great circle) of length $2\theta_{\text{par}}$ (Star B in Fig. 2).

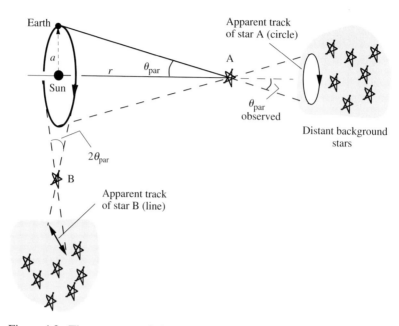

Figure 4.2. The geometry of trigonometric parallax. The angular position (celestial position) of a relatively nearby star changes relative to the distant background stars (taken to be at infinite distance) as the earth proceeds around the sun. The apparent tracks of two stars are shown, star A at the ecliptic pole with a circular track, and star B in the ecliptic plane with straight-line track. The parallax angle of $\theta_{\text{par}} = 1''$ occurs if the star is at a distance of 3.09×10^{16} m $= 3.26$ LY (light years); astronomers call this distance 1.0 parsec (pc).

A star at an intermediate position tracks out an ellipse; the semimajor axis of which is θ_{par}. At distances beyond $\sim$1000 light years (LY), the angular motion becomes too small to measure from ground-based telescopes. The Hipparcos satellite (1989–93) measured parallax angles with precisions of about 1 milliarcsecond (1 mas) for stars brighter than about 9th magnitude and thus could measure distances to about 3000 LY.

The definition of the unit of distance used traditionally by astronomers, the *parsec* (pc), derives from the geometry of Fig. 2. If the angle is precisely $\theta_{\text{par}} = 1.00''$, then the distance to the star is specified to be 1.0 *parsec* (pc), from "*par*allax" and "arc*sec*". This defines the parsec which turns out to be 3.26 LY $= 3.09 \times 10^{16}$ m. The parsec is another of those unhappy historical units; however astronomers use it extensively so it is worth knowing. Think: 1 LY $\approx 10^{16}$ m and 1 pc $\approx$ 3 LY.

$$1.00 \text{ LY} = 0.946 \times 10^{16} \text{ m} \qquad \text{(Light year)} \qquad (4.8)$$

$$1.00 \text{ pc} = 3.262 \text{ LY} = 3.086 \times 10^{16} \text{ m} \qquad \text{(Parsec)} \qquad (4.9)$$

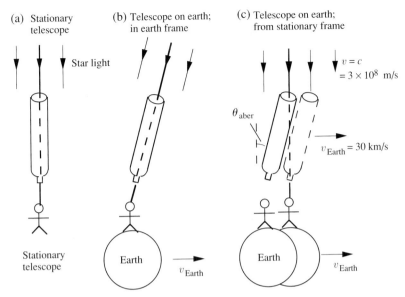

Figure 4.3. Stellar aberration. The apparent direction of an overhead star is modified due to the speed of the earth in its orbit about the sun. (a) Stationary telescope viewed from a stationary reference frame (b) Telescope in the (moving) reference frame of an observer on the earth. (c) Moving telescope from stationary reference frame.

The measurement of parallax angles is a standard method for the determination of distances to the closer stars (see Section 9.5).

Stellar aberration

There is also a small shift of star positions that arises from the *velocity* of the earth as it orbits the sun. (Parallax is a consequence of the varying *position* of the earth.) This effect is called *stellar aberration*. It should be calculated with special relativity, but since the earth is traveling at a speed much less than the speed of light, a classical derivation yields the correct magnitude of the effect. Consider starlight to be analogous to (vertically) falling rain as one runs through it. In the frame of reference of the runner, the rain does not fall precisely vertically; it appears to come from a slightly different (more forward) angle. It would be advisable for the runner to tilt her umbrella forward a bit. Similarly, starlight appears to come from a different angle, and the telescope must be tilted slightly forward to collect it (Fig. 3b).

The magnitude of the (maximum) aberration angle follows from the geometry of Fig. 3c which shows the situation from a stationary reference frame. The light falls vertically, and the telescope must be tilted at an angle such that light entering the top of the telescope will exit through the eyepiece when it reaches the bottom.

The required angle is given by $\tan \theta_{aber} = v_{earth}/c$, or to a very good approximation,

$$\theta_{aber,max} \approx \frac{v_{earth}}{c} = \frac{2.979 \times 10^4 \text{ m/s}}{2.998 \times 10^8 \text{ m/s}} = 0.994 \times 10^{-4} \text{ rad} = 20.5'' \quad (4.10)$$

The pointing direction of the telescope is the arrival direction of the light according to the moving observer. This results in an annual cyclic variation of the apparent positions of stars on the celestial sphere of up to 20.5″ per year.

The effect is maximal when the star is 90° from the earth's velocity vector, e.g., at the ecliptic pole, and the effect is absent when the star is 0° from the velocity vector. In the course of a year, a star at the ecliptic pole will thus track out a circle of radius 20.5″ on the celestial sphere, and a star on the ecliptic will move back and forth along a line (great circle) of length 41″. At intermediate positions, the star's track will be an ellipse. All stars in the same vicinity will suffer the same motion, so this effect can not be detected simply by taking photographs of a given region of the sky. It differs from parallax in this respect.

The direction and magnitude of aberration depend solely upon the time of year and the direction of the target star. It is easily calculated from appropriate formulae. These corrections are sufficiently small $\lesssim 20.5''$ that they traditionally have not been an issue in practical observing where the coordinates of a star are used primarily to point the telescope to the close vicinity of a star, say within ~1′. However, those who create catalogs of precise stellar positions and those who program computers for precise pointing of telescopes must include aberration in their algorithms.

Precession of the earth

Precession of the celestial equatorial coordinate system is due, as noted, to precession of the earth. The latter is rather like a spinning top that precesses about the vertical due to the torque of gravity.

Torque due to a ring of mass

In the case of the earth, the torque is due to the gravitational attraction of the sun, moon, and planets acting on the earth. These bodies could not exert a torque on the earth if it were perfectly spherical. However, the earth has an equatorial bulge; its radius is larger at the equator than at the poles. The sun, moon, and planets (especially Jupiter) are thus able to exert a torque on it. The bulge is due to the centrifugal force in the rotating frame of reference of the spinning earth; the centrifugal force increases with radius from the spin axis, and hence is largest at the equator.

Since the precession is very slow with a period of 25 770 yr, the sun might be viewed as circulating around the earth 25 770 times per precession period. If one runs this as a speeded up movie, the sun would appear as a ring of mass around the ecliptic (Fig. 4a). The moon's orbit lies only 5.1° from the ecliptic so its mass

(a)

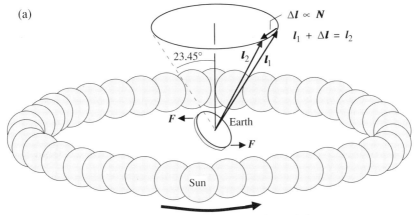

25 770 cycles of sun in one precession period

(b)

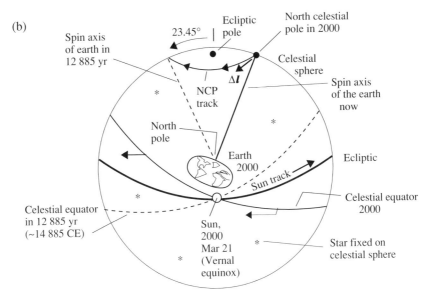

Figure 4.4. Precession of the earth's spin axis about the normal to the ecliptic plane with period 25 770 yr. (a) Time-averaged distribution of mass in the ecliptic plane due to the many cycles of the sun (and moon and Jupiter) which acts gravitationally on the earth's equatorial bulge to produce an instantaneous torque N and hence precessional motion. (b) Earth alignment relative to celestial sphere on 2000 Mar 21, with the ecliptic as the equator. The motions of the spin axis and the associated celestial equator (light solid line) are indicated with arrows. The vernal equinox moves to the left as the precession proceeds, i.e., in the direction of lower right ascension. The sun follows the ecliptic (heavy line); it is moving into the northern hemisphere on Mar 21. The situation after one-half precession cycle is shown with dashed lines.

also contributes to the ring. Similarly, Jupiter's inclination is only 1.3° so it too contributes to the ring of mass. The two small arrows labeled F on the earth in the figure represent the approximately horizontal forces that "pull" the bulge toward the ring of suns. Together these forces create a (vector) torque N that is directed more or less out of the paper. The resultant angular momentum change dl ("dee ell") would be parallel to the torque (from $N = \mathrm{d}l/\mathrm{d}t$). This causes the spin vector of the earth to precess as shown. The result is that the earth's spin axis traces out a circle on the celestial sphere with a half-cone angle of 23.45°. The center of the circle is in the direction of the ecliptic pole.

Take a closer look at the precessional motion. Figure 4b shows precession relative to the celestial sphere. The spin axis of the earth and the celestial equator in 2000 are shown. Recall that the celestial equator is the projection of the plane of the earth's equator onto the celestial sphere. The ecliptic is a fixed track among the stars because the earth's *orbit* is quite rigidly fixed in space. The sun's position on 2000 March 21 is shown; it is in the constellation Pisces and moving to the right on the ecliptic. Note that it is just crossing the celestial equator from south to north in its annual motion; it is at the vernal equinox, the zero of right ascension (in 2000).

As the earth begins to precess away from the 2000 position, the spin axis moves toward the reader (arrow Δl) and the celestial equator slides to the left (see arrows). The intersection of the celestial equator and the ecliptic (the vernal equinox) thus also moves to the left or the west, the *precession of the equinoxes*. This intersection is the zero of right ascension by definition so it too moves to the west. A given star in this region will thus find itself farther and farther east of the vernal equinox. Since right ascension increases to the east, the star will have an increasing right ascension from year to year. The declination will also change. For this reason, as noted in Section 3.2, one always must specify the epoch (year) of the coordinate system being used.

As the precession continues for ~12 885 yr, the spin axis will have precessed 180° to the position shown as the dashed line. The new celestial equator is also shown as a dashed line. An intersection of the celestial equator and the ecliptic is again at the same position among the stars, in Pisces, but when the sun passes this point it will be moving from north to south relative to the earth, i.e., the first day of fall in the northern hemisphere! Thus this position will be assigned right ascension $\alpha = 12$ h, and the $\alpha = 0$ position will be on the opposite side of the sphere.

Rate of precession

The rate in arcsec per year at which the vernal equinox slides around the ecliptic is

$$\text{Rate} = 360° \times 1/(25\,770) \times 3600 \quad \text{(Motion of the vernal} \qquad (4.11)$$
$$= 50.3''/\text{yr} \qquad\qquad\qquad\qquad \text{equinox)}$$

or 42′ in 50 yr, a sizable motion. The shifts in coordinates of a celestial object can be of this magnitude; they can also be much lower depending on the location of the star on the celestial sphere. The vernal equinox used to be in the constellation Aries; it was named the *first point of Aries* in the time of Hipparchus (~135 BCE). It is now in Pisces, and in about 600 yr, it will be in Aquarius, the *Age of Aquarius*!

Nutation

The earth also exhibits a small nutation (wobble) of 9″ with a period of 18.6 yr. This nutation is also due to gravitational effects of the moon and sun. The period is the same as that for the moon's orbit to precess one complete cycle with respect to the stars. See further discussion below (solar eclipses). The pole and equator of the equatorial coordinate system follow this motion also.

Calendar

What happens to the calendar and the seasons as the earth precesses? Consider a future date 12 885 yr from now. The sun will pass annually through Pisces as it always does, but as noted above it will be passing from north to south in earth coordinates. It thus will be the beginning of fall in the northern hemisphere; winter will be on the way. The location of the sun along the ecliptic (or equivalently, the location of the earth in its orbit) is thus not an indicator of the season because of precession. The season is indicated by the position of the sun in its track relative to the (precessing) vernal equinox. The right ascension of the sun will thus always indicate the season.

Since the adoption in 46 BCE of the *Julian calendar* by Julius Caesar, the calendar has been tied to the seasons. Thus March 21 should always take place at the beginning of spring when the sun crosses from south to north (at the vernal equinox), even though the vernal equinox keeps shifting relative to the stars. Stated otherwise, the Julian calendar is based on the *tropical year*, the time it takes the sun to travel from vernal equinox to vernal equinox.

The Roman calendar which preceded the Julian calendar was based on lunar cycles. The Julian calendar basically is the calendar now in use. It consists of a 365-d year interspersed with a *leap year* of 366 d every 4 yr, on the year numbers divisible evenly by 4. Thus, the Julian year is exactly 365.2500 d long, a standard Caesar adopted from the Egyptians. The Julian calendar was also set so that the vernal equinox fell on March 25 in 46 BCE, the traditional date at the time. This required a 3 month adjustment which caused great confusion (46 BCE contained 445 d).

The actual duration of a tropical year, 365.242 189 d, is slightly shorter than both the sidereal year (4) (see Fig. 4b) *and* also the Julian year. Thus, by the time of Pope

Gregory XIII, in 1582, the date of the vernal equinox had slipped back to March 11. Motivated by the drift of Easter Sunday later and later into the spring (Easter was referenced to the *calendar* date Mar. 21), Gregory instituted the *Gregorian calendar* which set March 21 to the vernal equinox. To keep it close to that date, he also adjusted the average length of the year by declaring that three out of four "century years" would *not* be leap years. In this scheme, 1700, 1800, and 1900 would not be leap years, but year numbers evenly divisible by 400, e.g., 1600 and 2000, would still be leap years as they were in the Julian calendar. This brings the calendar into good synchrony with the tropical year; the error is one day in ~3000 yr.

Catholic countries adopted the Gregorian calendar in 1582 but the British Isles and the United States did not do so until 1752 when parliament decreed that the day after Sept. 2 would be Sept. 14. This led to great consternation wherein some people demanded that their 11 d be returned to them. The czars of Russia never adopted the Gregorian calendar. Only after the 1917 revolution was it adopted by the Soviet Union (including Russia), in 1918. Astronomers still use the Julian year of 365.25 d $\times$ 86 400 s/d = 31 557 600 s and the Julian century of 36 525 d as sometimes useful time intervals.

This discussion traces the development of the Gregorian calendar, currently the international standard. Other important systems are (or were) the Mayan-Aztec, Hebrew, Asian Indian, Islamic, and Chinese calendars.

Zodiac

A great deal of attention is paid to the signs of the zodiac by astrologers. If you were born on December 7, as I was, your sign would be Sagittarius. The sun is supposedly located in the so-named constellation on that date. However, since the calendar follows the earth's precession, the sun is no longer in Sagittarius on Dec. 7; it is in the adjacent constellation of Ophiuchus. It was also there on the Dec. 7 (1930) when I was born. Thus, I can not understand why I am instructed to read the horoscope for Sagittarius. However, it doesn't really matter because there is no physical basis for the predictions therein.

Proper motion

The other important motion of the stars is that due to their diverse individual motions on the celestial sphere. Their transverse angular velocities relative to the solar system observer are known as *proper motions*, which we mentioned in Section 3.2. Since proper motions are angular motions, radial motion is not included. This projected angular velocity arises from the actual motions of the stars in inertial space relative to the barycenter (center of mass) of the (moving) solar system.

The effects of the earth's motion relative to the barycenter are subtracted from the quantities measured from the earth.

Proper motion is due in part to the movement of stars, including the sun, about the center of the Galaxy. The stars closer to the center of the Galaxy rotate with greater angular velocity than those farther out. Random motions of stars due to gravitational interactions also contribute to the proper motion of a given star.

Motion on celestial sphere

Proper motion is a rate of change of angular position, $\mu = d\theta/dt$ (radians/s) on the celestial sphere, relative to the distant galaxies. It is often given in the mixed units of milliarcsec per year (mas/yr) by astronomers. The direction of movement is given by specifying the rate of movement in the directions of right ascension α and declination δ, that is $(d\theta/dt)_\alpha$ and $(d\theta/dt)_\delta$. These angular displacements are *great-circle angles*, the angles you would obtain by measuring along two axes on a photograph of that small part of the sky using the appropriate scale factor for the camera to convert from mm to arcsec. The photograph is a projection onto a plane *tangent* to the celestial sphere at the position of the star.

One should be cautious about the relation between the great circle angles measured and the associated changes in right ascension and declination. In declination, there is a direct equivalence because declination is measured along the great-circle meridians. Thus a proper motion of $1''$/yr in declination would lead to a change in declination of $1.0''$ in one year, e.g., from $35°\,00'\,00''$ to $35°\,00'\,01''$.

In contrast, a proper motion of $1''$/yr in the direction of right ascension (α) could lead to a larger change in α, if specified in terms of angle, because the lines of constant α (the meridians) converge toward the celestial poles. For example, at declination $45°$, a great-circle angle of $1''$ would subtend $1''\sqrt{2} = 1.4''$ of right ascension. Since α is actually measured in h m s with 24 h corresponding to $360°$ on the equator, it follows that 1 h corresponds to $15°$ and that 1 s (of time) corresponds to $15''$ (of angle), and conversely $1''$ corresponds to $1/15$ s of time. Thus our $\sqrt{2} = 1.4''$ of right ascension (at $\delta = 45°$) would correspond to $\Delta\alpha = (1/15) \times \sqrt{2} = 0.094$ s of time. In one year the right ascension of the star in question might thus change from 23 h 00 m 00 s to 23 h 00 m 00.094 s. Alternatively, a catalog may give the proper motion in RA as a direct correction to α in seconds of time per year, s/yr.

The proper motion of each star is unique. Thus catalogs of precise positions must list the magnitude of this motion for each star in each coordinate, α and δ, on the celestial sphere. Only a few hundred stars have proper motions greater than $1''$/yr. The largest value $10.25''$/yr is for Barnard's star. Its close proximity to the sun, only 6 LY, leads to this large angular velocity. Measured proper motions range down to $\lesssim 1$ mas/yr.

Peculiar motion and local standard of rest

The velocity of a star in three-dimensional space relative to the barycenter of its neighbors is called its *peculiar motion*. It is likely due to gravitational interactions with other stars at some time in the past. Stars that approach close to one another experience mutual gravitational attraction and hence accelerations. The distances between stars are normally so great that their motions at present can be assumed to be straight lines over the mere ~100 yr of modern astronomy. The peculiar motions of stars in the solar neighborhood are measured with respect to the frame of reference that moves with their barycenter (center of mass), the *local standard of rest*. Their peculiar motions are typically of order 10 km/s with a few reaching 100 km/s. They contribute to the observed *angular* velocities relative to the celestial coordinate system, i.e., to the proper motions.

Solar motion

Measurements of the radial and transverse motions of stars in the vicinity of the sun, and in all directions about it, indicate that the sun is moving (in the local standard of rest) toward the bright star Vega in the constellation Lyra ($\alpha = +18$ h, $\delta = +30°$) at a speed of 19.5 km/s (4.1 AU/year). On the average, spectra of stars in this direction show a blue Doppler shift, and those in the opposite direction show a red shift. The celestial coordinate toward which the sun moves is called the *apex* of solar motion; the position from which it recedes is called the *antapex*. This motion is distinct from the general motion of the sun with its neighboring stars about the center of the Galaxy. See Section 9.5 ("Secular and statistical parallaxes") for more on this.

4.4 Lunar and planet motions – eclipses

The bodies in the solar system are quite close to us compared to the stars; thus they have substantial angular velocities relative to the earth. Their apparent positions on the celestial sphere change radically on an annual scale and by significant amounts day by day. The celestial motions of these bright objects have long fascinated humankind.

Eclipses of the sun and moon

The motions of the sun and moon on the celestial sphere lead to solar and lunar eclipses. The former occurs when the moon comes between the earth and sun and blocks, at least partially, the observer's view of the sun. The latter occurs when the moon is on the far side of the earth from the sun and enters the shadow of the earth.

"Orbits" of the moon and sun

The moon orbits the earth every 27.32 d (the *tropical month*, the time it takes between south-to-north crossings of the celestial equator, i.e., equinox to equinox. The sidereal period (relative to the stars) is nearly identical to this, within 10^{-4} d. Thus the moon changes its position on the celestial sphere a great amount each day, $360°/27.3$ d $= 13.2°$/d. This large night-to-night motion is ~ 25 moon diameters and hence is quite dramatic. Watch it move through the stars hour by hour or night to night. When the moon is at its closest to the sun (on the celestial sphere), the sun is far beyond it and illuminates only its far side. This is the *new moon*. We can see it a couple days later in the evening sky as a narrow sliver when it has moved away from the sun. The period from new moon to new moon, the *synodic month*, is somewhat longer, 29.53 d, because the moon has to move an additional distance each orbit to catch up to the sun which itself moves through the sky by $\sim 30°$ in a month.

The orbit of the moon is inclined $5°\,08'\,43''$ from the plane of the ecliptic (Fig. 5), and this value oscillates a small amount ($\pm 9'$ with a period of 173 d) due to torques arising from the earth's equatorial bulge and the sun. These torques (primarily that due to the sun) also cause the moon's orbital angular-momentum vector to precess about the normal to the ecliptic with a period of 18.6 yr. (The underlying physics is similar to that we used for the precession of the earth.)

The moon's orbit is also quite eccentric; the angular diameter of the moon decreases by about 12% from perigee (closest point to earth) to apogee (farthest point). Recall that the solar size varies 3% due to the earth-orbit eccentricity. The above mentioned torques cause the perigee to advance eastward with a period of 8.85 yr relative to stars. Referring to the line connecting the apogee and perigee, this motion is called the *advance of the line of apsides*.

For these reasons, the path of the moon through the stars, as viewed from the earth, gradually changes from month to month and year to year. Lunar eclipses of radio and x-ray sources that happen to be in the path of the moon have been used to obtain their precise celestial positions by observing the exact times of the occultation and reappearance of the star.

As it orbits the earth, the moon's track on the celestial sphere comes into the vicinity of the sun every 29.53 d. But since the moon's orbit (an approximate great circle) is tilted 5° from the sun's orbit (the ecliptic, also a great circle), the moon is usually well above or below the sun. However, the (apparent) annual solar motion around the ecliptic ensures that the sun will cross the moon's orbit twice a year, and an eclipse could take place.

This is illustrated in Fig. 5 where the great-circle tracks of the sun and moon on the celestial sphere are shown. As the sun moves along the ecliptic, it will pass through the intersections of the two orbital planes, called the *nodes* at ~ 6-month intervals.

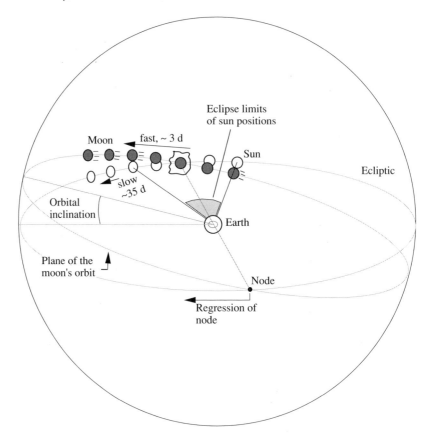

Figure 4.5. Tracks of moon and sun around the celestial sphere showing the eclipse limits. The orbits are inclined to one another by only 5° so an eclipse occurs on the earth each time the sun is within the ~35° region surrounding the node (intersection of the orbits). The sun (open circles) passes through the region slowly and the moon (dark circles) passes through rapidly. The nature of the eclipse depends on how far from the node the overlap occurs. The moon's orbit precesses with a period of 18.6 yr so that the node regresses 19.4° per year, leading to solar crossings at 173-d intervals. [Adapted from Littman and Wilcox, *Totality*, University of Hawaii Press 1991, p. 12]

The nodes will slowly drift westward along the ecliptic because of the 18.6-yr precession of the moon's orbit, called the *regression of the nodes*. This causes the interval between node crossings to be a bit shorter than 6 months, namely 173.3 d. At the nodes, the moon, in principle, could come into the line of sight to the sun. This would result in an *eclipse of the sun*.

Total and partial solar eclipses

A solar eclipse requires that both the sun and the moon arrive at the node at nearly the same time. This coincidence need not be precise because (*i*) the two orbits are tilted only 5° from one another, and (*ii*) the observer can be positioned anywhere north or south on the earth to bring the sun and moon into sufficient alignment (with the help of parallax) to yield at least a partial eclipse wherein the moon covers only a part of the sun. It thus turns out that the sun is susceptible to an eclipse (total or partial) somewhere on the earth for 30 to 37 d while it is in the vicinity of a node. Since the moon takes only 29.5 d to complete its cycle, it will always come through this region while the sun is sufficiently close to the node to be eclipsed. Thus an eclipse somewhere on the earth *must* occur every ∼173 d, the time between node crossings by the sun.

A *total solar eclipse* is the most dramatic because the sun's disk is completely blocked by the moon. This requires the earth–moon–sun alignment to be relatively precise. Also, the moon's angular diameter must be greater than that of the sun which is sometimes not the case because of the elliptical orbits of the moon and earth. *Partial solar eclipses* are more common. About 30% of solar eclipses are called "total" because the sun is totally eclipsed somewhere on the earth. In fact, a small zone of totality sweeps across the earth creating a "track" of totality. The sun will be partially eclipsed in adjacent regions. The other eclipses are either "partial" or "annular". In the former, no place on earth experiences totality. In the latter, the moon is directly in front of the sun but smaller in angular diameter, for some earth observers. Thus a thin ring of sun surface is seen surrounding the blackness of the moon.

The 18-year saros cycle

Since ancient times, it has been noted that a lunar eclipse will always be followed by a similar eclipse 6585.3 d later (18 yr 11.3 d if there are 4 leap years), and later it became known that solar eclipses also recurred with this same interval. Any solar eclipse will surely be followed 6585.3 d later by another of similar degree of occultation and duration. This is due to the approximate coincidences of multiples of four periodicities, each of which yield an interval near 6585 d as follows:

(*i*) 6585.32 d from 223 cycles of the moon's 29.530 59 d motion from new moon to new moon (synodic month). This establishes the exact interval between the two adjacent eclipses of the series (called the *saros*).

(*ii*) 6585.78 d from 19 annual cycles of the sun's 346.620 06-d motion from node to node (2 × 173.3 d = *eclipse year*). This places the sun at the time of an eclipse, see (*i*), very close (about 1/2 day or 1/2 degree) from where it was relative to the node at the time of previous eclipse. The eclipse will thus be of similar kind (partial or full) as the previous one; see Fig. 5.

(*iii*) 6585.54 d from 239 cycles of the moon's orbital period in its elliptical track from perigee to perigee (*anomalistic month* = 27.554 55 d); the perigee is the time the moon is closest to the earth in its eccentric orbit. This condition ensures that, at the time of the eclipse, see (*i*), the moon will be almost at the same place in its elliptical orbit and hence of about the same angular size viewed from the earth. This is another component that contributes to the similarity of two adjacent eclipses in the series. If one eclipse is a total eclipse, the next is likely be total also.

(*iv*) 6574.67 d from 18 cycles of the earth from perihelion to perihelion in its orbit about the sun (*anomalistic year* = 365.259 64 d). This is similar to (*iii*); it ensures the sun will be about the same angular size viewed from the earth as for the adjacent eclipse in the series. This too contributes to the similarity and duration of the two eclipses.

These numerical coincidences are not perfect so the eclipses will slowly evolve in character. A series of such eclipses (*saros*) will begin with a brief partial eclipse with the moon barely touching the sun, and observable only from the Arctic (or Antarctica). Gradually, the eclipses of the series move southward (or northward), become more and more complete until they become total or annular (or both, sequentially). Thereafter (possibly after $\sim$20 total or annular eclipses), the eclipses become partial again with gradually decreasing degree of occultation. Eventually, after about 75 cycles or $\sim$1350 yr from the start, the last minimal partial eclipse of the saros appears near the other pole of the earth. With this the saros dies; there are no more eclipses in the series. At any time, there are $\sim$38 saros in progress, one for every node crossing at 173.3 d intervals during the 18-yr period.

The total eclipse of 1919 was used to demonstrate that Einstein's prediction of the bending of starlight was quantitatively correct. Although there is no physical connection between the eclipses in a given saros cycle, much was made of the fact that the 1991 July 11 eclipse belonged to the same saros as the 1919 eclipse. Both were total eclipses of rather long duration (almost 7 min at the center of the shadow track) because the moon's angular diameter was large and sun's was small on those dates. That is, the moon was near perigee (closest point to the earth) and the earth near aphelion (most distant from the sun). At the times of other contemporary eclipses in this saros, e.g., 1919, 1937, 1955, 1973, 1991, and 2009, the earth and moon are at similar positions in their orbits, so these eclipses are all total and of similarly long duration.

Finally, one should note that the earth spin period (1.0 d) is not commensurate with the 6585.3 d saros period; the earth will have rotated 0.3 revolution beyond its original orientation at the time of the previous eclipse of the series. The eclipse track on the earth's surface is displaced westward $\sim$1/3 of the way around the earth, and a bit south or north, compared to the track of the previous eclipse in the saros. This and the limited width of the eclipse track made it more difficult to discover the 18-yr recurrence cycle of solar eclipses than was the case for lunar eclipses (see below).

Wonder and science

Total eclipses are a wondrous experience. During the eclipse, one can see with naked eye the solar corona extending several solar diameters beyond the (covered) solar disk. Beautiful red prominences can sometimes be seen with the naked eye. These are giant loops of hot gas emitting the red Balmer line of hydrogen. The shadow of totality on the earth at one instant is quite small, less than 300 km in diameter. It can be viewed only by those who are along its track across the earth's surface. Astronomers have a history of transporting equipment to remote places with great difficulty in order to carry out studies of the sun during a total solar eclipse.

Eclipses provided astronomers the opportunity to study the chromosphere and corona of the sun. For example, the discovery of highly ionized iron indicated that the corona is exceedingly hot, $>1 \times 10^6$ K. This was surprising since the photosphere is only 5800 K. The light we see from the corona is light from the *photosphere* of the sun, its visible surface, scattered toward us by either the energetic electrons of the ionized hot gas (the inner *K corona*), or by "dust" grains in the solar system (the outer *F corona*). Dust grains consist mostly of ice, silicates, and graphite and are also found in interstellar space.

The shadow of the total solar eclipse of 1991 July 11 passed directly over the mountain-top observatory of Mauna Kea, Hawaii. This allowed astronomers to use major telescopes for observations to study various aspects of solar energetics, e.g., the mechanism by which the corona becomes so hot, most likely, the release of magnetic energy by annihilation or twisting of magnetic fields. The relation between the various components of the solar surface and atmosphere that appear at radio, optical and x-ray wavelengths could also be studied.

A note about safety: one *must* use a safe sun filter when observing the disk of the sun before and after totality, by naked eye and especially with binoculars or telescope. But during totality, when the disk is completely covered, it is perfectly safe to use the naked eye with or without optical aids. The corona is roughly 10^6 times fainter than the solar disk. With a sun filter you would see nothing at all during totality!

Corona in x rays and visible light

An x-ray photo of the sun was obtained during the 1991 July 11 eclipse from a rocket launched at White Sands, New Mexico (USA) exactly when the sun was totally eclipsed in Hawaii (Fig. 6a). At this time in New Mexico, the (partial) eclipse had not quite started. The looming and approaching shadow of the moon is seen to the right at about 4 o'clock; the x rays can not penetrate the moon. The bright features in the figure are typically located at *active regions* where sunspots (in optical light)

(a)   (b)

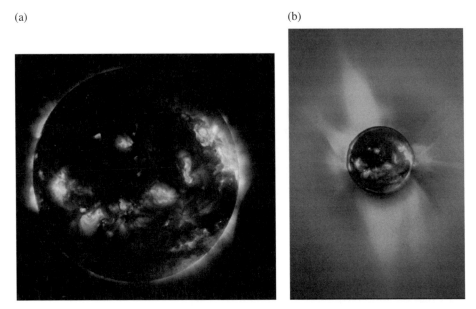

Figure 4.6. (a) X-ray image of sun taken from a rocket flown from White Sands, New Mexico at the time of the 1991 July 11 total eclipse in Hawaii, showing hot plasmas above active regions. The approaching moon shadow is seen at about 4 o'clock. (b) Composite of the x-ray image and optical images taken at the same time during totality with the CFH Telescope on Mauna Kea in Hawaii. It shows the x-ray features within the solar disk and optical features outside it. The extended corona is seen in photospheric light scattered toward the telescope by coronal electrons and dust. The outer ring is an artifact of the graded filter used to bring out the features of the outer (and fainter) portions of the corona. [L. Golub and S. Koutchmy/NASA/CFHT]

are present. The x-ray emission indicates the presence of extremely hot coronal plasmas ($\sim 10^6$ K), which are confined and guided by magnetic fields to form loops and other structures.

An optical photograph obtained during totality with the Canada–France–Hawaii Telescope at Mauna Kea is combined with the x-ray image in Fig. 6b. The optical contribution to Fig. 6 shows the extended corona reaching out to several solar radii and a small "prominence" at about 3 o'clock. This prominence was easily visible with the naked eye as red light from the hydrogen Balmer Hα line. The coronal configuration varies in appearance from eclipse to eclipse; in this case, it extends approximately north–south (up–down).

The optical exposure contributes nothing to the disk of the sun in Fig. 6b because the moon covers it; it is totally black. Note the black circle of the moon shadow surrounding the disk of the x-ray image in Fig. 6b. The x-ray contribution in contrast shows bright features across the face of the sun and around the edges just above the surface. Thus in this picture, one sees the corona, mostly transversely in the optical and mostly face-on in x rays, in a three-dimensional perspective reaching nearly down to the photosphere.

Lunar eclipses

Fifteen days before or after a solar eclipse, the moon has moved to the opposite side of the earth, almost directly opposite the sun. Since the earth–moon–sun alignment was quite precise just 15 d earlier, it is not unlikely that the moon will now enter into the shadow cast by the earth. This is known as a *lunar eclipse*. Lunar eclipses can be partial or total. At this time the moon is full, so the phenomenon can be quite dramatic. During a lunar eclipse, the moon does not completely disappear because sunlight scattered by the earth's atmosphere illuminates the lunar disk weakly. It may appear to have a dark reddish-brown color.

Observation of such an eclipse is not restricted to observers along a narrow path on the earth, as is the case for solar eclipses. Anyone on the moon side of the earth at the time of the eclipse can see the darkened moon. Thus for a given location on the earth, lunar eclipses are more probable than solar eclipses. It was these more frequent sightings of lunar eclipses that allowed the ancients (Chaldeans, over 2000 yr ago) to discover the 18 yr 11.3 d saros cycle, which could then be used to predict future lunar eclipses.

Planets

The motions of the sun and moon on the celestial sphere are both steadily eastward. In contrast the motions of the planets on the celestial sphere are quite complex. The planets all orbit the sun in the same direction as the earth (eastward viewed from the sun), but, viewed from the earth, they will move both eastward and at times westward (*retrograde motion*) due to the earth's motion. It is amazing that Kepler was able to deduce from such motions that a planet is actually moving in a simple elliptical orbit. The planetary motions are quite dramatic and easily noticeable by eye, more or less on a week-to-week time scale.

Sometimes two or three of the brighter planets will come close together in the sky. The novelty of this and its beauty, especially if the moon happens to be nearby, is a noteworthy occurrence. The orbits of the planets all lie within $3.5°$ of the ecliptic with the exception of Mercury ($7°$) and Pluto ($17°$). (One can argue that Pluto is not really a planet.) Thus the planets move along the sky very closely to the track of the sun, the ecliptic. Sometimes after sunset, one can see several planets (e.g., Venus, Mars, and Jupiter) spread along a great circle in the sky. They and the just-set sun nicely map out the ecliptic.

4.5 Measures of time

The study of celestial motions requires a definition of "time". It turns out that time keeping is not a simple process. There are many factors that contribute to this complexity. It was difficult in early times when clocks at different geographic locations could not be easily synchronized, and it is difficult in modern times as one pushes to obtain greater and greater accuracy, even with the aid of atomic clocks.

We humans instinctively conceptualize time as a smoothly running entity that can be agreed upon by all observers. This is the *Newtonian model* of time. That is, they can synchronize their watches (time and rate) and agree on the time of events at different locations. However, when high velocities or strong gravity are encountered, the comparisons are no longer so simple. General relativity (GR) gives a different model of time, taking into account the effects of speed and gravity. Special relativity takes into account the effect of speed.

In contrast to these models, there are *nature's clocks* which can be used to keep track of time. We have already seen that Caesar's model of time (his calendar; Section 3) did not track well nature's clock, the passing of the seasons. Manufactured clocks make use of nature's clocks. Traditionally they have been based on the natural frequency of the pendulum or that of the spring/mass oscillator. More recently the quartz crystal is commonly used to produce much more stable electrical oscillations. Each oscillation can be thought of as a tick of time, and time in "seconds" is defined in terms of a number of ticks. The SI standard of time is now based on oscillations of the cesium atom in *atomic clocks*.

Nature's astronomical clocks are the daily rotation period of the earth, the earth's annual rotation about the sun, and a rapidly rotating neutron star, seen as a regularly (up to a point) pulsing *radio pulsar*. A few of these pulsars exhibit an extremely high degree of stability that begins to rival that of atomic clocks, but they have not yet been used as a time standard.

One must keep in mind the distinction between a model of time and a clock of nature. One can tell time no better than the most stable clock one has. It may drift in rate, but there is no way to know it until a more stable clock is invented or discovered. In contrast, the model of time specifies ideally how time behaves, as a function of location, speed, and gravity. For example, the Newtonian model describes a time that is infinitely stable and everywhere the same. The story of time is one of the development and discovery of increasingly more stable clocks together with the development of models that successfully explain the subtleties of time keeping these improved clocks reveal.

Time according to the stars and sun

The most fundamental time keeping is based on the motions of the sun and stars as they pass overhead with seasonal variations.

Sidereal time

As the earth rotates under the sky, the zenith of a given observatory (e.g., Palomar Mountain) moves along the celestial sphere from west to east (Fig. 3.1). Thus, as we have described, a typical celestial object will rise in the east and move westward *relative* to Palomar Mountain. At any given time, the meridian of right ascension directly over the observatory is by definition the *sidereal time*, e.g., 15 h 25 m 35 s. A star at right ascension, $\alpha = 15$ h 25 m 35 s, crosses over the longitude meridian of Mt. Palomar at sidereal time 15 h 25 m 35 s (by definition). At this sidereal time, stars at $\alpha = 16$ h will not yet have transited the observer's meridian, and stars at $\alpha = 15$ h 00 m will have already transited it.

Sidereal time is strictly a *local time*. To set his sidereal clock, the observer need only look overhead to see what meridian of the celestial sphere is passing overhead; its right ascension is the time to which he would set the clock. If, at some instant, the observer compares his sidereal time to that of another observer at a different longitude, by radio signal for example, they would report different sidereal times.

Mean solar time

Solar time is another local time; it indicates the location of the observer's meridian with respect to the sun. By definition, the sun is overhead (or due south or north of the zenith) at noon, solar time. Solar time varies in rate relative to a fixed-rate universal time because of the eccentricity of the earth's orbit and the tilt of the earth's axis relative to the normal to the earth's orbital plane. *Mean solar time* averages out these variations. The difference between solar time and mean solar time kept by clocks varies by as much as 16 min during the year; the difference is known as the *equation of time*. Clock makers used to put a table of the time difference for each day of the year on their clocks.

Mean solar time still lacks something in convenience; it will differ in two nearby towns if one is west or east of the other because the sun will pass over one meridian before it passes over that of the other. To solve this problem, everyone in a given region (time zone) agrees to keep the time of an observer at the central meridian of the region. This is called *zone time,* e.g., Eastern Standard Time. There are 24 such time zones, each 15° wide, more or less. The historic observatory at Greenwich England lies at the zero of geographic longitude. The zone time of this region has been called *Greenwich Mean Time* (GMT). This system of time zones dates back to 1884.

Astronomers find it convenient to use the modern equivalent of GMT (Universal Coordinated Time UTC, see below) during their observations. Frequently, observations at different observatories must be compared, and comparison of data from different time zones can lead to confusion. Astronomers tend to use the date of the beginning of an observing night to label that night's data even if the data were actually taken after midnight. This can add more confusion. In contrast, the UTC time *and* date (at Greenwich) of each specific observation during the night is unambiguous.

Sidereal and solar (or zone) times are not synchronized to one another because the former is referenced to the stars and the latter to the sun which moves through the stars as viewed from the earth. Since the sun moves along the ecliptic, the lengths of the two types of days must differ slightly, by about 1 part in 365, or by about 4 min. Which is the longer? Since the sun moves along the celestial sphere in the same direction that the earth spins, a given point on the earth must move more than $360°$ (relative to the stars) in order to catch the sun. The solar day is thus longer than a sidereal day.

The mean solar day is divided into $24 \times 60 \times 60 = 86\,400$ UT seconds which are slightly elastic(!) due to small irregular variations in the earth spin rate (see below). In civil life, we keep track of time with fixed atomic or SI seconds. Thus,

$$
\begin{aligned}
1 \text{ Mean solar day} &= 24 \text{ h } 00 \text{ m } 00 \text{ s (solar time)} \\
&= 86\,400.00 \text{ s (solar time)} \\
&(\approx 86\,400 \text{ s atomic SI time})
\end{aligned}
\tag{4.12}
$$

The shorter sidereal day is

$$
\begin{aligned}
1 \text{ Sidereal day} &= 23 \text{ h } 56 \text{ m } 04.09 \text{ s (mean solar time)} \\
&= 86\,164.09 \text{ s (mean solar time)} \\
&(\approx 86\,164.09 \text{ s, atomic SI time})
\end{aligned}
\tag{4.13}
$$

which tells us that the sidereal day is shorter than the solar day by 235.91 s $\approx$ 4 min.

The sidereal day is referenced to the vernal equinox which moves slowly through the sky in the westward direction (Fig. 4). The day relative to the fixed stars is thus about 0.01 s longer than that given in (13). The relation (13) is valid even as the earth rotation rate varies irregularly because the mean solar day lengthens proportionally with the sidereal day.

The mean solar day is now defined such that, in about 1820, it would have been exactly 86 400 SI seconds in duration. It is now somewhat greater, by about 2 ms, due to the decreasing (on average) rotation rate of the earth. See discussion of UTC time and leap seconds below.

Universal and atomic times

Universal time (UT) and earth spin

The modern equivalent of GMT is *universal time*. There are two principal versions of universal time, UT (= UT1), and UTC. Here we discuss the former; the latter will come soon. The UT time is obtained from the overhead passages of stars rather than the passage of the sun, but it is continually adjusted (in principle) to closely approximate mean solar time at Greenwich such that the mean sun is on the meridian at noon and one day equals precisely 86 400 s. This time standard thus uses the earth spin as the clock.

Formally, UT is defined in terms of the sidereal time at Greenwich. Specifically, 0 h UT (midnight in Greenwich) on Jan. 1 is defined to occur at Greenwich mean sidereal time (GMST) ≈ 6.7 h. Examine Fig. 3.1 and note that on Jan. 1, which is 285 d after Mar. 21, the sun is about 18.7 h east of the equinox [(285 d/365 d) $\times$ 24 h $\approx$ 18.7 h], that is, $\alpha_\odot \approx 18.7$ h. At midnight in Greenwich on this date, the zenith will be on the side of the celestial sphere exactly opposite (in RA) to the sun, i.e., at 18.7 − 12 = 6.7 h. The sidereal time at Greenwich will thus be about 6.7 h at 0 h UT on Jan. 1, in accord with the definition.

Variations in the earth spin rate result in variations in the rate of time based on star transits, relative to an ideal Newtonian model. These variations can arise from tidal friction which slows the earth spin, varying angular momentum carried by oceans and the atmosphere which cause seasonal variations in the rotation rate and in the motion of the earth's spin axis relative to the earth's crust (*polar motion*). In large part these changes are unpredictable. UT has been corrected for polar motion (UT0 is the uncorrected version), but UT still slows and speeds up with the earth spin. The duration of a second or a minute actually grows and shrinks so that 86 400 s carries one from one mean solar transit to the next. For precise measurements, a variable second is not very useful. We use atomic time for that; see below. The precise relation between UT and GMST is presented in the next section.

Greenwich mean sidereal time (GMST) at 0 h UT

As stated above, one can set a UT clock by observing the stars passing over Greenwich at midnight on Jan. 1. When the appropriate meridian of the celestial sphere (sidereal time ~6 h 42 m = 6.7 h on Jan. 1) transits the zenith, it is exactly 0 h UT. The current relation used to set UT in terms of the star transits at Greenwich gives Greenwich mean sidereal time at 0 h UT in terms of the day of the year,

$$
\begin{aligned}
\text{GMST at } 0\,\text{h UT} = {} & 6\,\text{h }41\,\text{m }50.548\,41\,\text{s} \\
& + (8\,640\,184.812\,866\,T \qquad \text{(Definition of } 0\,\text{h UT)} \qquad (4.14) \\
& + 0.093\,104\,T^2 \\
& - 6.2 \times 10^{-6}\,T^3)\,\text{s}
\end{aligned}
$$

where the parameter T specifies the day for which one desires to find the GMST at 0 h UT. Recall that the part of the sky that is overhead at midnight in Greenwich changes with the seasons. Here T is measured in Julian centuries of 36 525 d from epoch J2000.0 (Jan. 1, 12 h UT, 2000). Specifically, $T = d/(36\,525)$ where d is the number of days from epoch J2000.0. The day count d is constrained to the values of $\pm 0.5, \pm 1.5, \pm 2.5 \ldots$ because it is applied only at 0 h UT. (We discuss epochs below.)

The first term on the right of (14) specifies the $\sim$6.7 h sidereal time discussed above. The second term is linear with time; in one (Julian) day, it advances the GMST at 0 h UT by 236.55 sidereal seconds to take into account the difference between the sidereal and solar days. (The difference between (12) and (13) is somewhat less, 235.91 s, because it is in solar seconds. The two kinds of seconds differ by one part in 365.) The numerical coefficient of this (second) term is the number of seconds of sidereal time that must be added for this correction in a Julian century. As an example, at 2000 Jan. 1, 0 h UT, the date is $d = -0.5$ d and this term subtracts 118.28 s (ST) ($=236.55/2$) from the first term as expected for the solar-day correction in 12 h of elapsed time. Twenty-four hours later, the correction is $+118.28$ s, the next day $+384.83$ s, and so on until in 1 yr the correction would correspond to 24 h bringing the GMST at 0 h UT back to $\sim$6.7 h ST. Finally, the two non-linear terms correct for polar motion; in one century ($T = 1$) the correction is $\sim$0.1 s.

Ephemeris second

Beginning in the late 1920s, it was realized that there were better clocks than the variable earth spin period. Accordingly, the highly stable orbital motion of the earth about the sun was adopted as a reference. Thus in 1958, the *ephemeris second* was defined as a fixed fraction of the "tropical year 1900", namely $1/(31\,556\,925.9747)$. Recall that the tropical year is the time required for the sun to travel from vernal equinox to vernal equinox on the celestial sphere. The ephemeris second was free of the vagaries of earth spin.

Atomic time (TAI)

Atomic clocks introduced a new level of accuracy to time keeping in the late 1940s. Comparisons of different atomic clocks indicated that these uncertainties were about 1 second in 3000 yr, or 0.3 ms in 1 year, or better. This means that the time of a clock tick after one year would have this uncertainty. This is 1 part in 10^{11} since there are $\sim 1 \times 10^{11}$ seconds in 3000 yr. There are several types of atomic clocks such as cesium beams, hydrogen masers, rubidium vapor cells, and mercury ion frequency standards. They have differing accuracies ranging up to 1 part in 10^{14} in a day. The fractional accuracies can differ for different time intervals.

The accuracy of atomic clocks permitted the demonstration by 1972 that the earth's spin period (the traditional clock) is lengthening at a variable rate averaging $\sim$1.7 ms per century, most of which is attributable to tidal action. This is not as small as it sounds because each day the small changes accumulate. After only one year at this slowdown rate, the earth clock would be $\sim$3.1 ms retarded. (After six months, the day would be 1.7 ms/200 longer than at the beginning. Take this as the average excess period during the year and multiply by 365.25 d in a year to obtain 3.1 ms.) Thus the earth clock is good to 1 part in $(1 \text{ yr}/3.1 \text{ ms}) = 1.0 \times 10^{10}$, or about 10 times worse than our rudimentary atomic clock with 1 part in 10^{11} accuracy. Thus atomic time became a better standard than the earth's spin.

In 1967, the *atomic second* (now called the *SI second*) was adopted as a fundamental unit of measurement,

$$1.0 \text{ atomic second} = 9\,192\,631\,770 \text{ cycles of Cs}^{133} \quad \text{(SI second)} \quad (4.15)$$

where the cycle count refers to the radiation from a ground-state transition between two hyperfine levels of Cs^{133}. The number of cycles was chosen to agree with the ephemeris second. *Atomic time* was defined to agree with UT at 1958.0. Currently, the atomic time standard is TAI (*Temps Atomique International*). It is based on the average of $\sim$150 atomic clocks in 30 countries. Note that UT and TAI run independently of one another; the one tied to the transits of the stars (i.e., earth spin) and the other to an atomic standard. TAI is stable to about 30 μs in a century, or 1 part in 10^{14}. Expected advances, e.g., with cold cesium in space, could yield stabilities 100 times better.

Universal coordinated time (UTC) and leap seconds

In 1972, *Universal Time Coordinated* (UTC) was adopted to bring together the UT and TAI systems. It is based on the atomic second but it is occasionally adjusted by the addition of an extra second, a *leap second*, to maintain it within 0.9 s of UT. In principle, the leap second could be subtracted, if necessary to maintain the 0.9 s difference.

Leap seconds are inserted into UTC when needed at midnight on June 30 or Dec. 31. Sometimes a Dec. 31 will last 86 401 s! Compare to the insertion of 3 months by Caesar in 46 BCE. This method allows UTC to reflect the changing spin period of the earth while maintaining the unit of 1 s as a very stable atomic unit of time.

One might ask why one adds an entire leap second every year or two when we argued above that the earth clock is retarded by only $\sim$3.1 ms in a year. Recall that this was the retardation relative to the tick rate at the beginning of the year. The leap-second corrections refer, though, to the date when the day was exactly 86 400 s long, which occurred in $\sim$1820, or $\sim$180 yr ago. Since that time, the day has lengthened by $\sim$1.7 ms/century $\times$ 1.8 centuries $\approx$ 3.1 ms. (Do not confuse this "length of day" with the 1-yr "clock retardation" value of 3.1 ms.) This would suggest that the day is now about 86 400.003 s in duration. In one year, after 365 rotations at this period, the discrepancy relative to the 86 400-s day would build up to 3.1 ms $\times$ 365.25 $\approx$ 1.1 s. Hence, even if the earth spin were to remain constant at a period of 86 400.003 s, the daily accumulation of 3-ms intervals would require a leap second every year or so to keep our watches in step with the sun's overhead transits.

The irregular variations of the earth clock are huge; annual fluctuations alone change the earth period by $\sim$1 ms or more. You can find a plot of the length of day excess over 86 400 SI seconds (LOD) at the US Naval Observatory "Earth Orientation" web site, http://maia.usno.navy.mil; click on "What is Earth Orientation?" The excess was 2 to 2.5 ms in $\sim$1993, but now as I write this in January 2003, it is fluctuating around 0.6 ms. The earth spin has been speeding up (!) over the past decade, but now is relatively steady. At the current 86 400.0006 s period, the earth clock would lose only 0.6 ms $\times$ 365 d = 0.22 s in a year compared to UTC. No wonder there have been no leap second insertions since 1999 Jan. 1 (through 2002).

Terrestrial time (TT)

Astronomers and others have a need for a continuously running time standard that is never adjusted for irregularities of the earth's rotation. A natural choice for this would be atomic time TAI. In fact, since 1972 this has been the standard for such purposes, except for a constant offset of 32.184 s required to match it to the previous (discarded) free-running standard, ephemeris time (ET) described above. This corrected time is known as *terrestrial time* (TT),

$$\text{TT} = \text{TAI} + 32.184\,\text{s} \qquad\qquad \text{(Terrestrial time)} \qquad (4.16)$$

For most purposes, the two time standards, ET and TT can be considered a single seamless time standard: ET before 1958 and TT from 1958 onward. It runs at the same rate as TAI. The TT day contains exactly 86 400 SI seconds.

An earlier name for TT was *Terrestrial Dynamic Time* (TDT). A dynamic time is one that is used as the independent variable in physical equations such as Newton's laws. Thus, in principle, an ideal TT could be obtained only from extended observations of the planetary motions in the solar system, and this time could in principle diverge from atomic time. Divergence could also arise from small inaccuracies in the rates of atomic clocks. At present, nevertheless, TT is defined to have the same rate as TAI, but with the offset given in (16). The nominal atomic clock is taken to be on the surface of the earth because experiments show that clock rates depend on their location in the earth's gravitational potential in accord with the GR model of time.

In practice, TT is a method of keeping track of the leap seconds inserted into UTC. This is important if one wants to calculate time intervals over several years as in the timing of pulsar pulses. The *Astronomical Almanac* published each year gives

Table 4.1. *TT and TAI offsets relative to UTC*[a]

Start (UTC)	End (UTC)	TAI − UTC (s)	TT − UTC (s)
1902 Jan. 1 0 h			~0[b]
1958 Jan. 1 0 h		~0	32.18[b]
1977 Jan. 1 0 h	1977 Dec. 31 24 h	16	48.184
1988 Jan. 1 0 h	1989 Dec. 31 24 h	24	56.184
1990 Jan. 1 0 h	1990 Dec. 31 24 h	25	57.184
1991 Jan. 1 0 h	1992 Jun. 30 24 h	26	58.184
1992 Jul. 1 0 h	1993 Jun. 30 24 h	27	59.184
1993 Jul. 1 0 h	1994 Jun. 30 24 h	28	60.184
1994 Jul. 1 0 h	1995 Dec. 31 24 h	29	61.184
1996 Jan. 1 0 h	1997 Jun. 30 24 h	30	62.184
1997 Jul. 1 0 h	1998 Dec. 31 24 h	31	63.184
1999 Jan. 1 0 h	>2002	32	64.184

[a] *Astronomical Almanac for 2002*, US Naval Obs. and Royal Greenwich Obs. (US Govt. Printing Office). All offsets from 1988–2002 are listed.
[b] ET − UT

a history of the changing offsets of TT and TAI from UTC. For example, during the 1989–1993 mission of the Hipparcos satellite, the TT offsets were 56–59 s and at the end of the century it was, and is at this writing, 64.184 s (Table 1). The irregular variation of the earth spin is evident from the irregular intervals between inserted leap seconds in Table 1. Table 1 shows that TAI was set to agree with UTC in 1958 and TT (actually ET) was set to agree with UTC in 1902. This is the origin of the 32.184 s offset in (16).

Keep in mind that an added leap second in UTC amounts to holding back the clock for one second before it becomes the next day. Thus TT times are increasingly greater than are UTC times, insofar as the leap seconds continue to be positive. One could imagine that someone, someplace, has a 24-h clock that reads TT time; i.e., it is never held back by the insertion of a leap second. It would be running ahead (fast) of a UTC clock by about 1 minute. If I were waiting to celebrate my birthday on 1990 Dec. 7, the TT clock would allow me to start whooping it up 57.184 s before the UTC clock strikes midnight.

If I wanted to know how many SI seconds (atomic time) I had lived, I would find the number of days since my date of birth, taking into account leap years (possibly by using Julian days; see discussion below). I would adjust this for partial days at each end of the interval and then multiply by 86 400 s/d. Finally I would add the difference in the two TT − UTC offsets at the ends of the interval, $(TT - UTC)_{now} - (TT - UTC)_{birth}$. The latter step takes into account the leap seconds that were inserted between the two dates.

TT is a *terrestrial* time scale. It is used for timing events occurring at the earth or distant events observed at earth and timed with clocks located on the earth's surface (mean sea level). For example, it is used for expressing the locations of earth orbiting satellites as a function of time.

Barycentric times

According to the GR model of time, a clock deep in a potential well will run more slowly than one less deep. Also, a clock moving at high speed relative to some stationary clocks runs more slowly than the "at rest" clocks it passes. The latter is the *time dilation* effect also encountered in special relativity. A clock on the earth, with its elliptical orbit, experiences both effects relative to a stationary clock far from the solar system. Each effect has a fixed and a cyclic component. The cyclic terms arise from the eccentricity of the earth's orbit; on an annual basis, both the speed and the depth in the solar potential change.

Barycentric Dynamical Time (TDB) is a *coordinate time* in general relativity; it marks the progress of "time" in the GR model, but is not necessarily the time kept by any particular real clock. It is defined in a coordinate system with spatial origin at the solar-system barycenter. It may be used as the independent variable in the equations of motion that represent the positions and velocities of solar-system bodies in general relativity. One may think of it as the time kept by a clock that is on the surface of a hypothetical earth that orbits the barycenter in a strictly circular orbit at constant velocity, at an orbital radius and velocity typical of the actual earth. It runs at the same rate as TT except that it is free of the cyclical effects of the earth's elliptical orbit, to which TT is subject. The annual periodic term in the difference TT – TDB is of amplitude only 1.7 ms, while other periodic terms due to the planets and moon contribute up to 20 μs.

In 1991 the International Astronomical Union defined *Barycentric Coordinate Time* (TCB), a coordinate time for another system with spatial origin at the barycenter of the solar system. This too can be the independent variable in the equations of motion. TCB is the time kept by a series of synchronized clocks that are at rest relative to the barycenter and far removed from it. They are in flat space where gravitational redshift and velocity effects are null. TCB can be called "far away time". At present, the TCB clock runs faster than TDB by a constant 49 s per century. This time is appropriate for keeping track of events taking place throughout the solar system.

TCB differs from TDB by the constant rate offsets of the two relativistic effects mentioned above, namely time dilation and solar gravitational potential. TCB time runs faster than the TDB clock for each effect. Recall that our hypothetical TDB clock is in an earth-like but circular orbit. It therefore runs slower than the "far away" TCB clocks by (*i*) 33 s per century because it is in the gravitational potential (or space distortion) of the sun's gravity, and (*ii*) 16 s per century because its orbital velocity is 29.8 km s^{-1}, or 10^{-4} times the speed of light, relative to the solar system barycenter. The latter effect is the relativistic time dilation, or equivalently, the transverse Doppler effect. The two effects together yield the above mentioned total rate difference of 49 s per century.

Julian date (JD)

The comparison of observations over many years is not simple because of the varying numbers of days in a year (due to leap years) and the different numbers of days in the several months. Accordingly, a continuously running counting system for days is used in astronomy; these are known as Julian dates (JD). The Julian date 0.0 was set prior to most dates one would encounter in astronomy, namely at noon on Jan. 1 in 4713 BC. It was defined by Justus Scaliger in 1582 at the time of the initiation of the Gregorian calendar and named after his father Julius Caesar Scaliger, not the Roman emperor. We discussed the several calendars in Section 3 above.

Julian days are counted as continuously running numbers which are now approaching 2.5 million. The Julian day beginning at noon on 2000 Jan. 1 is JD 2 451 545, while for Jan. 2 it is JD 2 451 546, etc. The Astronomical Almanac gives equivalences between Gregorian dates and Julian dates (JD). The beginning of each Julian day is now defined to be at noon at Greenwich, either 12 h UTC or 12 h TT. More precise times within a Julian day are indicated with decimal figures, not h m s. If one uses JD with precision of 1 minute or better, it is important to specify the units one is using, such as JD(TT), JD(TDB), or JD(UTC), because, for example, JD 2 451 545.000 00 (TT) and JD 2 451 545.000 00 (TDB) both occur about one minute before JD 2 451 545.000 00 (UTC).

Julian days can be grouped conveniently into centuries of 100 yr. Now the number of days in a Gregorian century will vary depending on the number of leap years in that given century. This motivates reference to the (Roman) calendar wherein each year has exactly 365.25 d (i.e., a leap year every 4 yr without fail) so that a *Julian century* always has 36 525 d. Thus J2100.0 will be exactly 36 525 d later than J2000.0. If further, the Julian day is defined with TT time wherein every day has exactly 86 400 s (no leap seconds), we have a system wherein every Julian century has 86 400 s/d $\times$ 36 525 d/C $=$ 3 155 760 000 s, and where each second is the atomic (or TT) second, also known as the *SI second*.

Epochs for coordinate systems

The equatorial coordinate system used for celestial measurements depends on the orientation of the earth, and this is a continuously changing function of time

(Section 3.2). The time chosen during some period (usually decades) for the specification of celestial coordinates in catalogs and communications between astronomers is called the *standard epoch*, traditionally expressed in years.

The standard epochs in use in the last century, B1900.0 and B1950.0 were based on the Besselian year which begins when the mean sun is at $\alpha = 18$ h 40 m. As noted in Section 3.2, B1950.0 occurred about 2 hours before the New Year of 1950.

The standard epoch in use today is J2000.0 (TDB); it is based on the Julian century/day/second system just described. The epoch J2000.000 was set to occur exactly at 2000 Jan. 1, 12 h (TDB), i.e., at JD 2 451 545.0 (TDB). All other epochs E are defined relative to this,

$$JD(TDB) = 2\,451\,545.0 \ (TDB) \tag{4.17}$$
$$+ [(E - 2000.00)\ 365.25] \qquad (\text{Defines epoch})$$

where E is the epoch in years, e.g., $E = 1991.25$ is the approximate mean epoch of the observations made by the Hipparcos satellite. From (17) one finds

$$\begin{aligned}
J1900.0 &= JD\ 2\,415\,020.0\ (TDB) = 1899\ Dec.\ 31,\ 12\ h\ TDB \\
J1991.25 &= JD\ 2\,448\,349.0625\ (TDB) = 1991\ Apr.\ 2,\ 13\ h\ 30\ m\ TDB \\
J2000.0 &\equiv JD\ 2\,451\,545.0\ (TDB) = 2000\ Jan.\ 1,\ 12\ h\ TDB \qquad (4.18)\\
J2100.0 &= JD\ 2\,488\,070.0\ (TDB) = 2100\ Jan.\ 1,\ 12\ h\ TDB \\
J2200.0 &= JD\ 2\,524\,595.0\ (TDB) = 2200\ Jan.\ 2,\ 12\ h\ TDB
\end{aligned}$$

Each of the century epochs in (17) will occur somewhat before UTC noon according to the TT – UTC offsets; see Table 1. Recall that TDB differs from TT by at most 1.7 ms. Although the definition of epoch is based on TDB time, one would do well to eliminate the possibility of confusion by writing J2000.0 (TDB). Sometimes one sees J2000.0 (TT) which is effectively the same thing, within 1.7 ms. One often sees simply J2000.

The two epochs J1900 and J2000 are separated by a Julian century, exactly 36 525 d; see (17). This leads to a one-day date shift in the Gregorian calendar date because it has no leap day added in 1900 February whereas the older Julian calendar does. (See Section 3 above, "Calendar".) Thus this Gregorian century was one day shorter than the Julian century, namely 36 524 d (to better match the tropical year). The next century yielded identical dates because 2000 was a leap year in both calendars, and the following century again differs because 2100 will not be a (Gregorian) leap year.

This system of Julian centuries thus gradually gets out of step with the Gregorian calendar just as Caesar's calendar got out of step with the seasons, by 3 d every 400 yr or 15 d in 2000 yr. Does this mean that astronomers have adopted again

Caesar's calendar for the epoch definition? Not really, because they do not give month names to it, nor do they live by it. Although they do use the length of the Julian century, they chose not to set the J2000.0 epoch to the extrapolation of the New Year from Caesar's Julian calendar, but rather defined J2000.0 to be on 2000 Jan. 1 of the *Gregorian* calendar, and set it at noon rather than at midnight, following Scaliger's convention for the Julian date.

In another 2000 yr, J4000.0 will occur on Jan. 16 12 h (TDB). On this date, at a rate of insertion of leap seconds into UTC of somewhat less than one per year, TT time will be advanced over UTC by, say, 1500 s more than today's ∼1 min offset. Thus J4000.0 should occur roughly 26 minutes before UTC noon on 4000 Jan. 16.

A *modified Julian date* (MJD) defined as $MJD = JD - 2\,400\,000.5$ is sometimes used. It starts at midnight in Greenwich rather than at noon. MJD is a smaller number than JD and thus is less cumbersome to use in plots and text. Again, if precision is required, one should specify MJD(UTC), MJD(TT) or MJD(TDB).

Signals from pulsars

The timing of signals from outer space has long been a fundamental part of astronomy. The discovery of radio pulsars extended this aspect of astronomy to very short time scales, to seconds and milliseconds. These rotating neutron stars emit a pulse of radio noise once each rotation. The pulse is probably due to acceleration of electrons along the magnetic field lines emerging from the pole of the star. The rotation rate of these stars can be very stable if they have minimal energy loss from magnetic dipole radiation. The pulsar PSR 1937 + 21 is particularly stable; it has a pulse period of 1.6 ms, and its stability rivals that of atomic time. The time standard does not now make use of the signals from such pulsars, but it may come to pass that certain pulsars will become an important time keeping standard.

Problems

4.2 Gravity

Problem 4.21. (a) If the entire mass of the sun were compressed into a sphere the size of the earth, what would be its density relative to that of the earth? (This is typical of a *white dwarf* star.) (b) If your scale indicates 80 kg when you weigh yourself on earth, what would it read if you weighed yourself with it on the surface of the white dwarf? (c) Could you pick up a penny? See Appendix for the masses and sizes of the earth and the sun. A penny weighs about 2 g. [Ans. (b) $\sim 10^7$ kg]

Introduction to Problems 4.22–25. The moon's interaction with the earth's oceans causes the earth to slow down and the moon to gain energy. Energy is dissipated by the movement of the tides across the earth, but angular momentum of the moon–earth system must be conserved. The moon thus moves into a higher orbit and gains angular momentum while the earth loses angular momentum. We explore this evolution in these problems. We neglect the effect of the sun's gravity and other factors affecting the earth spin and lunar orbit. The long-term lengthening of the earth's spin period due primarily to the tidal effect is $(dP/dt)_0 \approx 1.7$ ms per century. The current lengthening of the moon's orbital period is about $(dp/dt)_0 \approx 35.3$ ms per century corresponding to a changing earth–moon distance of 38.2 mm yr^{-1}. The current sidereal periods (relative to fixed stars) of the earth spin and moon orbit are respectively $P_0 = 86\,164$ s and $p_0 = 27.322$ d (1 d $= 86\,400$ s). The earth mass is $M = 6.0 \times 10^{24}$ kg and the ratio of masses (earth to moon) is $M/m = 81.3$. The earth radius is $R = 6.4 \times 10^6$ m and the mean moon–earth center-to-center distance is currently $r_0 = 3.8 \times 10^8$ m. The moment of inertia of the earth is $\sim MR^2/3$. In the following, let all angular momenta be about one axis and ignore moon spin. Take the earth to be a solid body (except for the tides). Use Newtonian expressions for circular orbits where $M \gg m$. Neglect any effects of the sun's gravitational field. Let Ω represent the angular velocity (rad/s or s^{-1}) of the earth spin and ω that of the moon's orbital motion.

Problem 4.22. Earth–moon: Spindown of the earth. (a) If the moon was once part of the earth (but was later somehow ejected while conserving system angular momentum), what would have been the rotation period of the earth before the ejection? (b) If the slowdown rate of the earth spin dP/dt is taken (improbably) to be constant at its current value since it had that period, how long would it take the earth to spin down to its current period? How does this compare to the age of the earth, 4.5×10^9 yr? Ignore the nature of the ejection process. (c) What does our simplistic assumption that $dP/dt = $ constant imply about the rate of change of the angular velocity of the earth $d\Omega/dt$ (rad s^{-2}), i.e., its angular acceleration, as a function of Ω? How does the torque on the earth spin depend on Ω in this approximation? Your answer could be compared to physical models of the torque exerted on the earth by lunar tides. [Ans. $\sim$4 h; $\sim 10^{10}$ yr; $\propto -\Omega^2$]

Problem 4.23. Earth–moon continued. Compare the relative slowing rates to theory. (a) Find an expression for the ratio of the angular acceleration of the earth spin to that of the moon orbit, $[(d\Omega/dt)/(d\omega/dt)]$, in terms of M, m, G, R, ω, under the sole assumption that angular momentum is conserved. Proceed as follows. Write an expression for the total system angular momentum L; include only earth spin and moon orbital motion. Use Kepler's third law to eliminate the orbital radius r of the moon, and finally set $dL/dt = 0$. Find the numerical value of the ratio for the current angular velocity of the moon. (b) Find the actual ratio observed today, from the data on the periods given in the "Introduction" above, and compare to the theoretical value just derived. Hint: What is the relation between dP/dt and $d\Omega/dt$? [Ans. $\sim$40, $\sim$40]

Problem 4.24. Earth–moon continued. Evolution of angular velocities under angular momentum conservation. (a) Find an expression for Ω in terms of M, R, G, ω, and L the total angular momentum (actually its component along the earth spin axis). Under the assumption of strict angular momentum conservation for all time at the current value $L = L_0$, create a plot of Ω vs. ω for the earth–moon system from $\omega \sim 10^{-7}$ s^{-1} to at least $\omega = 10^{-3}$ s^{-1} (i.e., $p = 2\pi/\omega = 1.7$ h). Program your calculator, calculate 3 points per decade of ω (or more if the function changes rapidly), and tabulate your results before plotting. Find analytically and/or empirically, and show on your plot, the position (ω, Ω) where (*i*) the rotations are synchronous, $\Omega = \omega$ (two places), (*ii*) the angular accelerations are equal (see previous problem), (*iii*) the angular momenta of the earth and moon are equal, (*iv*) the spins are those of today, and (*v*) the earth spin is zero, $\Omega = 0$. At each point label or tabulate the equivalent rotational periods in hours or days. (b) For the two cases where $\Omega = \omega$, find the distances to the moon from the earth center, in units of earth radius; i.e., find r/R for each case. Compare to today's values. [Ans. synchrony at $P \sim 5$ h and $\sim$50 d at $r/R \sim 2$ and 90; Ref. Counselman, *ApJ* **180**, 307, 1973]

Problem 4.25. Earth–moon continued. Energy evolution. (a) Find an expression for the total mechanical energy E of the system, including earth spin, moon kinetic, and moon potential energies, the latter going to zero at infinity, as a function of M, m, R, L, and ω where L is the total (conserved) angular momentum. Eliminate Ω with the expression for $\Omega(L)$ from the previous problem, 4.24a. (b) Plot log E vs. log ω for the actual values of M, m, R, and L_0 for values of log ω ranging from –7 to –2. Program your calculator to do this. (c) Demonstrate empirically with your calculator that the energy reaches an extremum (max. or min.) at the two points where $\Omega = \omega$ (see previous problem). (d) Demonstrate analytically that at an energy extremum the frequencies Ω and ω are equal. Hint: justify and use $dE(\Omega, \omega)/d\omega = 0$ and $dL(\Omega, \omega)/d\omega = 0$. (e) If the moon were initially at the inner-orbit extremum (with high $\omega = \Omega$), what are the possible scenarios for its evolution as it experiences tidal friction? (f) Qualitatively, what would be the expected effect on the earth–moon system of ocean tides raised by the sun's gravity after outer synchrony is reached? [Ans. (b) Sample point: when $\omega = 356 \times 10^{-6}$ s^{-1} at the inner, rapid synchrony point, $E \approx +4.2 \times 10^{30}$J]

4.3 Apparent motions of stars

Problem 4.31. Draw a horizon coordinate system for an observer in the northern hemisphere at latitude $+60°$. Show the tracks of five stars: a star that is always north of the observer, a star that never sets, a star that rises and sets in the north and another which rises and sets in the south, and a star never visible to the observer. Indicate with small arrows the range of track locations that satisfies each of these five conditions.

Problem 4.32. (Library problem) (a) Refer to the star charts of *Norton's 2000 Star Atlas* to find the boundaries of the constellation Aquarius. Determine from the rate of motion of the first point of Aries (the vernal equinox) along the ecliptic when the vernal equinox will move into the constellation Aquarius. (b) In what constellation is the sun on *your* birthday in ~2000? The Sun's daily coordinates are in the *Astronomical Almanac*, or you can deduce it from the "Index Maps" in Norton. (c) Is it possible on your birthday to see the stars of the astrological sign of the zodiac for your birthday? If not, could you go to the other side of the earth, to China, to see the stars of your sign on your birthday? If not, how might you manage to succeed in this endeavor? Does precession help? (d) Could you see or observe these stars easily at other times of the year? [Ans. Precession does help, even today!]

Problem 4.33. Make sketches of the apparent annual motion of a given star at distance 32.6 LY (=10 pc) due to parallax and also due to aberration as the earth circles the sun (assume a perfectly circular orbit) for the following cases: (a) a star lying on the ecliptic (at ecliptic colatitude $\theta = 90°$) at ecliptic longitude $\phi = 90°$. (Here, ϕ is measured from the vernal equinox in the same direction as right ascension α; i.e., our star is at $\alpha \approx 6$ h), (b) a star at the north ecliptic pole $\theta = 0°$, and (c) a star 30° from the ecliptic equator at $\theta = 60°$ and at $\phi = 90°$. In each case, sketch the stellar track as *viewed from the earth in the ecliptic coordinate system* indicating magnitudes and directions of angular displacements. Also indicate the star locations on the tracks for the dates Mar. 21, June 21, Sept. 21, and Dec. 21. (d) What are the practical consequences of parallax and aberration for an observer? What measurements would one make in order to construct the track? [Ans. Tracks are (a) lines, (b) circles, (c) ellipses]

Problem 4.34. Find, from the earth's precessional motion (Fig. 4b) and the current J2000.0 coordinates, the approximate J2100.0 coordinates for four hypothetical stars: (a) on the ecliptic at $\alpha = 0$ h, (b) on the ecliptic at $\alpha = 6$ h, (c) at the north celestial pole, (d) at the north ecliptic pole. In each case, first write down the J2000.0 celestial coordinates, α, δ. Obtain your answers by visualizing and sketching the motion of the celestial coordinate system in the locale of the star. Work solely in degrees (not h m s). Make use of flat space and small angle approximations. In each case, give the (approximate) great circle distance (in degrees) that corresponds to the shift of coordinates of the star. [Ans. the shifts are ~1.5°; ~1.5°; ~0.6°; ~0°]

4.4 Lunar and planet motions – eclipses

Problem 4.41. (a) Confirm that the four cyclic periods given in the text lead to similar eclipses at intervals of 18 y 11.3 d (or 10.3 d). Explain the role of each period in bringing this about. (b) Show that the length of the saros is about 1350 yr. Hint: consider that the sun is susceptible to an eclipse for ~35 d while it is near a node. (c) How do the shadow tracks of totality on the earth change location for sequential eclipses in a given saros? Find approximate magnitudes in km and directions of the shifts in the E–W and N–S directions. Hint: consider spin of the earth and the change of sun–moon alignments as the saros evolves. [Ans. (c) $\sim 15\,000$ km E–W; ~ 300 km N–S for shadows at low geographic (or ecliptic) latitudes.]

Problem 4.42. The moon is currently receding from the earth at the rate of about 38 mm/yr. About how long will it take for total eclipses to no longer occur because the moon is never large enough (in angle) to cover the sun? Because of the elliptical orbits of the earth and moon, the current largest angular diameter of the moon (as viewed from earth) is about $33'\,30''$ and the minimum sun diameter is about $31'\,32''$. The current minimum perigee distance of the moon is 3.564×10^8 m; assume that it moves away from the earth at the aforementioned 38 mm/year. [Ans. $\sim 10^9$ yr]

Problem 4.43 (challenging problem). Here we derive the average precession rate of the moon's orbit.

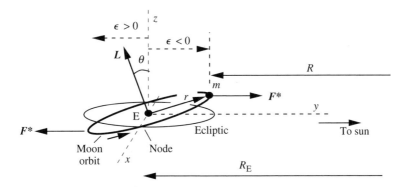

(a) Argue qualitatively from the sketch that the angular momentum vector of the moon's *orbit* should precess about the normal to the ecliptic due to the forces shown. (Neglect the effect of the earth's equatorial bulge.) Will the precession rate be the same at all times of the year? (b) Consider the radial forces (along the sun–earth line) on a test mass m near the earth (e.g., the moon) in the *non*-inertial (accelerating about the sun) *frame of the earth*. Specifically, show that the total (net) force F^* due to the sum of (i) the outward (away from sun) centrifugal force arising from the earth's orbital motion and (ii) the inward attractive gravitational force due to the sun (mass M) is

$$F^* = \frac{3GMm\epsilon}{R_E^3}$$

where R_E is the earth's distance from the sun (1 AU) and $R = R_E + \epsilon$ is the radial distance of the mass m from the sun. How does the force vary at different positions in the moon's orbit? (Hint: see sketch) What does this force have to do with the ocean tides? Why do we neglect the gravitational force due to the (assumed spherical) earth? (c) Demonstrate that, in general, the angular velocity of precession, i.e., the angular velocity of the node, is $\Omega = \tau/(L \sin \theta)$ where τ

and L are the magnitudes of the torque applied to, and angular momentum of, the moon with origin at the earth, respectively, and where θ is the angle between the angular momentum L and the ecliptic normal. (Make an appropriate vector diagram in spherical coordinates.) (d) For the special case where the moon and the sun are both $90°$ from the nodes, what is the torque on the moon about the earth center, and what is the implied angular velocity of precession? Useful values are: $\theta = 5.1°$; $m = 7.35 \times 10^{22}$ kg, $\omega_{\text{moon orbit}} = 2.7 \times 10^{-6}$ rad s^{-1}. [Ans. $\sim 80°$ yr^{-1}; this is a maximum value. Averaging over the lunar orbit and over all sun directions yields approximately the actual value of $19.4°$/yr]

4.5 *Measures of time*

Problem 4.51. Explain the following statement: "It is fortunate indeed that solar and sidereal clocks are not synchronized. If they were, we could never observe certain regions of the sky". Could this situation occur in another star–planet system? If it could, would space-borne telescopes orbiting the planet make more of the sky accessible for observations? Explain.

Problem 4.52. (a) Use (14) to find the precise (to 0.01 s) GMST on 1995 Dec. 7 at 0 h UT. It happens that Dec. 7 is the 341st day of 1995. (b) Repeat for your birthday at 0 h UT in 2005. To calculate the parameter T, watch out for leap years, or use the Julian-date table in the *Astronomical Almanac*. [Ans. (a) 5 h 01 m 11.0 s]

Problem 4.53. (a) What is the approximate *sidereal time* at the instant shown in Fig. 3.1 at Palomar Mountain and in Moscow? Palomar is at longitude $118°$ W and Moscow at $38°$ E. (b) What is the *local* sidereal time at my home on my birthday at 1800 h zone time (6 PM EST) on 1995 Dec. 7 to the nearest second (of sidereal *time*, not arcseconds)? I live in Belmont MA at longitude $71°$ $08'$ $17''$ W. In the preceding problem, or in the 1995 *Astronomical Almanac*, you will find that GMST at 0 h UT on Dec. 7 (i.e., at Greenwich at the beginning of the day) is 5 h 01 m 11.0 s. You should correct this for (i) clock time noting that Boston time is 5 h earlier than Greenwich UT and that the sidereal day is only 86 164.1 s long, and (ii) for longitude. (c) What is the sidereal time at *your* home at 6 PM on *your* birthday this year (as you read this)? [Ans. (b): $\sim$23 h]

Problem 4.54. How many days have elapsed between your birth and your 20th birthday? How many SI seconds (SI = TT)? Try using the Julian date table in the *Astronomical Almanac*.

Problem 4.55. (Some knowledge of relativity would be helpful for this one, but it is not necessary.) Demonstrate that a terrestrial (on earth) clock should run more slowly than Barycentric Coordinate Time (TCB) by 49 s per century. Take into account separately the two effects mentioned in the text: (a) the earth clock is in the gravitational potential well of sun, and (b) the earth is moving fast ($1.0 \times 10^{-4}c$) relative to the solar system barycenter. Hints: (a) a photon leaving an atom in a gravitational potential well at radius r from a non-rotating mass M with frequency v_r will arrive at infinite distance with the lesser frequency v such that

$$\frac{v}{v_r} = \left(1 - \frac{2GM}{c^2 r}\right)^{1/2}$$

An observer detecting it at infinity concludes therefore that the originating atom (clock) is running slower than his local clock. In the case of weak gravity, the fraction inside the parentheses is much less than unity. (b) Apply the relativistic time dilation ratio $\gamma = [1 - (v/c)^2]^{-1/2}$. Note that we mix two theories here, general and special relativity; in fact GR embraces both effects. [Ans. (a) + (b) $\approx$ 50 s]

5

Telescopes

What we learn in this chapter

Telescopes and **antennas** collect photons, and the detectors at their foci record the information content of the radiation, its **intensity** and **polarization** as a function of time, and also its **frequency distribution** and **direction of arrival**. There are several common **configurations of optical telescopes**. **Focal length** and **aperture** determine the **plate scale**, **sensitivity** and potential **resolution** of the telescope. **Non-focusing** instruments are used by **gamma-ray astronomers** while **x-ray astronomers** use both focusing and non-focusing systems. Telescope resolution may be limited by **diffraction**. The **point-spread function** describes the shape of the (single pixel) telescope **beam**. The resolution of large ground-based optical telescopes is severely limited by **non-planar wavefronts** caused by atmospheric turbulence. **Speckle interferometry and adaptive optics** are techniques for overcoming this limitation.

5.1 Introduction

The systems that extract information from faint signals about distant celestial bodies are the source of essentially all our astronomical knowledge. Telescopes collect and concentrate the radiation, and the instruments at their foci analyze one or more properties of the radiation. The systems used for the various frequency bands (e.g., radio, optical, and x-ray) differ dramatically from one another.

The faint signals must compete with background noise from the cosmos, the atmosphere, the earth's surface, and the detectors themselves. These noise sources differ with the frequency of the radiation. Advances in astronomy often follow from improved rejection of noise so that fainter signals can be detected. For example, improved focusing yields a smaller spot on the image plane (film or CCD), and the signal need only compete with the smaller amount of instrument noise occurring at this smaller region.

In this chapter, we present some characteristics of telescopes including their focusing properties. The diffraction phenomenon that can limit the resolution of telescopes is described. Current efforts to develop systems to remove the blurring due to the atmosphere are also discussed. In the following chapter we describe detectors that are placed at the foci of the telescopes.

5.2 Information content of radiation

All astronomical telescope and detector systems have the same purpose, namely, the study of incoming photons with the maximum possible sensitivity, and with the optimum frequency, timing, and angular resolution. One can not always attain the best possible performance in all these aspects at the same time.

A stellar object at a great ("infinite") distance appears to us as a "point" source; its angular size is smaller than is resolvable by our eye or instrument. Light rays may diverge isotropically from it, but at the great distance of our telescope, the small subset of rays impinging on it is effectively parallel. The beam of photons arriving at the earth is thus like rain falling everywhere parallel to itself. In terms of waves, the wavefronts are everywhere normal to the propagation direction. This signal from a point-like source is called a plane wave.

Telescopes capture the portion of the incoming energy that impinges on the telescope aperture. A larger telescope can collect more energy (rain) each second. The instruments on the telescope are used to determine the properties of the collected electromagnetic radiation (or incoming photons). The properties that can be measured are limited in number. They are:

(*i*) The rate (number per unit time) of arriving photons. This rate follows from the total power radiated by the source of the radiation, the average energy of the individual photons, and the distance to the source. (At radio frequencies one measures the amplitude of the electromagnetic wave in lieu of counting photons.) This rate can vary with time, for example from variable stars and pulsars. The former variations arise from periodic changes in radius and brightness of a star, while the latter arise from the rotation of a neutron star.

(*ii*) The arrival directions of the photons, or equivalently, the regions of the sky from which they originate. This allows one to describe the angular shape of the source on the sky. The photon numbers and energies from different directions determine, for example, the brightness distribution of a diffuse nebula.

(*iii*) The photon energy $h\nu$ (or equivalently the frequency or wavelength) of the radiation. This allows one to determine how the incoming radiation is distributed in frequency (the *spectrum*). For example, a concentration of energy at one frequency (a *spectral line*) would indicate the existence of a particular atom, such as hydrogen, undergoing a specific atomic transition. The existence of such transitions gives information about the temperatures and densities in the atmospheres of stars as well as their speeds from Doppler shifts of the frequency.

(*iv*) The polarization, i.e., the directions of the transverse electric vector E of the incoming electromagnetic wave. A predominance of vectors in one direction is indicative of polarized light. This can indicate, for example, that the emitting particles (electrons) are significantly influenced by magnetic fields in the emitting region or that the light has been scattered by dust grains in the interstellar medium.

The sensitivity or precision with which a given telescope–detector system is able to measure these quantities is crucial to understanding the data obtained with it. For example, the angular resolution is the capability of the system to distinguish (or resolve) two adjacent objects, expressed as the minimum separation angle. This is typically $1''$ for a ground-based optical telescope, about $0.05''$ for a large space-borne optical telescope (e.g., the Hubble Space Telescope), and better than $0.001''$ for several radio telescopes operating together from locations on different continents.

Similarly, frequency resolution is the ability to distinguish two spectral lines at closely adjacent frequencies. High-dispersion, *echelle-grating spectrometers* used on optical telescopes attain resolutions in wavelength of $\Delta\lambda \lesssim 0.10$ nm. Since the wavelength of optical radiation is $\sim$500 nm, the *resolution*, defined as $\lambda/\Delta\lambda$, is $\sim$5000. Timing resolution is the ability of the instrument to distinguish the arrival times of single photons (or groups of photons) that arrive at closely spaced times. Pulses of radio emission arrive from spinning neutron stars with separations as small as 1.6 ms.

5.3 Image formation

Telescopes and antennas are the light collectors of astronomy. They come in varying shapes and sizes that depend in part on the frequency of radiation they are designed to detect. Most systems concentrate the incoming radiation by means of *focusing*. Optical telescopes gather light with a lens or a reflecting surface (a mirror). Radio telescopes make use of reflecting metal surfaces. X-ray telescopes make use of the reflecting character of a smooth metal surface for x rays impinging on it at a low glancing ("grazing") angle, like a stone skipping on water. Some radio and x-ray detection systems and all gamma-ray systems do not focus the radiation. These are non-focusing systems.

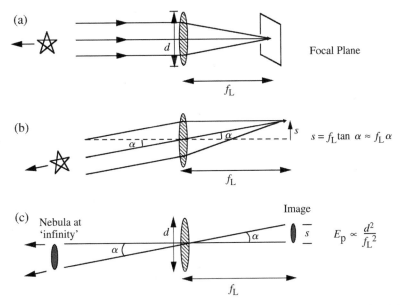

Figure 5.1. The focusing characteristics of an ideal thin lens. The focal length f_L and the aperture d are each measured in meters. (a) Parallel beam arriving along the lens axis, and focusing at distance f_L. (b) Off-axis parallel beam (small angle α) also converging at distance f_L, but displaced a distance $s = f_L \tan \alpha$ from the lens axis. (c) Extended source subtending angle α and depositing, in a fixed time, energy $E_p \propto (d/f_L)^2$ onto a single pixel of the image plane.

Focusing systems

Focal length and plate scale

The radiation from a very distant point-like star arrives at earth as a parallel beam of light. If the light impinges normally onto a thin (ideal) convex lens (Fig. 1a), a parallel bundle of rays will focus to an on-axis point image in the *focal plane*, a distance f_L (*focal length*) beyond the lens. If the parallel rays arrive at an angle α from the lens axis, they will also focus to a point in the focal plane (Fig. 1b), but at a distance s removed from the optical axis. The vertical position of the focus is defined by the ray that passes through the center of the lens; it will transit the (thin) lens without being deviated. The relation between these quantities is

$$s = f_L \tan \alpha \xrightarrow[\alpha \text{ small}]{} f_L\,\alpha \qquad \text{(m)} \qquad (5.1)$$

This geometry can also be applied to a properly figured concave mirror (or system of mirrors) that brings a parallel beam of light to a focus.

If there were two stars, one on axis and the other off axis as shown in Figs. 1a,b, they would be separated in the sky by the angle α and on the plate by the distance *s*. The relation (1) gives the star separation in the focal plane, or on a photographic plate exposed in the focal plane. A nebula with angular diameter α (Fig. 1c) would have this same diameter *s* on the photograph. A large focal length yields a large star separation or nebular image, and a small focal length yields a small image. A small focal length requires the lens or mirror to refract the rays more strongly (Fig. 1a), and this leads to difficulties of design and limitations in performance (e.g., depth and breadth of the well-focused region). On the other hand, such a telescope can fit inside a smaller, and hence cheaper, telescope building.

The *plate scale* describes the angle that is imaged onto unit length of the plate; it is simply the ratio of α and *s* which is the inverse of the focal length,

$$P_s = \frac{\alpha}{s} = \frac{1}{f_L} \qquad \text{(rad/m; plate scale)} \qquad (5.2)$$

The units are m^{-1} or, equivalently, radians/meter; radians have no dimension. A large plate scale means the image size *s* is small and vice versa. In practice, the plate scale is usually given in "arcsec per mm" ($''$/mm). The focal length of the Lick 3-m (diameter) telescope is $f_L = 15.2$ m giving a plate scale of $14''$/mm at the prime focus. A $1°$ piece of the sky would occupy a full $1/4$ m in the focal plane. A given telescope may offer a choice of several focal lengths, for example by changing the *secondary mirror* (see below).

Aperture and deposited energy

The rate of energy deposited on a single grain of film, or on the single pixel of a modern electronic imaging device, determines whether a given incident energy flux can be detected in a given time. A large telescope aperture (diameter *d*) will increase the energy flow onto the detector because a larger part of the incoming wavefront is intercepted and focused. For a perfect point source, perfect atmosphere, and perfect lens, all the collected photons from a source will be deposited onto the same grain of the film or the same pixel of an electronic detector. In this ideal case, the aperture alone determines the needed exposure; the focal length does not enter.

However, if the celestial source has a significant finite angular size, like the moon or a nebula, the energy will be deposited over a number of grains or pixels (Fig. 1c). If the image is spread over a large number of pixels because of a large focal length, a longer exposure is required to obtain a detectable signal in a given pixel. If, in contrast, the image is concentrated in a small region because of a short focal length, there is more energy deposited in a given pixel in a given time. In this case, the source image appears smaller but is more quickly detected.

The area of a circular image of diameter *s* is proportional to s^2. Thus, for a nebula of (fixed) angular size α, the energy deposited onto a single pixel for a fixed telescope mirror aperture is $E_p \propto s^{-2} \propto f_L^{-2}$ where we used $s \propto f_L$ from (1). Also, a larger aperture will allow more photons to be collected proportionally to the collecting area a of the mirror where $a \propto d^2$. Thus, for diffuse sources, the energy deposited per unit time onto a single pixel depends on both the aperture and the focal length,

$$\Rightarrow \quad E_p \propto \left(\frac{d}{f_L}\right)^2 \qquad \text{(Energy per unit time onto single pixel)} \qquad (5.3)$$

The ratio f_L/d is called the *focal ratio*,

$$\mathcal{R} \equiv \frac{f_L}{d} \qquad \text{(Focal ratio)} \qquad (5.4)$$

From (3) and (4) it is apparent that the focal ratio is an inverse measure of how fast energy is deposited on an element of the image plane. One refers to the "speed" of the optical system; it is proportional to the energy E_p deposited in a given time. Thus, speed $\propto E_p \propto \mathcal{R}^{-2}$. A greater speed means a photograph or other measurement may be carried out in less time.

The *focal ratio* is usually indicated with the notation "$f/\mathcal{R}$"; i.e., a focal ratio of 6 is written "$f/6$", when, in fact, it is the aperture d which is equal to $f_L/\mathcal{R}$; see (4). When you see "$f/6$", think "$\mathcal{R} = 6$, the focal length f_L is 6 times the aperture d". Amateur cameras usually have a focal ratio adjustment which can be varied from about $f/2$ to about $f/16$. This is accomplished by changing the aperture with an adjustable diaphragm; the focal length does not change. A "50-mm lens" refers to the focal length f_L. Zoom lenses change the focal length.

The speed of an optical system depends only on the ratio $\mathcal{R}$; it is independent of the specific camera used. A 1-m telescope with $f/6$ optics will be just as fast as a 4-m system with $f/6$ optics. Less light is collected by the smaller 1-m aperture, but the focal length is also shorter. This causes the energy to be concentrated onto a smaller area in the image plane. Thus the energy deposited per pixel remains the same for the two systems. Slower optics (greater $\mathcal{R}$) are sometimes desirable; the image is more spread out (greater magnification) and the angular resolution is improved. But it takes longer to get a good exposure.

A special kind of telescope design, the *Schmidt telescope* uses a refracting corrector plate in front of the principal (primary) mirror to produce high quality images over a large $5° \times 5°$ angular field. It features a short focal length. The focal ratio for the large Palomar Schmidt telescope is $\mathcal{R} = 2.5$. The short focal length yields low magnification causing a lot of energy to be focused onto each pixel. The Schmidt design is thus very fast. It was the ideal instrument to make the Palomar Observatory Sky Survey described in Section 3.4.

Telescope configurations

Optical light may be collected and focused by means of a transmitting lens that refracts the rays as shown in Figs. 1a and 2a. The disadvantage of a lens in astronomy is that the light must traverse the glass which can lead to imperfect focusing due for example to color dependence of the index of refraction (*chromatic aberration*). Also, the lenses become very heavy as they become larger. This makes difficult the precision positioning and support of the lens in a movable telescope structure as is required for a good focus. In contrast, a relatively lightweight mirror may be used to reflect the light to a focus (Fig. 2b). In this case it is only the shape of the mirror that must be precisely machined and maintained. All current major optical telescopes use reflective optics.

A *primary mirror* concentrates the light at a *primary focus* (Fig. 2b). For the very largest telescopes, instruments can be mounted at the primary focus and an astronomer can ride around in a little cage which surrounds the focus. This allows her to change photographic plates, all the time listening to classical music, seemingly suspended among the stars. We say "ride" because the telescope will point to many different positions in the sky on a typical night of observing. Unfortunately, modern electronic detectors generally make such rides unnecessary now. Most radio telescopes operate in the prime-focus configuration. They often feature a large metallic antenna dish that reflects the radiation to a detector at the prime focus.

Alternatively, a *secondary mirror* can direct the light through a hole in the large primary mirror to the *Cassegrain focus* (Fig. 2c). The secondary mirror blocks some of the light entering the telescope, but only a small fraction of it. The Cassegrain focus is convenient because large, heavy instruments can be mounted more easily on the back end of the telescope. The secondary mirror can be changed to modify the effective focal length in many telescopes. One practice is for the secondary-mirror structure to contain two mirrors (Fig. 2c). Rotation of the secondary mirror structure by 180° moves the alternate mirror into the beam.

The light from the primary mirror can be intercepted by a flat mirror which directs the light toward the side of the telescope to the *Newtonian focus* which can be examined with an eyepiece mounted on the side of the telescope (Fig. 2d). This scheme is often used for amateur telescopes. In large telescopes, the light can be directed by a series of mirrors to a temperature-controlled room below the telescope to the *Coudé focus* (not shown). There a large spectroscope disperses the light into its spectral colors with high resolution (large $\lambda/\Delta\lambda$).

Traditionally optical telescopes are mounted with two orthogonal axes of rotation, one polar (pointing to the celestial pole), called a *polar mount*; see "Why equatorial coordinates?" in Section 3.2. Radio telescopes and new larger optical telescopes

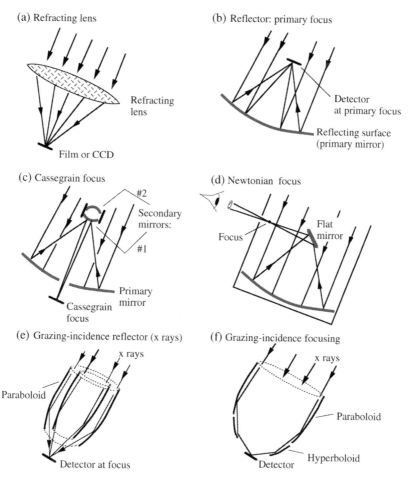

Figure 5.2. Focusing systems. (a) Refracting lens. (b,c,d) Reflecting telescope with primary focus, Cassegrain focus, and Newtonian focus, respectively. The choice of two secondary mirrors in a given telescope allows the telescope to operate with two different focal lengths. (e,f) Grazing incidence x-ray telescopes with one reflection and two reflections. Additional collecting area can be obtained by nesting the mirrors as shown in (e). The foci shown here are unrealistically close to the collectors.

are often mounted in the *altitude-azimuth* configuration where one axis or rotation is horizontal and the other vertical. This simplifies the mechanical design and makes it easier to keep the primary mirror in its optimal shape. It is easily distorted by thermal gradients and the varying forces of gravity as the telescope moves to new orientations. The newest optical telescopes have sophisticated systems for monitoring and adjusting the shape of the primary mirror so that it can be kept in its optimal shape throughout a night of observing.

Telescope designs also pay great attention to keeping thermal gradients and convection to a minimum. Unknown to astronomers for a century was the fact that thermal currents within the telescope building were a major source of poor image quality. This arose from solar heating of concrete structures that are slow to cool and to power dissipation in the building (heated offices, etc.) The Magellan/Baade telescope in Chile is obtaining images as good as 0.30″ (full-width at half maximum intensity) with good thermal and mirror-shape control.

The grazing incidence arrangement for x-ray astronomy ($E \lesssim 10$ keV) is shown in Fig. 2e. The x-rays reflect off a very shiny surface that looks like the inside of a cylinder but in reality has a parabolic shape in cross section, a paraboloid. This form will focus perfectly a point source at great distance that is on the axis of the paraboloid, but will not focus well objects that are off the axis. In the most elegant systems, the x rays reflect twice from the inner surface, first from a parabolic surface and then from a hyperbolic surface (Fig. 2f). This provides good off-axis focusing.

The grazing geometry yields a relatively small collecting area since the photons must strike the inclined surface of the mirror. To improve the collecting area, a number of mirrors can be nested within one another; a second such mirror is shown in Fig. 2e. Some x-ray systems emphasize large collecting area rather than high angular resolution. They have conical mirrors that approximate the parabolic shape and have dozens of such mirrors nested about a common axis.

Non-focusing systems

Electromagnetic radiation with photon energies above ∼10 keV (e.g., high-energy x rays and gamma rays) can not be focused with currently known techniques. Thus alternative non-focusing methods of constructing images are used. Although generally non-focusing systems are much less sensitive to faint sources, they are sometimes preferred even when focusing is available. For example, in the 1–10 keV band, the study of time variability of bright sources requires large collecting areas which are difficult to attain with x-ray focusing systems. The Rossi X-ray Timing Explorer (RXTE, launched 1995) uses such techniques to study black hole and neutron star systems.

Tubular and modulation collimators

Mechanical collimators may be used to restrict the regions of the sky that their detectors can "see". One type is simply a set of stacked tubes, like handfuls of soda straws (Fig. 3a). Since the radiation from a particular point on the sky impinges on the entire detector, the signal from the star must contend with background from the entire detector. The range of angles $\Delta\theta$ from which photons can reach a point on the detector through a tubular collimator is shown in Fig. 3a (inset); $\Delta\theta \approx d/h$.

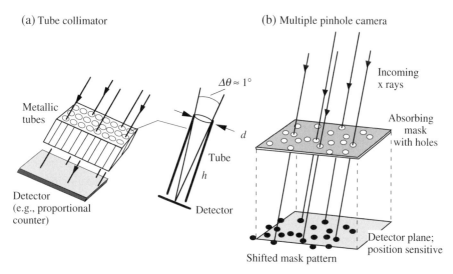

(a) Tube collimator

(b) Multiple pinhole camera

Figure 5.3. Non-focusing collimation. (a) Mechanical tubular collimation ("soda straws") and an expanded view of one of the tubes showing the angular extent of the field of view. (b) Operation of a multi-pinhole camera. The location of the shifted image of the mask pattern reveals the arrival direction of the photons.

This opening angle is comparable to the angular resolution. The angular resolution attainable with tubular collimators is typically $\sim 1°$. The detector used with such collimators is likely to be a *proportional counter* or a *crystal scintillator* (Sections 6.2 and 6.4).

Tubular collimators make possible x-ray or gamma-ray systems with larger apertures than focusing telescopes, albeit at the expense of angular resolution. Timing studies of bright x-ray sources are made possible by the large aperture because large numbers of detected photons are needed to provide the required statistical precision in short time intervals. The detectors need not be *position sensitive*; the detector output does not distinguish where on the detector the photon landed. It is the mechanical collimator that defines the arrival direction of the incoming rays.

Multiple pinhole collimator

If one uses a detector that locates the position at which a photon strikes its surface, such as a piece of film, or a CCD detector (Section 6.3), other arrangements become possible. For example, a mask with randomly placed "pinholes" may be placed above the detector. A point source will then project a pattern of pinhole images onto the detector (Fig. 3b). The sideward displacement of the pattern on the detector surface directly indicates the direction from which the photons arrived, i.e., the location of the source on the sky. This is an extension of the simple pinhole camera from one to many pinholes. The multiple pinholes provide more effective area, but they introduce more interference between the patterns if there are two or more sources in the field of view (FOV). A typical system would have a large FOV, say 30°. The angular resolution is much less, say 10', because it derives from the (small) pinhole diameter and the (large) distance between the mask and detector.

The existence and location of multiple point sources in the FOV of such a system can be deduced by searching the data mathematically for the multiple mask patterns. At each point x,y on the detector surface, there is an accumulated or recorded number of photons. Subtract a constant from all the values to force the average over all x,y to be zero. That constant will, of course, be the average of the original numbers. The result is a function that varies with x,y and takes on positive and negative values from pixel to pixel. Call this the response, $R(x,y)$; it is the real data. Now, compare this to the response $R_p(\alpha,\delta,x,y)$ expected at x,y for a single hypothetical point source at some arbitrary sky position α,δ, and similarly adjusted to zero average. This is a trial function. This comparison is accomplished with the *cross-correlation function* (CCF), $C(\alpha,\delta)$,

$$\Rightarrow \qquad C(\alpha,\delta) \equiv \iint_{\substack{\text{All } x,y \text{ in} \\ \text{plane of detector}}} R(x,y)R_p(\alpha,\delta,x,y)\,\mathrm{d}x\,\mathrm{d}y \qquad (5.5)$$

(Cross-correlation function, CCF)

If one had chosen the response function for the position α,δ where a source really exists, and if that source were the only one in the field of view, the two functions R and R_p would be identical at all points x,y, except for statistical fluctuations in the real data. When one is positive, the other would also be positive, and similarly for the negative excursions. Thus, at all points, their product would be positive, and the sum (integral) of the product over the detector plane would then yield a large positive number. If on the other hand, there is no source at the chosen position, the two functions (trial and real) will differ more or less randomly with respect to one another. The products will be both randomly positive and negative, and the result of the integration will be approximately zero.

Calculation of $C(\alpha,\delta)$ for all trial positions (α,δ) in the field of view would thus yield only one position (resolution element) with a high value. All other positions would yield near-zero correlations. A sky map could be constructed with the values of $C(\alpha,\delta)$. It would show a high spot at the position α,δ of the actual source.

This technique can be used even if several celestial sources are in the field of view; the CCF map will reveal them. Each will appear as a bump in the map roughly proportional to its intensity. The net result is a pseudo or reconstructed image of the sky. The map may contain *side lobes* and other bumps or features that are not in the real sky because the mask patterns are not perfectly random. Thus one must

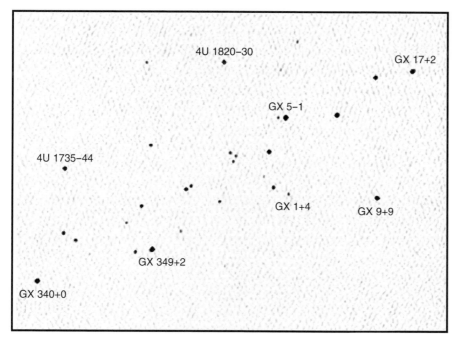

Figure 5.4. Reconstructed "image" of x-ray sources in the galactic center region, obtained with the wide field camera (WFC) multiple-pinhole instrument on the Italian–Dutch BeppoSAX satellite (launched 1996). The ~30 x-ray sources above threshold lie along about 40° of the galactic plane. The GX source names are in galactic coordinates (degrees) while the 4U names are in celestial coordinates. [From J. In't Zand, *Proc. 4th Integral Workshop; ESA-SP*, astro-ph/0104299]

take care not to believe everything one sees in such a map, or in data of any kind. (See Section 7.3 for map cleaning algorithms.)

A French experiment Sigma on the Russian satellite Granat and the Dutch Wide Field Camera (WFC) on the Italian BeppoSAX satellite were pinhole cameras that imaged the x-ray sky at photon energies of 2–30 keV. Since many of these sources are variable in intensity, successive exposures permitted the intensities to be monitored on extended time scales. A reconstructed image of the galactic center region from the WFC is shown in Fig. 4.

Although such systems can yield sky maps including a number of discrete sources, they do not provide true imaging. Unlike focusing systems, the signal from a given point source illuminates the entire counter and hence must contend with the background from the entire detector as well as with the flux from all the other sources in the field of view. These both contribute to the background noise. Such systems generally yield angular resolutions down to a few arcminutes whereas focusing systems yield resolutions of $\lesssim 1''$ in the optical and x ray.

Some real telescopes

Most radio astronomy is ground based, e.g., the new (in 2000) steerable-dish 100-m diameter Byrd Green Bank Telescope in West Virginia, the huge (300-m) fixed antenna dish in Arecibo, Puerto Rico, and the cooperative world-wide collection of diverse telescopes that work in concert to obtain extremely high angular resolutions, called *very long baseline interferometry* (VLBI; Section 7.4). Radio astronomy from space can extend our knowledge of the radio sky to frequencies below the ionospheric cutoff at ~10 MHz. Ideally this would be carried out with telescopes on the far side of the moon to eliminate radio noise from the earth. Observations from orbiting satellites can improve the angular precision of VLBI measurements. The Japanese HALCA satellite, launched in 1997, carried out VLBI in conjunction with ground-based telescopes, and a successor mission is being planned.

At microwave frequencies, the Cosmic Background Explorer (COBE) satellite (1989–93) studied the residual radiation from the hot early universe and established with greatly increased confidence that it did in fact originate in the early universe. It also found tiny fluctuations in its brightness that are most likely the origins of galaxy formation. Recent followup studies of these fluctuations with, for example, the Wilkinson Microwave Anisotropy Probe (WMAP, 2001) yield improved values for parameters that describe the universe, such as the Hubble constant. The Submillimeter Wave Astronomy Satellite (SWAS, 1998) studies star formation; it maps the location of spectral lines from molecules such as H_2O and O_2 in *giant molecular clouds* and *dark cloud cores* at frequencies near 500 GHz. The Submillimeter Array is an array of eight 6-m telescopes being built on Mauna Kea in Hawaii that will reach to 900 GHz with angular resolutions as fine as 0.1''.

In the infrared, the ground-based all-sky survey at wavelength 2 μm (2MASS) has greatly expanded our knowledge of the IR sky. The orbiting Infrared Astronomical Satellite (IRAS, 1983) had previously surveyed the sky at 12, 25, 60, and 100 μm, and studies of these sources were the objective of the European Infrared Space Observatory (ISO, 1995–1998). Infrared radiation passes through interstellar clouds of gas and dust much more easily than does optical radiation. Thus IR astronomers can peer into dusty regions where stars are currently being formed. A major infrared satellite mission is the Space Infrared Telescope Facility (SIRTF, scheduled for 2003).

At optical wavelengths, a number of large (~10-m diameter) ground-based optical telescopes with new capabilities are now operational or coming on line, e.g., the Keck twin telescopes on Hawaii, the Gemini twin telescopes in Hawaii and Chile, the twin Magellan telescopes in Chile, the European quadruple Very Large Telescope in Chile, and the Japanese Subaru Telescope in Hawaii. In space, the

Hubble Space Telescope (HST, 1990), being above the atmosphere, provides resolution of about 0.05″. The more finely focused images enable it to detect fainter and more distant objects than heretofore possible. *Speckle interferometry* and *adaptive optics* are new techniques coming into use to overcome the atmospheric blurring for ground-based telescopes, at least for some observations (see below). The use of multiple telescopes at one site enables optical astronomers to use *optical interferometry* to greatly improve angular resolutions (Section 7.4), at least for bright sources.

The orbiting International Ultraviolet Explorer (IUE, 1978) was sensitive over the range 3.8–10.8 eV (320–115 nm). The HST, although primarily an optical telescope, is sensitive from the near-IR up into the UV band: $E = 1.1$–10.8 eV ($\lambda = 1100$–115 nm). The important Lyman-alpha transition of hydrogen at 10.2 eV (see Fig. 10.1) is accessible to the HST. Spectroscopy in the far UV is being carried out by the Far Ultraviolet Spectroscopic Explorer (FUSE, 1999, 10.3–13.8 eV, 120–90 nm) which studies the spectral lines of atoms in the diffuse gases of the interstellar medium, supernova remnants, and other galaxies. At higher photon energies, beyond the 13.6 ionization energy of hydrogen which makes the ISM quite opaque, the Extreme Ultraviolet Explorer (EUVE, 1992–2000) studied nearby isolated hot white dwarfs, white dwarfs in accreting binary systems (cataclysmic variables), and the local hot interstellar medium itself.

The Uhuru satellite (1970) was the first orbiting satellite dedicated solely to celestial x-ray astronomy; it demonstrated the existence of neutron stars in binary stellar systems. It was followed by a succession of primarily US, European, and Japanese satellite missions that extended the field into a major branch of astronomy with imaging, spectroscopic, and timing capabilities. Current or recent missions include the German Roentgen satellite ROSAT (1990), the Japanese ASCA (1993), the US Rossi X-ray Timing Explorer (1995), and the Italian and Dutch BeppoSAX (1996). More recently launched (1999) are the powerful US Chandra and the European XMM–Newton observatories. They are large x-ray telescopes with reflective optics that focus x rays, yielding images and spectroscopy of distant objects with unprecedented sensitivity.

Gamma-ray astronomy has been carried out from space with a series of satellites leading up to the recent Compton Gamma-Ray Observatory (CGRO) which studied *gamma-ray bursts* (GRB) extensively. The recently launched Integral and the forthcoming SWIFT and GLAST missions will carry on the field. A major breakthrough in gamma-ray astronomy has been the determination that the explosive GRB take place in distant galaxies and hence are the most energetic explosions known, except for the Big Bang origin of the universe.

At energies beyond about 10^{12} eV, the field is called TeV astronomy or VHE (Very High Energy) astronomy. The fluxes of such energetic gamma rays are so low that detections by a satellite would be rare. However, the gamma rays of these energies develop in the atmosphere into a cascade of electrons and lower energy gamma rays, known as an *extensive air shower* (EAS; Section 12.3). The electrons emit light in the atmosphere (Cerenkov light; Section 12.2) that can be seen from ground level with big crudely-focusing (by optical-astronomy standards) light-collecting telescopes. Some dozen sources have been detected, primarily galactic pulsars, supernova remnants and jets in extragalactic *blazars* (the nuclei of very active galaxies). This astronomy has been carried out with ground-based telescopes, e.g., in Arizona and the Canary Islands at the Whipple and HEGRA observatories, respectively. Although the number of detected sources is few and the angular resolution modest, the processes studied are among the most energetic in astronomy. New more powerful facilities coming on line over the next few years are HESS (European), CANGAROO III (Australian), and VERITAS (US).

Gamma-ray astronomy above $\sim 10^{14}$ eV is carried out through detections of electrons in the gamma-ray initiated EAS that reach ground level. To date, no verified signals from point sources have been detected in this energy range. The high background due to EAS initiated by the more numerous incident protons mitigates against such detections.

5.4 Antenna beams

Meaning of a "beam"

The concept of an *antenna beam* is intrinsic to all astronomy. The beam is simply the portion of the sky observed by the detector at a given time (Fig. 5). For example, in a non-focusing detection system, mechanical collimators might restrict the *field-of-view* to a circular region on the sky of 0.7° radius. The detector would be said to have a 0.7° beam (half width) or 1.4° (full width) that views $\sim \pi (0.7)^2 = 1.5$ deg^2 of the sky.

A parabolic radio antenna is a classic example of a focusing system. If this antenna were broadcasting (rather than receiving), the power would be emitted more or less into a cone of angles, the antenna beam, with the power per unit solid angle at a maximum on the view axis and falling off at increasing angles from it. The power would not be emitted in a perfectly parallel beam (i.e., to a point at infinity). This is due to the phenomenon of diffraction that arises from the limited diameter of the antenna.

The same antenna in the receiving mode receives radiation from this same cone of angles; any celestial source within it would be detected, with efficiency (sensitivity) depending on the source location relative to the view axis. If several point-like sources lie in this region, they would be confused or "unresolved" (Fig. 5a). The

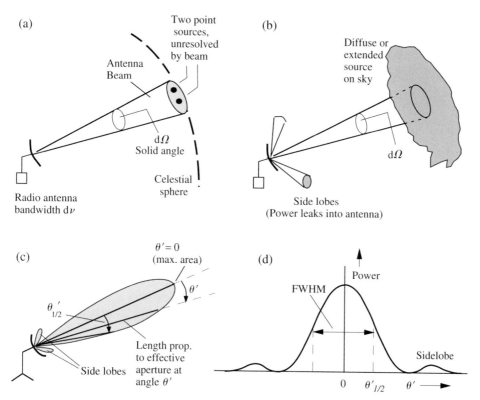

Figure 5.5. Antenna observing (a) two adjacent point sources that are not separated (resolved) by the beam and (b) a diffuse source that has greater angular extent than the beam. The beam includes side lobes wherein small amounts of power from unwanted directions can enter the receiver. (c) Power diagram; the beam is drawn so that the radial distance from the origin (antenna) to the solid line represents the effective sensitivity or area of the telescope in that direction. (d) Power received by the telescope as a function of the angular displacement of the source θ' from the telescope center line. The full width between the half-power points is the full width at half maximum (FWHM) beam size, or equivalently, the half power beam width (HPBW).

angular resolution of the telescope system is comparable to the angular size of the beam.

Each small portion (resolution element) of the film or CCD in a camera can be thought of as a detector that views, say, a $1'' \times 1''$ portion of the sky. Adjacent elements view adjacent portions of the sky. Thus a camera or focusing telescope is in effect a multiple-beam instrument. Such an imaging system is able to record the signal coming from different sky positions simultaneously, whereas a single-beam system must study adjacent portions of the sky sequentially. Examples of single-beam systems are the parabolic radio antenna with a single detector at the focus and an optical telescope with a single small hole in the focal plane. In the latter case, all the light from the star of interest passes through the hole and the photon number may be measured with an electronic device called a photomultiplier tube which we describe in Section 6.2; see Fig. 6.1.

Point spread function

When a beam has a small angular size, closely spaced sources can be better resolved. Even if only one source is in the region, a narrow beam gives less contamination from background radiation from directions adjacent to the source. On the other hand, a broader beam is more efficient if one is searching a large portion of sky for sources or if one is studying diffuse radiation from the sky. However, it might confuse or wash-out bright spots in the emission pattern.

At the focus of a telescope, the beam size is the portion of sky that one single pixel in the focal plane can "see". Blurring by the telescope, the atmosphere, or by diffraction means that a faint point-like source will appear as a blurred (enlarged) image in the focal plane. Pixels in this region thus detect photons from the source even though they are not at the exact image position of the point source. Each pixel thus sees a larger portion of the sky than it would in the absence of blurring; the beam of each pixel, and hence of the telescope, is thus increased by any blurring (defocusing). In an optical telescope at ground level, this blurring may be primarily due to variable refraction in the atmosphere. (Diffraction is more important only for the smallest telescopes; see below.) Typically atmospheric blurring of $1''$ means the telescope beam size is effectively $1''$.

The analysis of images of the sky requires knowledge of the response of the telescope to a point source, in particular the x, y distribution of deposited energy in the image plane. This is a two-dimensional function $f(x, y)$, the *point spread function* (psf). This multi-pixel image maps the single-pixel beam shape.

A typical psf will be peaked in the center with the deposited energy falling off with distance. For modestly bright objects recorded on photographic film, the film saturates; it reaches maximum blackness at the center of the image. But light scattering on the film surface enlarges the exposed region. Thus on photographs of the sky, the brighter stars appear bigger, not brighter. The psf in this case is flat and broad; the effective beam is larger for the brighter sources.

The number of photons collected (in a given time) is a strong function of the angular position of the source relative to the center of the beam as noted above. The effective area of a telescope is different for different parts of the antenna beam; an object directly on the axis of a radio telescope will deliver more of its energy to the focus than will an off-center one. Sometimes, a source far off to the side can be weakly detected if the beam has undesirable *side lobes* (Figs. 5b,c). A strong source in a side lobe will be indistinguishable from a weak source in the main beam.

An antenna beam can be drawn as shown in Fig. 5c. The radial distance from the origin (antenna) to the lobe boundary at a chosen angle is proportional to the effective area of the antenna for radiation arriving from that angle. Typically, the greatest efficiency is in the forward direction.

The quoted angular width of the beam depends on how the edge is defined. A definition often used is the *full width half maximum* (FWHM; Fig. 5d). If $\theta_{1/2}'$ is the angle from beam center at which the power is reduced to $1/2$ its maximum value, the FWHM angle is $2\theta_{1/2}'$. Radio astronomers refer to this as the *half-power beam width* response (HPBW). Although the power received at $\theta_{1/2}'$ is $1/2$ the power in the center of the beam, the total power *enclosed* within this angle can be substantially greater than (or less than) 50% of the total power over all angles.

A good telescope beam will have a highly peaked response function which will include $\gtrsim$90% of the received power within the FWHM limits, but a poor beam can have a response with large "wings" that result in only a small portion of the power falling within the FWHM angles. Another useful definition of beam width is the (half or full) angle that encloses 90% of the power. A beam with large wings would have a large value of this angle and vice versa.

Diffraction

One reason a telescope beam may not be as narrow as one might wish is an interference phenomenon known as *diffraction*. If a parallel beam from an infinitely distant source is incident upon an antenna, there is interference between the different parts of the incoming wavefront, called *wavelets* in the Huygen method of describing wave propagation. If the telescope has a limited diameter (i.e., it is not infinitely large), the interference will produce a blurred (enlarged) image of size that depends upon the diameter of the telescope.

Fraunhofer diffraction

A formal derivation of diffraction sums the effect of wavelets originating at each imaginary segment of an aperture such as that shown in Fig. 6a. The aperture could be the aperture of the primary mirror of a telescope. The wavelets are in phase at the aperture if they originate in a plane wave. They interfere with one another as they propagate downward from the aperture. The resultant propagation directions deviate from the vertical because the blocked parts of the original wavefront are missing.

Fraunhofer diffraction is the special case of diffraction where the distance to the image plane is large compared to the diameter d of the aperture. A segment of a plane wave of light leaving the aperture at a given angle θ (Fig. 6b) thus will illuminate one part of the distant image plane, and the light leaving at a different angle will illuminate another portion. In this manner, two stars will appear as two spots on the distant focal plane. Alternatively, one could insert a thin lens just below

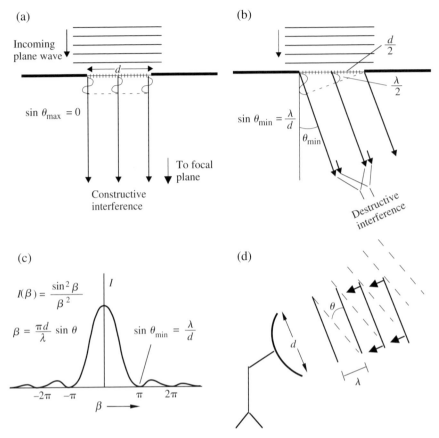

Figure 5.6. Fraunhofer diffraction of an incoming parallel beam by a single slit viewed from its short end. The slit is imagined to consist of the multiple segments shown. The image plane is taken to be at a very large (infinite) distance downward. (a) Forward direction ($\theta = 0$) wherein the wavelets from all segments are in phase giving maximum intensity. (b) Direction θ defined by $\sin \theta = \lambda/d$, in which the summed wavelets yield zero intensity, the first minimum of the response. (c) The one-dimensional response function, intensity vs. $\beta(\theta)$. (d) The half angle of diffraction smearing $\theta_{min} \approx \lambda/d$ represents an uncertainty in the arrival direction that is equivalent to the wavefront arriving at the telescope aperture with one edge lagging the other by one wavelength.

the aperture to bring the rays at each angle to a focus on a nearby focal plane. This simply moves the image plane closer and is the arrangement of a typical telescope.

   In either of these cases (with and without a lens), for each angle θ, one can examine the interference of the wavelets from imaginary segments of the aperture. The simplest geometry to consider is an aperture which is a long, narrow slit (running into and out of the paper) of width d. Figures 6a,b show the narrow dimension of the slit. At the angle $\sin \theta = \lambda/d$ (Fig. 6b), the rays from the left and center of the aperture are perfectly out of phase because the path length difference is exactly $\lambda/2$. The rays from the segments just to the right of these (shown as short arrows in Fig. 6b) also are perfectly out of phase with each other for the same reason. In fact, each pair of segments in the left half of the slit has a partner in the right half which exactly cancels the first. Thus, one would expect no light at this particular angle,

$$\sin\theta_{\min} = \frac{\lambda}{d} \qquad \text{(Angle of destructive interference;} \qquad (5.6)$$
$$\text{slit interference)}$$

In contrast, the rays directed straight ahead at $\theta = 0$ (Fig. 6a) are all in constructive interference.

A proper summation of the phases and amplitudes of all the wavelets over the narrow slit for each angle θ leads to an intensity pattern of the form,

$$I(\beta) = \frac{\sin^2\beta}{\beta^2} \qquad \left(\beta = \frac{\pi d}{\lambda}\sin\theta\right) \qquad \text{(Fraunhofer} \qquad (5.7)$$
$$\text{slit diffraction)}$$

This function (Fig. 6c) has a maximum value of 1.0 at $\beta = 0$, or $\theta = 0$. It has minima with zero value when $\sin\beta = 0$, i.e., at $\beta = \pm n\pi$, where n is an integer ≥ 1. The first minimum at $\beta = \pi$ corresponds to the angle $\sin\theta_{\min} = \lambda/d$ shown in Fig. 6b. The second minimum is at $\beta = 2\pi$, or $\sin\theta_{\min} = 2\lambda/d$. Thus the minima will generally be separated by $\Delta\theta \approx \lambda/d$ for small angles, $\theta \ll 1$ so that $\sin\theta \approx \theta$. The two central minima are an exception; they are separated by twice this amount.

The circular aperture of a telescope mirror yields a circular diffraction pattern as expected from the symmetry. Again, a proper summation of the wavelets must be carried out to obtain the radial variation of brightness. The result is similar in form to Fig. 6c. It turns out that the angular radius $\theta_{\min}$ of the first minimum (dark ring), which encircles most of the light, for $\lambda \ll d$, is

$$\theta_{\min} \approx 1.22\frac{\lambda}{d} \qquad \text{(rad; circular aperture; } \lambda \ll d) \qquad (5.8)$$

where λ is the wavelength of the incoming radiation and d is the diameter of the mirror or antenna. A sketch of the appearance is presented in Fig. 7. This is called the *Airy diffraction pattern*, and the inner part, within $\theta_{\min}$ is called the Airy disk.

The diffraction phenomenon is equivalent to saying that the antenna can not distinguish the arrival directions that lie within angles $\theta_{\min} \approx \lambda/d$ of the source directions; the two sources can not be resolved if they are closer together than this angle. (Even so, with sufficient signal, the response will be broadened or asymmetric, possibly revealing the presence of the second source.) As shown in Fig. 6d, the angle $\theta_{\min}$ corresponds to a shift of phase of only one wavelength across

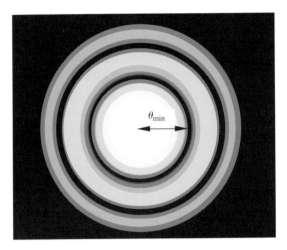

Figure 5.7. Approximate appearance of the Fraunhofer diffraction pattern from a parallel beam imaged through a circular aperture. The radius of the first minimum (black) is θ_{min} where $\sin \theta_{min} \approx 1.22\ \lambda/d$ for $\lambda \ll d$. The inner bright circle is called the Airy disk.

the face of the telescope. This is an easy way to remember and rederive, approximately, the expression (6) or (8). One can think: "the telescope can not distinguish incident angles that yield less than 1 wavelength phase shift across the face of the telescope. The uncertainty in arrival direction is thus $\sim\lambda/d$.

Radio resolution

Diffraction is the primary factor that limits the resolution of radio telescope beams. For example, a 16-m dish antenna observing at $\nu = 100$ MHz ($\lambda = c/\nu = 3$m) has a diffraction pattern, or half beam width, of

$$\theta_{min} \approx 1.22\frac{\lambda}{d} = 1.22\frac{3}{16} = 0.23 \text{ rad} = 13° \qquad \text{(radio)} \qquad (5.9)$$

which is terrible! Observations at 5 GHz ($\lambda = 0.06$ m) yield resolution improved by a factor of 50, or $\sim$1/4 degree, comparable to the angular radius of the moon,

$$\theta_{min} \approx 1.22\frac{0.06}{16} = 4.6 \times 10^{-3} \text{ rad} = 0.26° \qquad \text{(radio)} \qquad (5.10)$$

This is still very poor resolution by optical standards.

Radio astronomers, nevertheless, obtain spectacular angular resolution by using widely separated telescopes working in concert (*interferometry*), to $\lesssim 0.001''$ as noted above. The separation between two such telescopes effectively increases the telescope size, d (see Chapter 7).

Optical resolution

At optical wavelengths, the diffraction limit, or beam size, is much smaller because the wavelengths are much smaller. For $\lambda = 500$ nm and a small telescope of aperture $d = 0.40$ m,

$$\theta_{\min} \approx 1.22 \frac{500 \times 10^{-9}}{0.40} = 1.5 \times 10^{-6} \text{ rad} = 0.3'' \quad \text{(optical;} \qquad (5.11)$$
$$d = 0.4 \text{ m)}$$

Larger telescopes do not improve the situation, as would be the case if diffraction were the only limiting factor. Atmospheric turbulence limits the resolution, or *seeing* to about $1''$ or at best $\sim 0.3''$. Thus, the best possible resolution can be obtained if the telescope is only ~ 0.4 m in diameter and if the atmospheric turbulence is minimal. For telescopes in space, the atmosphere is no longer an issue, and larger diameters do help. The 2.4-m Hubble Space Telescope (HST) was designed to approach a theoretical resolution of $0.05''$ (at $\lambda = 500$ nm). In principle, the new generation of ground-based telescopes of ~ 10-m diameter have a diffraction limit of $0.013''$, but resolution at this level can not be attained without additional measures described below.

X-ray resolution

Present x-ray grazing incidence telescopes are not limited by diffraction because the wavelengths are 1000 times smaller than optical wavelengths. Rather, they are limited at present by the quality of the metallic reflecting surfaces (mirrors) to an angular resolution of about $1''$. The mechanical collimators discussed above are another factor of 100 removed from this limit. Thus diffraction plays no limiting role in x-ray astronomy imaging today.

5.5 Resolution enhancement

Turbulence in the atmosphere limits the resolution that can be obtained by an optical telescope. There are major efforts underway to develop systems that will correct for this effect. Here we describe the nature of the atmospheric blurring, an *a posteriori* method of correcting for the effect known as *speckle interferometry*, and an active method wherein the telescope optics are continuously changed to compensate for the atmospheric irregularities known as *adaptive optics*.

Isophase patches and speckles

Consider a distant point-like star. In the ideal case of no atmosphere, the electromagnetic signal would be a plane wave upon its arrival at the telescope (Fig. 8a). Its image in the focal plane would be an Airy disk of angular size $\sim \lambda/d$ where d

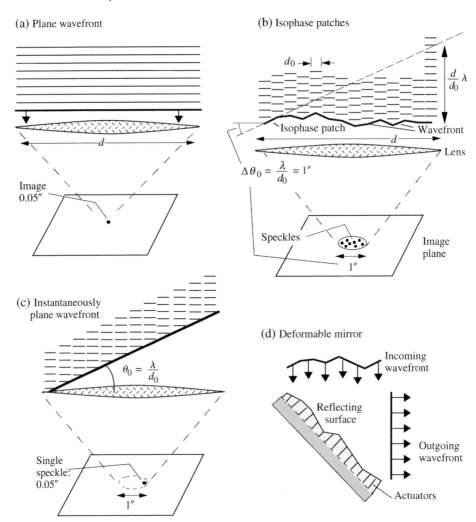

(a) Plane wavefront

(b) Isophase patches

$d_0 \rightarrow| \ |\leftarrow$

$\frac{d}{d_0} \lambda$

Isophase patch   Wavefront

$\Delta\theta_0 = \frac{\lambda}{d_0} = 1''$ d Lens

Image
0.05''

Speckles   Image
plane

$1''$

(c) Instantaneously
plane wavefront

$\theta_0 = \frac{\lambda}{d_0}$

(d) Deformable mirror

Incoming
wavefront

Reflecting
surface

Outgoing
wavefront

Single
speckle:
0.05''

$1''$

Actuators

Figure 5.8. Isophase patches arriving at lenses and mirrors. (a) Plane-parallel wavefronts yield a small diffraction-limited Airy disk characteristic of the entire lens diameter. (b) Turbulent cells in the atmosphere lead to a non-planar wavefront. The individual isophase patches independently image to large Airy disks, and the interference between the multiple wavefronts leads to speckles in the image plane. Each speckle is a mini-image of the source. (c) Improbable situation wherein phase delays at one instant of time form a tilted planar wavefront and thereby produce a single small speckle (image). (d) Deformable mirror that uses actuators to reshape the reflecting surface every ~ 1 ms. A non-planar wave becomes planar.

is the diameter of the telescope mirror. As noted, a 2.4-m telescope would yield a disk of $\sim 0.05''$. (We use angular units to describe the linear image size.)

In fact, the atmosphere has density enhancements (*turbulent cells*) of sizes $d_0 \approx 0.1$ m that delay portions of the wavefront en route to the telescope. These cells are carried rapidly across the telescope line of sight by high-altitude winds of speeds ~5 m/s. The result is that a distorted (non-planar) wavefront arrives at the telescope (Fig. 8b), and the shape of this varies rapidly with time. The figure shows the front at a given instant. Segments of the wavefront comparable to the ~0.1 m size of the turbulent cells will be nearly planar; they are called *isophase patches*. There would be about 24 such patches across the diameter of a 2.4-m telescope, and ~500 across the entire two-dimensional aperture.

Each of these isophase patches acts like a plane wave from the distant source and, in passing through the telescope, creates its own slightly offset image of the star. The offset angle will be $\lesssim \lambda/d_0$ (Fig. 8b); if it were larger it would not be an "isophase" patch. The overall image thus consists of many interfering plane waves arriving at the focal plane over distances comparable to the angle $\lambda/d_0 \approx 1''$. This overall image size is much larger than the ideal of $0.05''$ for our 2.4-m telescope.

The overall image size may also be found as follows. Since an isophase patch is only 0.1 m in size, it uses only 0.1 m of the telescope aperture. The image would be the Airy disk expected for a 0.1-m telescope, namely $\lambda/d_0 \approx 1''$, compare to (11). The overall image is thus a summation of 500 slightly displaced overlapping (and interfering) $1''$ Airy disks. This again leads to an overall image size of $1-2''$ at any given instant of time.

Inversely, the size of the isophase patches may be obtained from the detected image size ($\sim 1''$). From the diffraction relation (6),

$$\Delta\theta_0 \approx \lambda/d_0 \approx 1'' \qquad \text{(Overall image size)} \qquad (5.12)$$

For $\lambda = 500$ nm, this expression gives the isophase patch size we have already adopted,

$$d_0 \approx 0.1 \text{ m} \qquad \text{(Size of isophase patch)} \qquad (5.13)$$

The ~500 overlapping Airy disks lead to pronounced interference in the focal plane. This leads to bright and dim patches known as *speckles* across the $1-2''$ images, reminiscent of the light patterns seen on the bottom of a swimming pool. The size of the speckles may be obtained by noting that the wavefront of the leftmost isophase patch (dashed line) lags the wavefront of the rightmost patch by d/d_0 wavelengths (Fig 8b),

$$d\,\Delta\theta_0 = d\frac{\lambda}{d_0} = \frac{d}{d_0}\lambda \qquad \text{(Max. path length difference)} \qquad (5.14)$$

where d is the diameter of the telescope lens or mirror. This leads to the possibility of as many as d/d_0 interference maxima and minima along a given line in the image plane.

The $1''$ image of size λ/d_0 is thus broken up into d/d_0 speckles (at most) along one dimension, each of which has equivalent angular width of

➡ $$\text{Speckle angular size} = \frac{\Delta\theta_0}{d/d_0} = \frac{\lambda/d_0}{d/d_0} = \frac{\lambda}{d} \tag{5.15}$$

This is the diffraction limit of the telescope, namely $0.05''$ for $d = 2.4$ m, or $0.013''$ for $d = 10$ m. Each speckle thus has the small size of the Airy disk expected in the absence of atmospheric turbulence!

There are thus ~500 spatial elements across the two-dimensional image at the focus of our 2.4-m telescope. Only some of these will be bright (constructive interference); other elements will be dark (destructive interference). In general the bright spots (speckles) will be located randomly in the image plane and they will vary with time in location, brightness and number.

Each speckle can be interpreted as an image of the point source. If the source consists solely of two closely spaced stars, the light from each would pass through the same atmosphere cells and hence be refracted similarly. Each speckle would thus become a double speckle, or an "image" of the double star. If the source has an arbitrary shape, each speckle will have that shape, but with $0.05''$ resolution. Or more correctly stated, each speckle will be a convolution of the source shape and the $0.05''$ Airy disk.

An extreme (and improbable) example of the speckle phenomenon is shown in Fig. 8c. In this case, the phase delays, for an instant of time, happen to yield a plane wave with a fixed angle, $\theta_0 = \lambda/d_0$ across the entire telescope aperture. The wavefront momentarily mimics a plane wave from an offset source. The telescope lens will thus focus this wave perfectly with the full resolution of the lens, $\Delta\theta = \lambda/d \approx 0.05''$ for $d = 2.4$ m, but at a slightly offset position. Thus there is only one speckle, and it is *the* image.

This actually approximates the effect for a telescope that is comparable to or smaller than the isopatch size (0.1 m) because the light entering the telescope at a given time passes through only one isopatch. However, averaged over time, the image appears blurred because the atmospheric density enhancements are carried by winds across the telescope aperture. At speeds of order 5 m/s, they will move the distance of one cell in a time of about (0.1 m)/(5 m/s) = 20 ms. Thus the single image of our small telescope will move around rapidly within the overall image in this characteristic time. The 20 ms time is not much faster than the temporal resolution of the eye, so one can actually see the star image bouncing around when looking through a small telescope.

For large telescopes there will be many speckles of varying brightnesses. As the atmosphere moves, the speckle pattern continually changes, which again leads to a general fuzziness when integrated over time. A long-duration image of the binary star system HR 4689, taken with a 2.34-m telescope, is shown in Fig. 9a; its

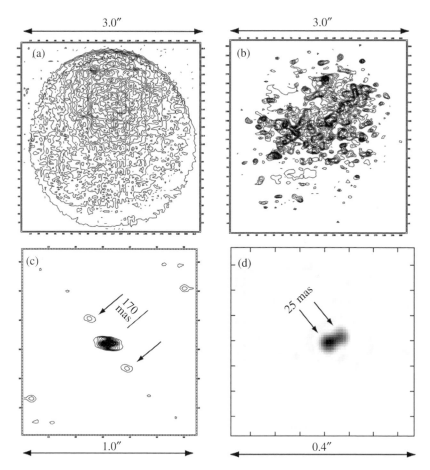

Figure 5.9. (a) Sum of 128 specklegrams of binary star HR4689, obtained at the 2.34-m Vainu Bappu Telescope, Kavalur, India. (b) One of the 128 specklegrams. (c) Reconstruction with autocorrelation function of the 128 specklegrams, with expanded scale. The primary star is the central peak; its partner is one of the two secondary peaks shown with arrows, separated only 170 milliarcseconds (mas) from the primary star. (d) Reconstruction from 2000 specklegrams of binary star system 41 Dra with only 25 mas separation. Arrows mark the two stars. North is up and east is to the left in these figures. [Adapted from: (a) K. Saha, "Speckle imaging: a boon for astronomical observations", *Proc. Conf. Young Astrophysicsts of Today's India*, 2001; (b,c) K. Saha and D. Maitra, *Indian J. Phys.*, **75B**, 391 (2001); (d) Balega *et al.*, *Astron. Lett.* **23**, 172 (1997)]

extent is almost 3″. In fact, it is the summation of 128 short (∼10 ms) exposures, one of which is shown in Fig. 9b. In the latter *specklegram*, one can easily see the individual speckles, each of which is a diffraction limited (0.067″) mini-image of the binary star system. One can not see the binary in this image because the number of photons in a given speckle is so small. However, the information is embedded in the thousands of speckles in the 128 exposures.

Speckle interferometry

In the technique known as *speckle interferometry*, a series of short exposures (e.g., 10 ms) is used to capture the speckles. Keeping in mind that each speckle is a diffraction limited image of the source, the quality of this image may be enhanced by superimposing lots of the speckles in one exposure and summing the counts in them. This improves the strength and the precision of the result. One further improves the quality by carrying out this summation on each of the hundreds or thousands of short exposures and superimposing and summing the results into a single diffraction-limited image. Alternatively, one may choose to use only the brightest speckle in each exposure for the superimposition/summation process.

The resultant high-resolution image of the source can reveal significant structure, such as a finite-sized circular disk in the case of a nearby star or a double image in the case of a binary star system. A necessary condition for the retrieval of an image is that multiple photons (at least two) contribute to each speckle; otherwise the desired angular information is not contained in the individual speckles and is thus not retrievable. The speckle technique can therefore be used only for the brighter celestial objects.

We show in Fig. 9c a reconstructed image of HR 4689 derived from the 128 exposures similar to Fig. 9b. It was obtained with an *autocorrelation* method. The brighter of the two stars is at the location of the strong central peak. The fainter star in the binary is represented by one of the two fainter peaks to the upper left (NE) or lower right (SW), marked with arrows. The fainter star is found to lie only 0.17″ from the brighter star. The width of the individual peaks is only about 0.067″, the diffraction limit for a 2.34-m telescope. This is a remarkable result from observations where the overall image size is almost 3″ (Fig. 9a). The ambiguity in the location of the fainter star in Fig. 9c is a natural consequence of the autocorrelation method.

An autocorrelation tests the correlation of the image with itself for all possible x-y (left/right and up/down) offsets between the two (identical) images. See (5) for definition of the one-dimensional correlation function. The plotted intensities in Fig. 9c are the values of the autocorrelations at the various offsets; position on the plot represents the x-y offset. The autocorrelation value will be greatest when the offset is zero because every speckle will be matched with itself and the correlation will be perfect. This is the central peak in Fig. 9c. The smaller side peaks arise at the two offsets where one of the two stars in the image overlaps the other.

Finally, we show in Fig. 9d a reconstruction by a different method of the binary 41 Dra, which reveals a stellar separation of only 0.025″.

Speckle interferometry is an *a posteriori* method of correcting for atmospheric turbulence; the images are constructed after the observations. In contrast it is possible to actively correct the wavefront to its original planar form, as we now discuss.

Adaptive optics

Under certain circumstances it is possible to actively control the optics of a telescope during observations to counteract atmospheric turbulence. This is known as *adaptive optics*. Such systems must also detect the state of the atmosphere so that the proper corrections may be applied.

Deformable mirrors

In an adaptive-optics system, the non-planar wavefront is corrected in real time prior to its arrival at the image plane. If successful, the telescope would then yield point-like images with the theoretical resolution of the telescope. Active wavefront correction has been accomplished with the use of a deformable mirror placed in the light beam after it is collimated by the telescope optics.

The instantaneous shape of the wavefront is independently measured (as described below), and this is used to change the mirror shape to correct for the atmospheric defocusing. These corrections must be applied every few milliseconds to keep up with the effect of the turbulence. A typical mirror might have diameter 0.1 m and several hundred tiny electrically driven pistons (actuators) which adjust the height of the reflecting surface by up to several micrometers in several microseconds.

The function of the deformable mirror is simple to understand. If the wavefront in a given region is delayed, the mirror surface in that region is moved forward to shorten the optical path relative to other positions (Fig. 8d). This can be done anywhere in the optical path where the "rays" from the various portions of the initial wavefront are separated, i.e., they are not at a focus.

The system may also contain a separate movable flat mirror to move the image centroid in two dimensions to keep it at a single position in the focal plane. This reduces the excursions required of the actuators in the deformable mirror. In a simpler system, the flat mirror is used alone to track the brightest speckles, which can provide improved images on its own.

Sensing the wavefront shape

Such a system requires knowledge of the instantaneous state of the atmosphere on time scales of ~ 1 ms. This is done by sensing the instantaneous shape of the wavefront each millisecond. The faint stars studied by many astronomers do not provide sufficient light for this. It is necessary therefore to use a bright reference star. Unfortunately, this star must lie within a few arcseconds of the real star if the same portion of the atmosphere is to be sampled. This is a very unlikely occurrence. Thus a bright artificial star is required.

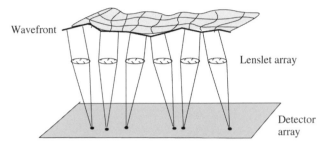

Figure 5.10. The Hartmann wavefront sensor. A two-dimensional array of lenslets records the displacement of the images (black dots) each of which represents the tilt of a small segment of the wavefront (dark line). Only the front row of lenslets is shown. [Adapted from W. Wild and R. Fugate, *Sky & Telescope*, June 1994, p. 24; courtesy MIT Lincoln Laboratory]

One technique is to use a ground-based laser. Its beam may be projected into the sky with either the main telescope or a smaller auxiliary telescope. The beam may be used to illuminate the thin layer of sodium gas found near the top of the atmosphere, at 90-km altitude in the mesosphere. In this case, the laser frequency is set to the frequency of the sodium line so that it excites the sodium atoms which radiate and thus serve as an artificial star. The laser does not radiate continuously; it sends brief flashes during which time the normal star is not studied. This avoids contamination of the true starlight by laser light. The flash from the artificial laser "star" is detected and analyzed for the wavefront distortions. The appropriate error signals are then sent to the deformable mirror.

The sensing and analysis of the wavefront shape may be carried out in a number of different ways. One of these is the Hartmann wavefront sensor. A mosaic of small "lenslets" sample the wavefront and project the image of the point source on a portion of the detector plane (Fig. 10). The locations of the several images yields the tilts of the sampled portions of the wavefront. Together, the array of lenslets samples the tilts of all the isophase patches. The shape of the wavefront can be determined from these tilts. Other methods for sampling the wavefront make use of interference techniques.

Unfortunately, the physical location of the laser star is not fixed in time; it wanders back and forth a bit due to the turbulence cells encountered by the laser beam on its upward path. The data from the laser star can thus not be used to control the flat movable mirror that adjusts the image centroid. The light from a real star must be used for this purpose. A relatively faint star can be used if the laser star is used to adjust the deformable mirror at the same time.

Complete system

An entire detection and correction system is shown in Fig. 11. The real star and the laser star are observed with the same optics. After passing the correction mirrors, the

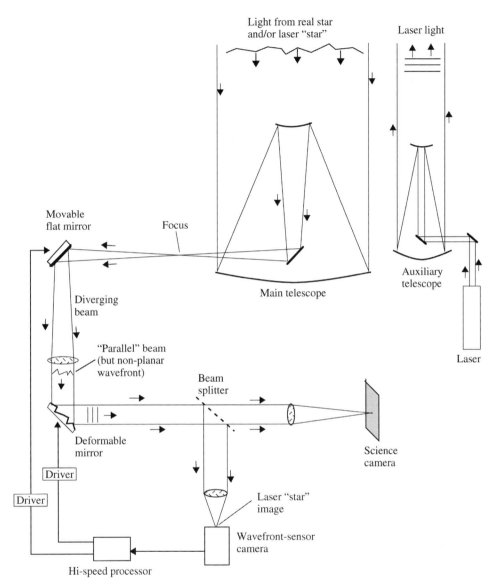

Figure 5.11. Complete adaptive optics system. A small telescope directs laser light to a point high in the atmosphere to create an artificial laser "star". The light from the real star, or intermittently from the laser star, arrives at the main telescope and is directed to a movable flat mirror which can tip and tilt to correct for centroid displacement of the image. The light is then directed to a deformable mirror which corrects for non-planarity of the wavefront. (The light from the laser-star is used for this.) The light beam is then split with a portion directed to a wavefront sensor, e.g., the Hartmann sensor (Fig. 10), the output of which is used to control the two correcting mirrors. The balance of the light is directed to the science camera.

beam is split into a wavefront-sensing beam and a science beam. The former is
sent to the wavefront-sensing hardware and analysis software which detect deviations from a planar wave with proper centroid location. Corrections are then
applied to the two mirrors. The modified beam is then sensed; it should be more
planar, but the atmosphere will have changed somewhat. The process continues
in this closed feedback-loop mode. The desired result is an image that closely
approximates that expected from the telescope in the absence of atmospheric
turbulence.

Adaptive optics systems are now producing results. The spectacular image of
the galactic center (Fig. 12) taken with the 3.6-m CFHT telescope in Hawaii is an
example. The smallest star images are only 0.13″ FWHM in size and about 1000
stars are visible in the original image.

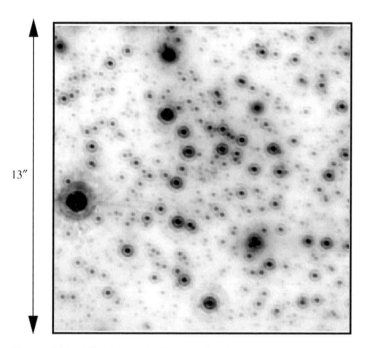

13″

Figure 5.12. Adaptive optics image of galactic center region in the near infrared
(K band at 2.2 mm). Its extent is 13″ × 13″ and the point spread function (faint
image size) is only 0.13″. The brighter stars are deliberately over exposed to bring
out the fainter stars of which up to 25 per square arcsec can be seen on the original
image. The faintest stars have K ≈ 19. The diffraction rings about the brighter
stars indicate that the intrinsic resolution of the 3.6-m CFHT telescope has been
reached. A 14.5 mag star 20″ removed was used for wavefront sensing. [From
Rigaut *et al.*, *PASP* **110**, 152 (1998), Fig. 12, with permission]

Problems

5.3 *Image formation*

Problem 5.31. (a) Confirm that the Lick 3-m telescope has plate scale of $\sim14''$/mm, given its focal length of 15.2 m. What distance on the plate would correspond to a great circle angle of $1°$? (b) What is the focal ratio $\mathscr{R}$ of this telescope? (c) A detectable image of a distant star (i.e., a point source) can be obtained on film at the focus of the telescope in an exposure of duration T if the flux density (W m^{-2}) at the telescope is $\mathscr{F}$. The circular image size is $1''$ in radius due to variable atmospheric refraction. How long does it take to detect a uniformly bright nebula of angular radius $2'$ if its total flux is $1000\mathscr{F}$? (d) What would be the required exposure time if the telescope were 1/4 the size of the Lick 3-m in all respects, i.e., $d = 0.75$ m and $f_{\mathrm{L}} = 3.8$ m? [Ans. ~0.2 m; ~5; $\sim15T$; —]

5.4 *Antenna beams*

Problem 5.41. Why is it desirable to have ground-based optical telescopes larger than the 0.40-m aperture which provides the best-possible resolution for a traditional telescope?

Problem 5.42. (a) What is the diffraction-limited resolution in arcsec of the Arecibo radio telescope "dish" of diameter 300 m operating at a wavelength of 20 mm? (b) If instead one had two such telescopes separated from one another by 10 km, what resolution (arcsec) might be attained, at the same wavelength? Repeat if the telescopes are separated by 7500 km (e.g., on different continents). (c) What is the theoretical diffraction-limited resolution for a 1-m diameter x-ray telescope for 1 keV photons (in arcsec)? [Ans. $\sim15''$; $\sim0.5''$, $\sim10^{-3''}$; $\sim10^{-3.5''}$]

5.5 *Resolution enhancement*

Problem 5.51. Describe the nature of the speckle pattern as a function of time that one might expect from a large telescope (e.g., 2.4-m diameter) and compare this to that expected for a small telescope of diameter only 50 mm. In each case, how many speckles might be expected and how might they appear to move according to an observer watching the pattern through an eyepiece? The turbulent cells in the atmosphere are typically 0.1 m in diameter.

6

Detectors and statistics

What we learn in this chapter

The **detectors** at the foci of telescopes may be **position-insensitive** such as the classic **photomultiplier** and the simple **proportional counter**. **Position-sensitive** detectors at the focus of a telescope provide an overall field of view that includes many beams (resolution elements). The **charge-coupled device** is widely used in **optical** and **x-ray** astronomy for this purpose. Its **internal structure** and **operation** reveal its strengths and weaknesses. Gamma-ray astronomers use **plastic** and **crystal scintillators** and **spark chambers** or their equivalent. Examples are the **EGRET** and **BATSE** instruments that were in orbit during the 1990s. The precision of a detected signal is limited by **statistical** and **systematic errors**. Knowledge of **basic statistical theory** enables one to assess the significance and meaning of one's data. Aspects of this are the character of **statistical fluctuations** (the **Poisson** and **normal distributions**), **background subtraction** with **error propagation**, and comparison of data to a **model** with a **least squares fit** and the **chi square test**.

6.1 Introduction

At the focal plane of a telescope, an image is formed. It can be viewed directly by eye in two ways, on a piece of frosted glass placed in the focal plane or through an eyepiece. In the latter case, the focal point of the eyepiece is placed at the image so as to create a parallel beam of small extent (pupil sized) that can be refocused by the eye. For most astronomy, detectors are placed in the focal plane to absorb the photons and convert them to electrical signals. The classical detector at optical wavelengths was photographic emulsion on a glass plate, but now it has become the *charge-coupled device* (CCD) which is now found in digital cameras.

The detectors used in astronomy are a diverse set. The detection apparatus of radio astronomy bears absolutely no resemblance to that of optical or gamma-ray astronomy. Despite this, the characteristics of the radiation (frequency, arrival direction, etc.) that one attempts to measure are similar for all wavelength bands. The operations of detectors is a rich field worthy of a textbook in itself.

In general, the objective of modern astronomical instruments is to acquire data with high efficiency in a form that can be analyzed by computer. For most purposes film is no longer considered convenient or useful except for a few specialized uses such as the wide-field Schmidt-telescope sky surveys. Even these are scanned photoelectrically and converted to digital form. Modern instruments can record individual photons with high efficiency and can rapidly extract positional, spectral, and temporal information. This capability often comes from the simultaneous acquisition and analysis of many positional and spectral channels of data, often in real time as the data are collected. In fact computers are often incorporated directly into the physical instrument itself.

In this chapter, selected examples of detection systems are presented, not in an attempt to be comprehensive, but rather to illustrate the types of measurement processes that take place before data appear in an astronomer's computer. If one understands one's detector down to the quantum/atomic basics, one can carry out measurements with greater confidence and awareness of possible shortfalls. We then present some basic statistical theory that underlies all astronomical measurements.

6.2 Position-insensitive detectors

Position-insensitive (non-imaging) detectors have played a significant role in astronomy. They do not record the location of an impinging photon. Two such detectors are briefly described here, the photomultiplier tube and the basic proportional counter. The former is for optical radiation and the latter for x rays. Each converts the incoming photon to an electron via the *photoelectric effect*. Multiplication of electron number within the detector then yields sufficient charge to be recorded by conventional electronics.

Photomultiplier and photometry

A classic example of a position-insensitive detector is the *photomultiplier tube*, which was long used to measure precisely the brightness of stars in different wavebands within the optical band, the visual (V), blue (B), and ultraviolet (U). The process of making such measurements is known as *photometry*. In this use, the tube records continuously all the light passing through a small hole in the focal plane of an optical telescope (Fig. 1a). The photomultiplier has been mostly replaced by the CCD for this purpose but is still widely used in high-energy astronomy.

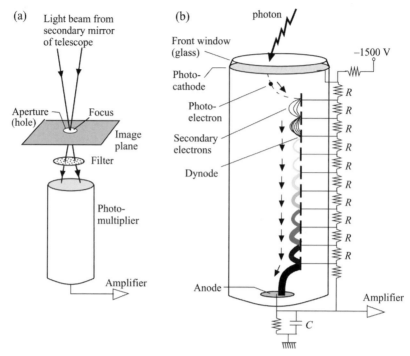

Figure 6.1. (a) Classic arrangement for quantitative measurement of flux of light from a star, known as photometry. CCDs are now used for photometry. (b) Operation of a photomultiplier. An incident optical photon interacts with the photocathode, ejecting a photoelectron which is replicated manyfold by the dynodes. The resultant large charge of electrons is collected at the anode. In real tubes, the dynodes are arranged to optimize the collection of electrons.

A photon entering the tube (Fig. 1b) impinges on *photocathode* material deposited on the inside surface of the front glass. This results in the ejection of an electron (*photoelectron*) by means of the *photoelectric effect* (Section 10.2). The ejected electron is then accelerated by an electric field toward a terminal called a *dynode* which it strikes with high energy. The dynode material responds to the collision by emitting several low-energy electrons. The emitted electrons are then accelerated toward another dynode and the multiplication is repeated. After some 10 steps, the original electron has become several thousand electrons. This charge is then collected on an anode/capacitor and the associated voltage pulse is large enough to be recorded. Each incident photon is thus recorded. The number of recorded photons in some time interval is the measure of the intensity of the incident light.

If the light passes through a color filter that transmits only blue light (B) before reaching the photomultiplier tube (Fig. 1a), the flux in that color band is measured precisely. Similar measurements in the ultraviolet (U) and in the visual or yellow (V)

are the basis of the a standard system of flux measurements known as the *UBV* system; for more on this system, see Section 8.3. Photomultipliers are also used to detect fluorescent and Cerenkov light from *extensive air showers*. These are cascades of charged particles and photons in the atmosphere that arise from a single very energetic cosmic ray particle, usually a proton, that enters the solar system and interacts with an atomic nucleus in the earth's atmosphere. In TeV astronomy, the Cerenkov light is collected with large mirrors that focus it onto photomultiplier tubes.

Proportional counter

X rays may be detected in large *proportional counters*. The detector consists of a metal box of inert gas such as argon or xenon with a fine wire (~25 μm diameter) running down its center. The wire, or *anode*, is held at high positive voltage, +1500 V, relative to the box so there is an electric field permeating the detector; any free electrons are attracted to the anode. One side of the gas-tight box consists of a thin entrance "window" of a material of low atomic number, often ~50-μm beryllium (Be), through which the x rays can enter without being absorbed (Fig. 2a). The basic proportional counter does not record the arrival position of the x ray but it does record the arrival time of an x ray with its approximate energy (or frequency).

An incoming x ray will pass through the counter gas without loss of energy until at some point it interacts catastrophically with an atom via the photoelectric effect in which it disappears and a *photoelectron* is ejected. This interaction takes place on a probabilistic basis; there is an average distance (*mean free path*) the x ray will travel in the gas before interacting, but one can not predict where in the gas any given x ray will interact.

The mean free path of the interaction increases with energy; high-energy x rays (*hard x rays*) will penetrate more material, on average, than will low-energy x rays (*soft x rays*). The degree of absorption is also a strong function of the atomic number Z of the absorbing material. X rays above energy ~2 keV will easily penetrate the Be window ($Z = 4$) of the detector and enter the gas. The relatively high atomic number of the gas, $Z = 18$ and $Z = 54$ for argon and xenon respectively, leads to significant absorption in the gas. At sufficiently high x-ray energies, the gas becomes quite transparent to the x rays and most of them will not be recorded. This occurs at ~15 keV for a counter of 10-mm depth filled with argon at one atmosphere pressure. One can reach ~60 keV with a counter of depth 35 mm filled with xenon at one atmosphere.

Absorption of the x ray in the counter gas results in the creation of ion pairs in the detector which can be recorded electrically. The details of the detection process in the gas are as follows for an incident 6.0 keV x ray in argon gas. The photoelectron

(a) Basic proportional counter

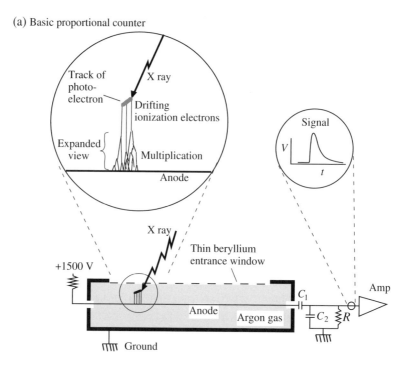

(b) Position-sensitive proportional counter

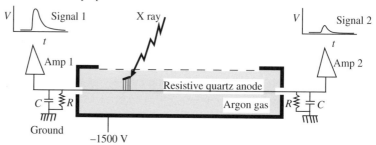

Figure 6.2. Proportional counter. (a) The incident x ray penetrates the thin window and ejects a photoelectron from an argon atom. Secondary electrons may also carry some of the energy. The electrons travel about 1 mm or less and create ion pairs (electrons and positive ions) along their tracks. One such track is shown as the short dark line. The electrons are attracted to the anode, near which they multiply by ionizing other argon atoms. (b) Position-sensitive proportional counter. The anode is resistive and the deposited charge is fed to preamplifiers at each end. The ratio of detected voltages yields the position along the wire of the deposited charge. The anode may be kept at ground potential to avoid large high-voltage capacitors, e.g., C_1 in (a); in this case the body is run at high negative voltage (don't touch!).

is most probably ejected from the ground state of the argon atom, known as the K shell or $n = 1$ state. It takes 3.2 keV of energy to ionize the atom from this state, that is, for the ejected electron to overcome the potential energy of the bound state. Thus the photoelectron will emerge from the atom with only $6.0 - 3.2 = 2.8$ keV of kinetic energy. It is this photoelectron that gives rise to the ion pairs.

In contrast to the catastrophic interactions of photons, a charged particle travers-ing material such as a gas will lose energy gradually through many ionizations of the atoms along its track. Its electric field ejects outer shell electrons from the gas atoms with which it has near encounters, thus creating a track of ions and electrons. An electron in argon gas loses about 25 eV of energy for each electron–ion pair it produces. It dissipates all its energy and comes to a stop in 1 mm or less. In our case, it will have created about $2800/25 = 112$ ion pairs, with statistical fluctuations of $\pm\sqrt{112}$ (see discussion of statistics below in Section 3). These ion pairs represent only 2.8 keV of the original 6.0 keV energy. Thus, so far, less than half of the x-ray energy is in a form that can be recorded by the detector.

Most or all of the missing energy may also be recovered through the relaxation of the originally disturbed atom with the missing K shell electron back to its neutral state. First, it fills its vacated ground state, probably with an electron from the $n = 2$ (L) shell. This transition to a lower energy state results in the release of a Kα x ray (*fluorescence*) of energy 3/4 the K shell ionization energy, or 2.4 keV, or it may eject an electron from the L shell (*Auger effect*). In the former case, the fluorescent x ray will likely interact with another gas atom, ejecting a photoelectron from the $n = 2$ or higher state.

Thus, in either case, another electron is traveling through the gas, and it produces more ion pairs. The Auger electron will not have the full 2.4 keV because, as before, it had to expend energy climbing out of the electric potential of its parent atom, namely $1/4 \times 3.2 = 0.8$ keV for the $n = 2$ state of argon. The remaining kinetic energy of the electron, 1.6 keV, creates $\sim$64 ion pairs along its track and these are recorded along with the first $\sim$112 ion pairs.

There is still missing energy, and it is stored in the atom or atoms that have L shell, or higher shell, vacancies. The involved atoms continue to relax with the emission of electrons until most or all of the original 6.0 keV has been converted into about (6 keV)/(25 eV) = 240 ion pairs. All of this happens extremely rapidly, in nanoseconds. The ion pairs from all these interactions are now waiting to be recorded. The recording process takes microseconds.

The electric field in the detector causes the electrons to drift inward toward the central anode while the ions drift outward toward the walls of the counter. The electrons will suffer many elastic collisions with gas atoms as they proceed toward the anode. The electric field strength increases rapidly as the distance to the anode decreases and is extremely strong in the immediate vicinity of the anode. When the electrons are close to the anode, they gain sufficient energy between collisions to ionize the atoms with which they collide. Each such inelastic collision produces another electron, thus doubling the number of electrons. Subsequent collisions continue the doubling until the number of electrons has been increased by a factor of several thousand when they strike the anode. This multiplication creates sufficient charge for registration by electronic circuits. The deposited charge is recorded electronically as a voltage or current and then as a digital number.

This entire process is fairly linear throughout: a more energetic x ray will result in more energy being deposited in the gas as ion–electron pairs, and the multiplication by the electric field is by an approximately fixed factor if the detector voltage is not too great. Thus the recorded charge will be a rough measure of the energy of the incident x ray; thus the name "proportional counter".

Statistical fluctuations in the number of ion pairs lead to fluctuations in the recorded charge (*pulse height*) for a given x-ray energy. If one plots the distribution of recorded pulse heights for incident 6.0 keV x rays, the result will be a peak at the pulse height corresponding to 6.0 keV, with a full width at half maximum (FWHM) of ~20% of 6.0 keV. The proportional counter is a low resolution spectroscope.

If the voltage is too high, each x ray will lead to a breakdown in the counter. It thus behaves as a *Geiger counter* which counts the x rays (or other charged particles) but does not reveal their energies because all the voltage pulses are more or less identical, independent of the energy of the detected particle or photon.

Sometimes, in the proportional counter, the secondary 2.4 keV fluorescent x ray, discussed above for argon gas, will occasionally escape from the detector volume without being absorbed. Its energy is thus lost. This leads to an *escape peak* in the pulse height distribution. For incident 6.0-keV x rays in argon, this would appear as a secondary peak at $6.0 - 2.4 = 3.6$ keV. The strength of the peak depends on the counter geometry as well as on the *fluorescent yield* of the element in question, that is, the probability that a vacancy in the K shell will be filled via the emission of an x ray, in contrast to the direct emission of an electron via the Auger effect. In heavy elements, such as xenon, fluorescence dominates. The escape peak can be substantial in this case. In argon, the Auger effect dominates, and the escape peak is much less pronounced.

Proportional counters are very efficient detectors in that they count a high proportion of the incident x rays, until at high energies the gas becomes transparent. They can be made with rather large areas so they can collect x rays from a celestial source at a high rate, and they can be made to respond to energies as high as ~60 keV. They thus were the backbone of early x-ray detectors and as well as of the recent Rossi X-ray Timing Explorer, despite their relatively high backgrounds compared to focusing systems (Section 5.3).

6.3 Position-sensitive detectors

The sky has been recorded on photographic plates for more than a century. These position-sensitive detectors record the position of photons incident upon them. Today astronomers usually use electronic devices of greater efficiency for this purpose.

Position-sensitive proportional counters

Proportional counters can be constructed to provide the approximate location of the incident x ray, or more specifically, the position of the photoelectric conversion in the gas. This is indicated by the location on the anode of the deposited cloud of charge. This location can be determined by making the anode wire resistive and measuring the charge reaching each end of the anode as shown in Fig. 2b. Quartz wires with carbon coatings may be used for this purpose. The relative amounts of charge collected at the two ends of the anode indicate the location of the deposited charge. If the charge is deposited 1/4 of the way along the wire, the ratio of the detected voltage pulses will be inversely proportional to the wire lengths (or resistances) to the two ends, namely 3:1.

This technique can provide the location of the incident x ray along one dimension to better than a millimeter in counters of size about 60 mm. Techniques that make use of many parallel (and/or orthogonal) anode wires can yield two-dimensional positional information. Such detectors are used in x-ray astronomy and also in high-energy particle physics experiments. Imaging proportional counters are known for their high detection efficiency, but their spatial (positional) resolution ($\lesssim 1$ mm) is only modest.

Charge-coupled device

In optical astronomy, the *charge-coupled device* (CCD) has almost entirely taken the place of the photographic plate. This is a solid state device that can record the integrated intensity of light falling on it as a function of position on its surface (as does photographic film). The surface of the CCD is divided into rows and columns of *pixels*. A typical size might be 2048 pixels $\times$ 2048 pixels each of size 15 μm $\times$ 15 μm giving an overall size of 30 mm $\times$ 30 mm. When used to record images of the sky, the CCD is placed in the focal plane of a telescope, and a portion of the sky is focused onto it. In a spectroscopic mode, the light is dispersed into its component frequencies (colors) and then focused onto the CCD. In both cases, the usage is similar to that of photographic film.

CCDs are also in use in x-ray astronomy. In this case, they must be modified to take into account the greater penetrating power of x rays. X rays would mostly pass straight through the sensitive region of an optical CCD without interacting, and hence would not be recorded.

Structure of a CCD

The structure of a small portion of a CCD is shown in the side view of Fig. 3a. It consists mostly of a silicon substrate of depth ~260 μm (lower two shaded regions) doped with impurities to make it a *p-type semiconductor*. The impurity is phosphorus which results in a fixed negative charge throughout the material; the charges are not free to move. The uppermost part (~2 μm; upper shaded region) of the silicon substrate is doped with boron (rather than phosphorus) which makes it an *n-type semiconductor* (fixed positive charges). The charge density may be as much as 10^2 times higher than that of the p-type silicon.

Each type of material in isolation contains free charges to balance the fixed charges. There are positive *holes* in the p-type and negative electrons in the n-type. When the two materials are placed together in a CCD, negative and positive free charges diffuse across the boundary neutralizing each other thus creating a region *depleted* of free charges near and on both sides of the boundary. In addition, prior to an exposure to the sky, the n-type material is swept completely free of any remaining free electrons by imposed electric fields (see below). The electrons (black dots) shown in the figure arise from subsequent exposure to light as described below.

The fixed positive charges of the n-type region, being of high charge density and no longer neutralized by electrons, serve to repel the free holes remaining in the p-type. This forces them toward the bottom of the p-type material and enlarges (vertically) the depleted region (no free charges) in the p-type material to a depth of about 5–10 μm. The positive holes pile up in the lower part of the silicon, creating a field that opposes that of the n-type charges. This limits the depth of the depletion region to the quoted 5–10 μm which is typical of an optical CCD.

The depleted regions have very low conductivity (because of the absence of free charges) and can thus support an electric field. The undepleted region below ~10 μm contains free charges and the electric field therein is negligible. The electric field in the p-type depleted region is directed downward due mostly to the proximity of the strongly positive-charged n-type region. This yields an upward force on any free electron created in the region. Such an electron would thus be attracted toward and into the positive n-type region. There it would encounter a minimum of potential energy due to the strong positive charges, and it would become "trapped".

Above the substrate, one finds an insulator (silicon dioxide) and, above this, *electrodes* made of *polysilicon*, i.e., containing impurities that make them conductors.

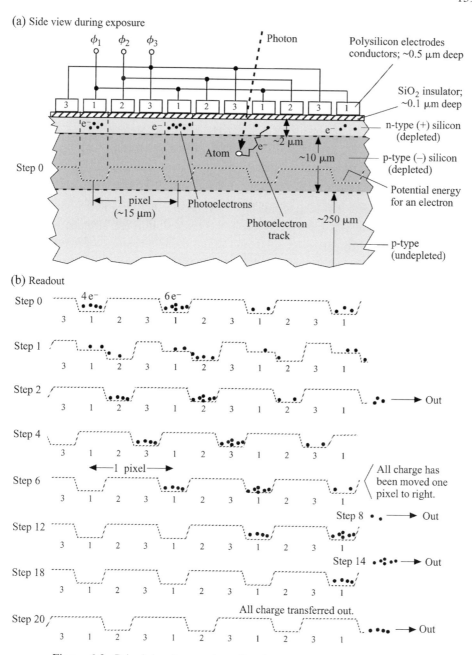

Figure 6.3. Principle of operation of a charge-coupled device (CCD) for visible light. (a) Cross section view. A photon creates a photoelectron in the silicon which is then collected in a low-potential-energy region for each pixel. (b) Readout scheme. After sufficient exposure and the detection of numerous photons, the charges are read out by manipulation of the voltages on the three sets of electrodes.

Every third electrode (numbered "1") is connected to a common bus so these can all be set to a given potential, ϕ_1. Similarly, all the "2" electrodes can be set to ϕ_2, and the "3" electrodes to ϕ_3. These electrodes can modify the potential in the n-type material such that it varies in the left–right direction.

Exposure to light

Now consider what happens when a CCD is exposed to light from the top. The photons will mostly pass through the thin electrode and insulator and will be absorbed in the depleted layers (Fig. 3a), in both the n-type and p-type silicon. A photon impinging on the silicon lattice will usually eject an electron from an atom (photoelectric effect). If this electron has more than 1.1 eV, it will find itself in an energy state known as the *conduction band* (still within the silicon), where it is essentially free. If the electron is created in the depleted region of the substrate as shown, the gradient of the potential (electric field) will drive it in the vertical direction to the minimum of potential energy near the center of the n-type region.

During the exposure to light, one set of the electrodes, labeled "1" in Fig. 3a, is set to relatively high (positive) potential, $\phi_1 \sim +5$ V compared to the lower potentials of the other two, ϕ_2 and ϕ_3. This creates a variation of potential in the left–right direction; the potential in the n-type layer under the "1" electrodes is raised, or the potential energy for an electron in these locations is lowered (dashed line). This causes the photoelectrons to collect at the locations of the "1" electrodes (but still in the n-type layer) as shown by the black dots in the figure. The pixel size is the width of three electrodes.

As the exposure continues, more photons arrive, more photoelectrons are created, and the charge trapped in the potential wells increases proportionally to the amount of light received at each of the $\sim 10^6$ pixel locations. Thus, an image of electric charges is gradually built up.

Readout of the image

After an appropriate time (seconds or minutes), the exposure is stopped, and the charge in each pixel is carried electrically out of the detector. This is accomplished by varying the potentials of the three sets of electrodes (Fig. 3b). Consider Step 0 to be the state of the three electrodes during the exposure, wherein the charge is collected at electrodes "1". In Step 1, the positive voltage of electrode "1" is dropped to an intermediate level (2.5 V) thus raising the potential energy for an electron. At the same time the potential energy at electrodes "2" is lowered. This causes the accumulated photoelectrons to move into well "2". Further raising of the potential energy of "1" (Step 2) traps the electrons in the wells at electrodes "2". All charges have now been moved one electrode position to the right, and the rightmost three electrons have been carried off the chip.

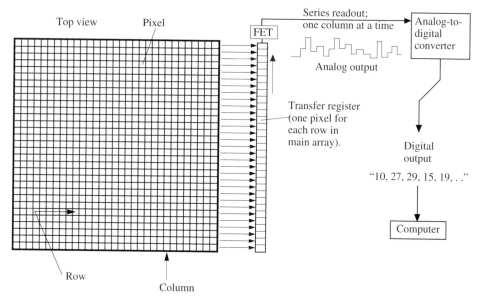

Figure 6.4. Readout method for the entire array. The charges in the pixels of the array are all shifted to the right by one pixel. The charges in the rightmost column are thereby moved into a transfer register and are then read out in the upward direction to a field-effect transistor (FET). This yields an analog signal with amplitude proportional to the charge in the pixel being read out. These analog numbers are converted to digital form with an analog-to-digital converter.

Two more steps of the voltages move the charges to the "3" position, and two more to the next "1" position, which is a full pixel from the starting position. It took six steps to reach this point. Further stepping eventually moves the charges in this row completely out of the image region (like a conveyor belt). The n-type region has been swept free of all photoelectrons.

The entire detector is read out as shown in Fig. 4. There are numerous rows and columns; only a portion of one row is shown in Fig. 3. The stepping shown in Fig. 3 is carried out for the entire detector synchronously. Thus after six steps, every pixel of information in Fig. 4 will have been moved one pixel to the right, and the contents of the rightmost column of pixels will have moved out of the detector array into a *transfer register* which, physically, is just another column of pixels in the CCD chip.

Before the main array is stepped farther to the right, the set of charges in the transfer register must be cleared. The charges in this register are thus stepped by electrodes to the top of the register in the same manner as the charge was stepped sideward in the main array. Thus one pixel of charge after another exits the top end of the transfer register. This sequence of charges is captured and amplified by a *field-effect transistor* (FET) to produce a serial analog output. The height (voltage) of each step of the output represents the amount of integrated light that had impinged on the pixel from which the charge originated. Finally, this series of analog voltages is fed into an analog-to-digital (A/D) converter which converts each step into a digital number that can be manipulated by a computer. This completes the readout of the rightmost column of the main array.

The main array is then stepped another six steps to again move all the charges one pixel to the right, this moves another column of pixels into the transfer register, and again these are serially moved into the FET and thence to the A/D converter. This continues until all columns of the CCD chip have been read out. The computer now holds the digital numbers that represent the entire image. It could then generate an image on a monitor that should appear similar to the image initially focused on the CCD.

The CCD itself is now almost ready for another exposure. Before starting the exposure, the entire readout process is repeated several times ("flushed") to clear the n-type region of any residual electrons. These might arise from a cosmic ray that had deposited a large amount of charge during the first exposure, which was not completely transferred out. Otherwise a residual image could appear in the next exposure. This sweeping ensures that the n-type region is completely depleted of electrons which in turn ensures that the depletion region in the p-type silicon is as deep as possible. This allows detection of the photons that happen to interact more deeply in the silicon; recall that only in a depleted region can an electric field be supported. The electric field drives the photoelectrons up to the n-type region where they can be collected and eventually read out.

This entire readout procedure may be visualized as a college band marching to the right (in Fig. 4) off a football field after its half-time performance. Consider each player to be the charge from one pixel. Just after the players in the front row of the band cross into the sidelines, the entire band halts and marches in place while the front-row players execute a "left face" and march one behind the other toward the end of the field. When the last player in the front row has moved past the leftmost person in the second row, the entire band takes a step forward and again halts while the second row does a left face and follows the first row. In this way, the entire band eventually clears the field.

Utility in optical astronomy

The CCD detectors are a powerful device for astronomy. They are sensitive across the infrared, optical, ultraviolet, and even x-ray bands. For optical photons they are ~70% efficient which means that 70% of the photons are converted to electrons. For comparison, the effective efficiency of the eye and film are each ~1%. Furthermore the CCD output is very linear over dynamic ranges of up to 10^5, i.e., the amount of charge collected is directly proportional to the integrated flux of light incident upon the pixel. In contrast, film is notoriously non-linear and limited in dynamic range to ≲100. Nevertheless, the spatial resolution of fine-grain films still exceeds that of CCDs, even though the latter have pixels as small as ∼15 μm.

The users of CCDs must contend with systematic and statistical errors as one must with all physical detectors. The FET introduces noise into the measured charge values. The noise in current CCDs corresponds to ≲5 electrons for each measured number. It is thus important to make the exposure sufficiently long so that substantially more than 5 photoelectrons (from the incoming photons) are accumulated in each pixel. Otherwise, the readouts will be mostly noise. If 25 photons are detected, the statistical noise in the signal ($\sim\sqrt{25} = 5$; see Section 5) will equal the readout noise. Longer exposures increase the signal and also the statistical noise from the signal. This improves the signal-to-noise ratio because the noise grows more slowly than the signal; see Section 5 below.

The readout of the entire CCD can take a fairly long time (∼40 s for 2048 × 2048 pixels). During this time, one is not accumulating photons. Astronomers usually like to record every last photon the telescope collects, so longer exposures with fewer readouts are preferred.

On the other hand, long exposures will be contaminated by *cosmic rays*, energetic charged particles traversing the atmosphere, which are continuously passing through the CCD chip. Such a particle will ionize the atoms in the chip and thus will deposit large amounts of charge in one or more pixels. These appear in the final image as a huge blob. Frequent readouts minimize the number of these in any given image. Also, if one is studying the time variability of a source, one might prefer to make many short exposures. Finally, one must read out the CCD before any of the pixels record so many photons that the wells are nearing saturation.

The individual pixels in CCDs are not perfectly uniform in sensitivity. This effect may be measured by means of "flat-field" exposures to the sky at dusk or dawn when the diffuse sky brightness overwhelms that from individual stars. Any non-uniformity in the resultant image is a direct measure of the relative sensitivity of the several pixels. This allows one to correct exposures of star fields for this effect.

Infrared, optical, ultraviolet, and x-ray photons may be recorded with CCDs. One should note that the CCD does not give information about the energy of the individual photons (except for x rays). Each photon gives rise to only a single electron. Thus the light must be analyzed for color by making exposures with different filters (e.g., U, B, V), or by focusing the output of a spectroscope onto the CCD. In the latter case, as noted above, the different frequencies are imaged to different locations on the CCD.

Adaptive imaging

An adaptation of the optical CCD could play an important role in removing the jitter of images due to atmospheric turbulence (Section 5.5). In this scheme, the CCD would replace the movable flat mirror (Fig. 5.11) which corrects for the motion of the brightest speckle or of the image centroid.

This CCD would be constructed with additional or modified electrodes that allow the accumulated charge to be moved pixel by pixel in two dimensions upon command. That is, one could shift charge two pixels to the right and one pixel upward, and moments later move three left and two down. One would use the information from a wavefront sensor to command these motions. In this way, the brightest speckle from a given star would always fall on top of the charge already collected from the brightest speckles in previous exposures, and sharper images should be forthcoming.

Utility in x-ray astronomy

In the case of x rays, an individual photon absorbed in the silicon creates a photo-electron. As in the proportional counter, the relaxation of the affected atom yields additional photons which produce additional photoelectrons. The total energy of all the photoelectrons is comparable to that of the incident x ray. These photoelectrons quickly lose their energy by ionizing atoms in the substrate; it takes 3.65 eV to create a single electron by ionization in silicon.

The number of ionization electrons created is thus roughly proportional to the energy of the incident photon. An x ray of 4-keV energy would result in $\sim$1000 secondary electrons. This leads to significantly better energy resolution than a proportional counter which produces far fewer ionization electrons for a given x-ray energy. These electrons from a given x ray will enter the potential well of the pixel. If the array is read out before a second x ray is likely to impinge on the same pixel, the measured charge in that pixel will represent the approximate energy of that single incident x ray.

In this way, a succession of CCD readouts can produce an x-ray image with knowledge of the spectral (energy) distribution of the photons at each location, albeit with only moderate spectral resolution. This is in contrast to optical CCDs where, in a given readout, the charge accumulated in each pixel represents the number of photons collected at that position; there is no spectral information.

Another important consideration in x-ray detectors is the depth of the depletion region as noted above. X rays can penetrate a substantial distance into the substrate. If a photoelectron is created in the deeper undepleted region, it will not be carried up to the n-type region. A good x-ray CCD detector requires the fabrication of silicon substrates that can support a deep depletion region.

The depth of the depletion region in the p-type material can be enlarged if the silicon of the chip is exceptionally pure and the amount of added phosphorus is low. This leads to fewer free holes and less opposition to the hole-clearing action of the electric field from the n-type layer. The depletion layer may be $\sim$50 μm deep in such chips. Such CCD chips are now in common use at the focal planes of x-ray observatories, e.g., Chandra and Newton.

Extremely high spectral resolution is obtained in such missions by the insertion of diffraction transmission gratings into the x-ray beam before it reaches the focal plane. This disperses the x-ray beam so that different x-ray energies are focused to different positions on the CCD.

6.4 Gamma-ray instruments

We describe here two gamma-ray experiments that were carried on the Compton Gamma Ray Observatory (CGRO;1991–2000), one rather complicated and the other quite simple. Both were quite effective for their respective objectives.

EGRET experiment

A very different type of position-sensitive device is the *spark chamber* used in gamma-ray astronomy experiments at energies above ~20 MeV. Figure 5a is a simplified view of the *Energetic Gamma-Ray Experiment Telescope* (EGRET) which orbited on the CGRO. It consisted of several types of detectors and was sensitive to gamma rays from energy 20 MeV to 30 GeV. It yielded both the arrival directions and the energies of the detected gamma rays. Its operation will be described briefly here. It demonstrated that certain active nuclei of distant galaxies known as *blazars* were prodigious emitters of gamma rays. It also produced an all-sky map of the diffuse emission of gamma rays produced in the interstellar medium of the Galaxy by energetic cosmic rays.

A forthcoming mission, GLAST, will extend these studies. It features a large silicon–tantalum detector wherein the silicon layers play the role of the spark chambers seen in Fig. 5a and described below.

Detector subsystems

Plastic scintillator anticoincidence

The EGRET detection system includes a *plastic scintillator* which forms a dome surrounding the upper portions of the system. Its purpose is to help discriminate against the energetic cosmic ray particles (mostly protons) that are continuously traversing the detector; Fig. 5a. When such a particle traverses the scintillator, it creates a track of ionization that results in a faint flash of light.

The scintillator consists of a clear plastic (polystyrene) whose atoms emit ultraviolet light as a result of having been ionized by the pulse of electric field from a passing charged particle. An additive to the plastic causes the UV light to be

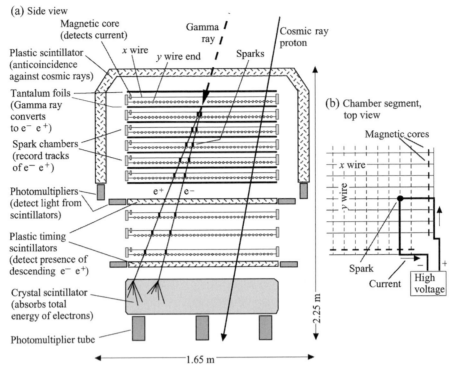

Figure 6.5. (a) Simplified view of the Energetic Gamma-Ray Experiment Telescope (EGRET) carried on the Compton gamma-ray observatory. It made use of scintillators and 36 spark-chamber modules (of which only 9 are shown) to detect and measure gamma rays from 20 MeV to 30 GeV while at the same time rejecting the much more numerous charged cosmic ray particles. Measurement of the tracks and total energy of the converted electron–positron pair yield the energy and arrival direction of the gamma ray. (b) Top view of wires in a portion of a spark chamber module. The spark is indicated by the filled circle, and the flow of current by the dark lines. Only one spark is shown.

absorbed and re-emitted as optical light. The scintillator is totally enclosed in a light-tight covering to exclude stray light. The inside of the covering is highly reflective so that the light rays reflect back and forth in the plastic with small absorption until they happen to strike one or more of the photomultipliers at the bottom edges of the scintillator (Fig. 5a). These convert the light to a voltage pulse which informs the digital logic that a cosmic ray particle traversed the experiment.

In contrast, a gamma ray is not charged and normally would not disturb the atoms of the plastic; it will be absorbed catastrophically in material of higher atomic number within the detector. One can thus distinguish a charged cosmic ray particle from a gamma ray by the presence or absence of a signal from the plastic scintillator. The absence of a such a pulse at the same time as other indications of radiation in the detector is called an *anticoincidence* event. It could well be a desired gamma ray.

Spark chamber detection of electron–positron pair

The gamma rays are detected in a bank of 36 spark-chamber modules stacked vertically on one another (Fig. 5a), They are interspersed with foils of tantalum, a material of high atomic number ($Z = 73$). The foils have a high cross section for the conversion of a gamma ray to an *electron–positron pair*, often called simply an *electron pair*. This interaction occurs when the gamma ray interacts with the Coulomb field of the nucleus of a high-Z atom in a process called *pair production*. The electrons created in this interaction take up most of the gamma-ray energy. Thus they have substantial kinetic energies and tend to travel in the same direction as the gamma ray as shown in the figure. The role of the spark chambers is to record the tracks of these particles.

Each spark chamber module is a large flat enclosure containing gas and two parallel planes of parallel and closely spaced wires (wire separation ∼1 mm). The two planes are separated by about 3 mm in the vertical (z) direction, and the planes are rotated such that the wires in one plane run at right angles to the wires in the other plane; the wires in the two planes run in the x and y directions respectively (Fig. 5a,b). It is like a window screen, except that the two planes (x and y) are separated vertically by about 3 mm.

The passage of the electron creates a track of ionization in the gas of each chamber it traverses. If a high voltage pulse is immediately applied between the top (x) and bottom (y) planes, a spark will occur at the x,y location of the ionization track. The high voltage must be applied soon but it can be delayed briefly until the detection logic indicates the desired type of event has occurred. One wishes to avoid unnecessary firings because it takes some time for the system to recover from the application of the high voltage.

The discharge thus takes place between the two planes, from one wire in the x plane to one in the y plane (dark marks in Fig. 5a and dark circle in Fig. 5b). A surge of current flows in each of these two wires (dark lines in top view, Fig. 5b). To record the x,y location of the spark, one must record which two wires carried the current. This is done with the aid of *discrete magnetic cores*, one of which is placed on the end of each wire (Fig. 5b).

Each core is shaped like a small doughnut and the wire passes through the hole. A current passing through a core will force it into a "+" state wherein the north magnetic pole is in the direction of current flow. If, before the spark, all of the cores in the chamber are placed in the opposite "−" state, then only those cores on the current carrying wires will be found in the "+" state after the spark.

After the event, the magnetic states of all the cores are pulsed with a voltage pulse to reset them. Those that had been flipped by the spark respond differently (electrically) from the others. In this manner, the circuitry registers which x wire and which y wire carried the spark current, and this defines the x, y position of the spark in the module in question. The x, y locations from the several modules permit one to follow the track of the electron or positron, or both, through the several detector layers and to locate the track or tracks in three dimensions.

The presence of two electron tracks means that each spark chamber module must produce a spark at two locations simultaneously. This requires a very uniform spacing between the two planes of wires so one spark will not short out the high voltage before the other spark can develop. A fast-rising and high-voltage pulse also helps insure that the two sparks will occur. The two tracks produce two x coordinates and two y coordinates which yields four x–y locations of which only two are correct. The resolution of this ambiguity by means of a simple rearrangement of the modules is left to the reader.

Timing scintillation detectors (up–down discrimination)

The electrons also pass through two plastic "timing" scintillators which are above and below the lower group of spark chambers (Fig. 5a). These detectors operate on the same principle as the anticoincidence scintillator; traversal by the electron pair results in an electrical pulse from the photomultipliers. In this case the times of the pulses are precisely recorded and compared. The sequence of times from the two scintillators indicates whether the particles were moving upward or downward. This allows one to eliminate upward moving cosmic rays. This discrimination requires very fast electronics because the particles are highly relativistic, traveling close to the speed of light, or 0.3 m per ns.

The passage of the electrons through the timing scintillators in the correct direction, together with the absence of any pulse in the anticoincidence scintillator, indicates the likely presence of a downward-moving gamma ray that converted to an electron–positron pair. The electronic logic then immediately directs the spark-chamber circuitry to apply high voltage to the chambers. This reveals the tracks of the electron pair as described above.

Energies and arrival directions

The electrons finally pass into a *crystal scintillator*. This is a high-Z material, *sodium iodide* (NaI), that causes the electrons to undergo multiple interactions until they have given up their entire kinetic energies to ionization. The recombinations of the ions and electrons cause it to scintillate, to emit light. The light is collected by photomultipliers. The combined output of all the photomultiplier tubes viewing this scintillator is approximately proportional to the total energy lost by the original positron and electron in the crystal scintillator.

The total energy loss of the two particles in their traversals through the chambers, the tantalum, and finally in the NaI should approximate the initial energy of the gamma ray. From this information and from the tracks of the two particles, one can deduce the arrival direction of the gamma ray and its total energy. The arrival direction can be obtained to about 10′ and the energy determined to about 15%. Note that the location of the gamma ray in the chamber is *not* important; it is the arrival direction that indicates from whence on the sky the gamma ray came.

BATSE experiment

The Burst and Transient Source Experiment (BATSE) also flew on CGRO. Its appeal for us here is the power inherent in its simplicity. Its primary objective was to study celestial flashes of gamma rays (*gamma-ray bursts*, GRB) that last only for a few minutes and that had been known but unexplained since 1967. The GRBs arrive at a rate of about one per day at completely random times with each coming from a different direction in the sky. This made them extremely hard to study; one never knew from where or when the next would suddenly appear; and then it would be gone. It was not even known if they came from our solar neighborhood, farther out in the Galaxy, or from intergalactic space.

The BATSE was designed to observe the entire sky surrounding the spacecraft (except that part blocked by the earth) in order to detect as many GRBs as possible. The objective was to study the intensity and spectral time profiles of many individual GRBs as well as the distribution of arrival directions and brightnesses of the class as a whole. It was sensitive in the energy range 20 keV to 2 MeV, much lower than that of the EGRET experiment (> 20 MeV). In its 9 years in space, it detected about 3000 bursts.

BATSE consisted of eight simple detectors that looked out from the eight corners of the spacecraft (think of a cube; Fig. 6a). Each consisted of a large (0.2 m^2) slab of NaI scintillator, sufficiently thick so that the gamma rays would interact within it, creating an avalanche of ionizing particles. The ionization releases optical photons which are detected with three photomultiplier tubes, giving an electrical pulse whenever a gamma ray is detected. The pulse amplitude represents the energy of the gamma ray. Penetrating cosmic ray protons were eliminated by a large sheet of plastic scintillator in front of the NaI. There was metallic shielding in the back to minimize gamma-ray detections from the rear. There was no collimation so the field of view of each was almost 2π sr, or 50% of the sky (Fig. 6b), with the eight fields of view overlapping substantially. Together they viewed the entire 4π sr solid angle surrounding the spacecraft, and typically up to four detectors would detect any given GRB.

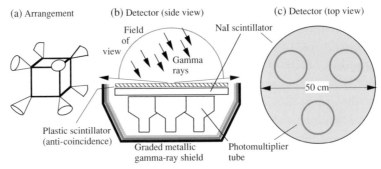

Figure 6.6. The Burst and Transient Source Experiment (BATSE) on the Compton gamma-ray observatory. (a) Eight detectors, each with a $\sim 2\pi$ sr field of view (much larger than shown), view outward from the eight corners of the spacecraft. (b) Side view of one of the detectors with field of view shown. Gamma rays can enter from any direction in the upper hemisphere. (c) Top view of one of the detectors.

A GRB consists of a flash of gamma rays. The number of gamma rays detected in a given time interval, say 0.1 s, indicates the intensity of the flash during that interval. The arrival direction of the gamma rays is determined by the relative numbers of detected gamma rays in the several detectors. A GRB arriving from a distant point arrives as a swarm of gamma rays in planes normal to the direction of propagation (like a plane wave). This flash of gamma rays is detected with several of the detectors with different signal strengths depending upon its arrival direction.

If the arrival direction is normal to one of the detectors, the effective collecting area is large and many gamma rays are absorbed, giving a large gamma-ray count. If the arrival direction is slanted, at an angle removed from the detector normal, the projected area of the detector is less, and fewer gamma rays are absorbed and hence counted. The signals from the several detectors taken together thus give a rough measure of the arrival direction, often within several degrees. They also give a measure of the *number flux density* of the gamma-ray burst (gamma rays m^{-2} s^{-1}). Summed over the duration of the burst, one measures the *fluence* (gamma rays m^{-2}).

The BATSE demonstrated that the GRB arrival directions are highly isotropic; suggesting they were from extragalactic space. It then played a major supportive role in the identification of *afterglows* of GRBs. These fading glows of radiation from the location of some GRBs are visible at radio, visible and x-ray wavelengths. They linger for days in the optical and even longer in the radio.

The Italian–Dutch BeppoSAX satellite discovered the afterglow phenomenon. It carried a multi-pinhole x-ray camera with a wide field of view that provided celestial positions of GRBs that were accurate to a few arcminutes. This allowed the sensitive focusing x-ray telescope on the satellite to find and study the fading x-ray afterglow. Ground-based optical and radio telescopes could do likewise, given the original arcmin position. Optical spectra of the afterglows provided redshifts that established the extragalactic origin of the GRB.

The GRBs are thus extremely luminous, liberating more energy than the entire rest mass energy of a neutron star. This makes them the most energetic explosions known in the universe, other than the big bang. They could result from the implosion of a massive stellar core into a black hole (*hypernova*), or from the final moments of two compact stars in a binary system, such as a black hole and a neutron star. As the two objects spiral into one another, radiating their energy away in gravitational waves, they merge into an even more massive black hole with a cataclysmic release of energy (see Section 12.4).

6.5 Statistics of measurements

In all scientific studies, the precision with which a quantity is measured is all important. For example, consider the detection or discovery of a weak source. If the uncertainty in the measured intensity is large, one might not be convinced the source was even detected; it could have been a perfectly normal fluctuation in the background noise. Another example is the comparison of two intensities. Did that quasar change its optical brightness since the measurement of 6 months ago? If the quasar in fact changed its intensity by 3% but the two measurements were each made with only 5% accuracy, true variability could not be claimed. To detect a 3% difference with confidence, one should require each measurement to be accurate to significantly less than "±1%".

In measurements governed by counting statistics, one must define what one means by an error or accuracy because there is no absolute limit to the possible fluctuations in a measured number. The measured number can, in principle, always be found to be bigger than some quoted limit, such as ±1%, if one is willing to wait long enough, possibly many lifetimes. Usually, one quotes a *one standard deviation limit* $\pm\sigma$ within which about 2/3 of the measurements will fall.

A quantity that varies with time or frequency can be compared to theoretical models such as a Gaussian peak or a blackbody spectrum. Here again the uncertainties in the data must be understood. Statistical fluctuations could either mask a possible agreement or they could make the data falsely appear to agree with the model.

The understanding of a few basic principles of statistics is often sufficient. If one must resort to complex statistical arguments to "prove" that a measured effect is real, it may, in fact, *not* be real. In this case, it is often wiser to return to the telescope for additional measurements so that one can make (or disprove) the case with the basic statistical arguments.

Here we present some of the concepts underlying the assignment of error bars to measured numbers and to values derived from them. We then present the least squares method of fitting data to a model and also the χ^2 (chi-square) method of evaluating such a fit.

Instrumental noise

Every instrument has its own characteristic background noise. In the absence of any photons impinging on it, apparent spurious signals will be produced. For example, as noted above, cosmic ray particles will pass through a CCD leaving an image in one or more pixels that could be mistaken for a stellar image. Also, the readout process produces a noise of its own. Cosmic rays passing through proportional counters produce pulses much like those due to x rays, and radio amplifiers exhibit quantum shot noise.

Instrument designers incorporate features to reduce or eliminate these effects as much as possible. For example, *anticoincidence* schemes may be incorporated to identify signals from the most penetrating cosmic ray particles, and the noise in many solid state detectors is reduced by cooling the detectors to liquid gas temperatures. However, residual instrument background always remains at some level. The observer copes with this by measuring the background as carefully as possible and then by subtracting it from the on-source data. However, the measurement of the instrument noise has its own error, and the result of the subtraction will still carry an uncertainty (error) associated with the background.

Some residual instrumental noise is *statistical* in nature (e.g., that due to randomly arriving cosmic rays) and it can be measured as precisely as desired, given sufficient time. Other noise is *systematic* in nature; it does not behave in a predictable manner and hence can not easily be eliminated by better measurement. Examples are the gain change in an amplifier due to unmeasured temperature changes, the aging of detector or electronic components, and the varying transparency of thin clouds passing in front of the celestial source under study. It is very important that the sources of, and estimates of, both the statistical and systematic errors in a given measurement be stated when presenting a result.

Statistical fluctuations – "noise"

Statistical noise is a term applied to the inherent randomness of certain types of events. Consider the detection of photons from a steady source, one that is not pulsing or flaring. Although there is an average rate of arrival of photons, the actual number detected in a limited time interval will fluctuate from interval to interval. The nature of these fluctuations can be quantified.

Poisson distribution

Consider a steady celestial source of constant luminosity, that produces, on average, 100 counts in a detector every second. The 100 photons do not arrive with equal intervals, 10 ms between each count, like the ticks of a clock. Instead, they arrive randomly; each photon arrives at a time completely uncorrelated with the others. (This is not the case for a pulsing source.) The average rate of arrival is governed by a fixed *probability* of an event occurring in some fixed interval of time. In our case, there is a 10% chance of an event occurring in every millisecond, but no certainty that the event will indeed arrive in any specific millisecond. This randomness leads to a variation in the number of counts detected in successive 1-s intervals, e.g., ... 105, 98, 87, 96, 103, 97, 101, ...

A distribution of counts $N(x)$ can be obtained from many such measurements; one measures the number of times N a given value x occurs. For example, $N(95)$ is the number of times the value 95 is measured. For a random process such as photon arrival times, this distribution is well known theoretically as the Poisson distribution. Our experimental distribution would not match this function exactly because each value $N(x)$ would itself have statistical fluctuations. We will find that longer observations with more accumulated counts improve the agreement with the theoretical function.

The *Poisson* distribution (not derived here) gives the probability P_x of detecting an integer number x of events,

$$P_x = \frac{m^x e^{-m}}{x!} \qquad\qquad \text{(Poisson distribution)} \qquad (6.1)$$

when the average (mean) number of events over a large number of tries is m. For example, if the mean (average) value of the measurements is $m = 10.3$, the probability of detecting $x = 6$ photons in a single observation is $P_x = 0.056$. If, instead, the mean is $m = 6.0$, the probability of detecting $x = 6$ is greater, $P_x = 0.161$, but still smaller than one might have guessed. This tells us that it is not particularly likely that one will actually detect the mean number. In these cases, the counts were accumulated over some fixed time interval.

The Poisson distribution is valid for discrete independent events that occur randomly (equal probability of occurrence per unit time) with a low probability of occurrence in a differential time interval dt. The distributions P_x vs. x for $m = 3.0$, 6.0 and 10.3 are shown in Fig. 7.

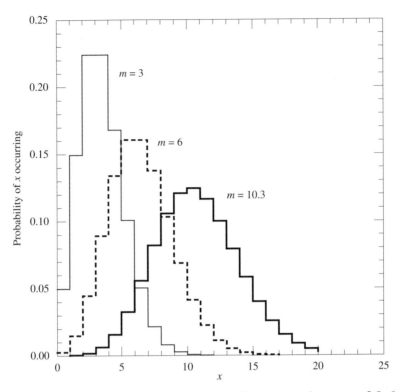

Figure 6.7. The Poisson distribution for small mean numbers, $m = 3.0$, 6.0 and 10.3. The ordinate gives the probability of the value x occurring, for the given mean value. Note the asymmetry of the histograms.

The probabilities of detecting $x = 0, 1, 2, 3, 4, \ldots 9$ for six values of m are given in Table 1. Note that the distribution in x is not symmetric about $x = m$; see also Fig. 7. The distribution is not the classic "bell curve" described below, particularly at low values of m. The distribution (1) is normalized to unity such that,

$$\sum_{x=0}^{x=\infty} P_x = 1 \qquad (6.2)$$

as it should if P_x is a probability. The sum (2) indicates that the sum of all probabilities for finding different values of x must equal unity. You can check that the summation yields unity, approximately, by adding the probabilities in a given row of the table. Finally, note that the probability of obtaining zero events is significant even if the mean m is as high as 3 or 4. You might expect two cars a minute to pass your house, on average, but in 13.5% of the 1-minute intervals, no cars will pass.

Table 6.1. *Sample values of Poisson function* P_x

x:	0	1	2	3	4	5	6	7[a]	8	9
$m = 1$	0.368	0.368	0.184	0.061	0.015	0.003	0.001	7E–5	9E–6	1E–6
$m = 2$	0.135	0.271	0.271	0.180	0.090	0.036	0.012	0.003	0.001	2E–4
$m = 3$	0.050	0.149	0.224	0.224	0.168	0.101	0.050	0.022	0.008	0.003
$m = 4$[b]	0.018	0.073	0.147	0.195	0.195	0.156	0.104	0.060	0.030	0.013
$m = 6$[c]	0.002	0.015	0.045	0.089	0.134	0.161	0.161	0.138	0.103	0.069
$m = 10$[d]	5E–5	5E–4	0.002	0.008	0.019	0.038	0.063	0.090	0.113	0.125

[a] The notation 7E-5 indicates 7×10^{-5}.
[b] The values of P_x for $m = 4$ at $x = 10$ and 11 are 0.005 and 0.002 respectively.
[c] The values of P_x for $m = 6$ at $x = 10$–14 are 0.041, 0.023, 0.011, 0.005, 0.002.
[d] The values of P_x for $m = 10$ at $x = 10$–18 are: 0.125, 0.114, 0.095, 0.073, 0.052, 0.035, 0.022, 0.013, 0.007.

Normal distribution

One can describe a continuous and symmetrical *normal distribution*, also known as the *Gaussian distribution,* even though Gauss's derivation (1809) was preceded by LaPlace's (1774) and DeMoivre's (1735). It gives the differential probability dP_x of finding the value x within the differential interval dx,

$$dP_x = \frac{1}{\sigma_w \sqrt{2\pi}} \exp\left[-\frac{(x - m)^2}{2\sigma_w^2}\right] dx. \quad \text{(Normal or Gaussian} \quad (6.3)$$
$$\text{distribution)}$$

This expression is characterized by two parameters, m (the mean which is the true value of the quantity being measured) and σ_w (a width parameter for now), whereas the Poisson distribution is described with the single parameter m. We introduce (3) solely as a mathematical function but its physical utility will soon become clear. We will find below that σ_w is also the standard deviation σ of the distribution. The subscript "w" is used here to distinguish temporarily the two quantities.

The normal distribution is the well known *bell curve* of probability; two examples are shown in Fig. 8 (smooth curves). It is symmetric about m and can extend to negative values of x. The coefficient $(2\pi)^{-1/2}$ is chosen so that this distribution is also normalized; that is, the integral of all probabilities is unity,

$$\int_{x=-\infty}^{x=\infty} dP_x = 1 \quad (6.4)$$

where the integral limits extend to all possible values of x.

The quantity σ_w in (3) describes a characteristic width of the distribution; at $x = m \pm \sigma_w$, the function has fallen to $e^{-0.5} = 0.601$ of its maximum value. It falls

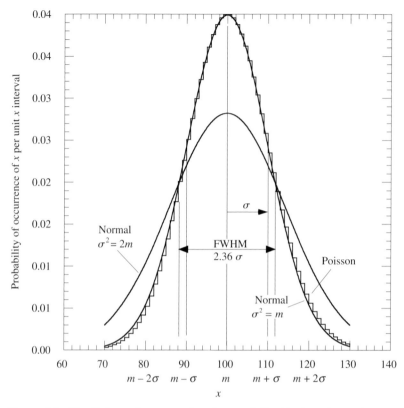

Figure 6.8. The Poisson (step curve) and normal distributions (smooth curves) for the mean value $m = 100$. The normal distribution is given for two values of the width parameter σ_w which is shown in the text to be equal to the standard deviation σ. The Poisson distribution approximates well the normal distribution if the latter has $\sigma = \sqrt{m}$. Note the slight asymmetry of the Poisson distribution relative to the normal distribution. The standard deviation and full width half maximum widths are shown for the higher normal peak; the two normal curves happen to cross at the FWHM point.

to $e^{-1} = 0.37$ at $x - m = \sqrt{2}\,\sigma_w$, i.e., when x is removed from m by $\sqrt{2}\,\sigma_w$. These widths indicate the spread of the function about the mean. The full width at half maximum (FWHM) can be shown to be $2.36\sigma_w$ (see Fig. 8).

One can show by integration of (3) that 68% of the area under the distribution falls within $m \pm \sigma_w$. That is, the probability of a measurement falling *within* $1\sigma_w$ of the mean is 0.68, or equivalently the probability of falling *outside* these limits is 0.32 (Table 2). The probability that the result of a measurement will fall outside $m \pm 2\sigma_w$ is 0.046 or 4.6%, or that it will fall outside $m \pm 3\sigma_w$ is 0.0027 or 0.27%. It is thus rather unlikely that a single measurement would yield a fluctuation of $3\sigma_w$ from the mean. It is extremely unlikely (one chance in two million) that a $5\sigma_w$

Table 6.2. *Normal distribution probabilities*

$\left(\dfrac{x_0}{\sigma}\right)^a$	Area (shaded) at $\|x - m\| > x_0$ [b]	$\left(\dfrac{x_0}{\sigma}\right)^a$	Area (shaded) at $\|x - m\| > x_0$ [b]
0	1.00	2.5	0.012 4
0.5	0.617	3.0	0.002 70
1.0	0.317	3.5	4.65×10^{-4}
1.2	0.230	4.0	6.34×10^{-5}
1.4	0.162	5.0	5.73×10^{-7}
1.6	0.110	6.0	2.0×10^{-9}
1.8	0.0719	7.0	2.6×10^{-12}
2.0	0.0455		

[a] Ratio of deviation x_0 to standard deviation σ. The standard deviation σ is equal to σ_w, the width parameter of the distribution.
[b] Probability of occurrence of deviation greater than $\pm x_0$.

fluctuation would occur. Such a fluctuation in a single measurement (trial) would be a miracle!

For large values of the mean m, the Poisson distribution approaches in shape the central part of the normal distribution if the width parameter σ_w of the latter is set to $\sigma_w = m^{1/2}$. (We do not demonstrate this formally.) Poisson and normal distributions are shown in Fig. 8 for a large mean value, $m = 100$. The normal curve with $\sigma_w = m^{1/2}$ approximates well the Poisson distribution. (In the figure we set $\sigma_w = \sigma$.) The normal distribution thus describes the arrival of random events for large m when $\sigma_w = m^{1/2}$. This then is one example of the physical utility of the normal distribution.

Variance and standard deviation

The width of a *measured* distribution indicates the range of values obtained from a set of individual measurements of x. This width may be characterized formally as the *root-mean-square deviation*, that is, as the square root of the average of the square of the deviations from the mean value. This parameter σ is called the *standard deviation*; its square σ^2 is called the *variance*,

$$\sigma^2 \equiv \frac{1}{n}\sum_{i=1}^{i=n}(x_i - m)^2 \qquad \text{(Definition of variance)} \qquad (6.5)$$

where m is the mean, n is the number of independent measurements of x, and the x_i are the individual measured values. In this case, m is the true mean, a value that can be obtained only with an infinite amount of data.

In practice the average value x_{av} of the n measured numbers may be the best approximation of m that is available. In this case, since m is not independently obtained, the divisor n in (5) is replaced with $n - 1$,

$$\sigma^2 \equiv \frac{1}{n-1} \sum_{i=1}^{i=n} (x_i - x_{av})^2 \qquad \text{(Practical variance)} \qquad (6.6)$$

The two expressions become equal for large n. The variance can be evaluated for any given experimental distribution whether it is random or not. One simply substitutes the raw data points into (5) or (6).

The variance (5) of a theoretical distribution such as the Poisson distribution (1) can also be calculated. For this case, it is useful to rewrite (5) in terms of the probability P_x used in the theoretical expressions. The summation would include subsets of terms wherein there were n_j occurrences of the same value x_j. The sum of this subset divided by n can be written as $(x_j - m)^2 n_j / n$. The overall summation (5) can be rewritten as a sum of such terms. That is, the summation will be over x_j or simply x, rather than the trial number i. Each term of (5) will thus contain the quotient n_j/n which is the probability P_x that the value x occurs. Thus (5) becomes,

$$\sigma^2 = \sum_{x=-\infty}^{x=\infty} (x - m)^2 P_x \qquad (6.7)$$

The variance of the Poisson distribution is obtained from substitution of (1) into (7). Subsequent (difficult) evaluation yields

$$\sigma^2 = \sum_{x=0}^{x=\infty} \frac{(x-m)^2 \, m^x e^{-m}}{x!} = m \qquad \begin{array}{c}\text{(Variance of} \\ \text{Poisson distribution)}\end{array} \qquad (6.8)$$

The standard deviation σ of the Poisson distribution is simply the square root of the mean number of occurrences(!),

$$\sigma = m^{1/2} \qquad (6.9)$$

If 100 photons are expected to arrive at a pixel of a CCD during an exposure of 1 s, based on the average rate of many prior trials, the standard deviation for the single measurement is $\sigma = \sqrt{100} = 10$. In subsequent measurements, the values will fluctuate by ± 10 or even ± 30 about the 100-count mean.

The uncertainty σ relative to the mean value in this case is $\sigma/m = 10/100$ or 10%. It is "a 10% measurement". Now, at this same rate, in 100 s, one expects

10 000 counts. In this case, the standard deviation for a single measurement is $\sqrt{10\,000} = 100$, and the relative uncertainty is much less, $100/10\,000$ or 1%. The accumulation of more counts leads to higher absolute fluctuations and uncertainty, but the relative or fractional uncertainty is reduced. The latter observation determines the average rate, namely 100 counts/s (if previously unknown) to $\sim 1\%$. Longer observations determine rates more precisely.

If only a single measurement is made with $m = 100$ and only 90 counts are detected, the true value is not known, but it probably lies within 10–20 counts of the measured value. One often bases the "error" on the measured, not the true, number, in this case $\sigma = \sqrt{90} = 9.5$. Here one is adopting the measured value as an approximation of the true mean.

The variance of the normal distribution is obtained through substitution of (3) into the integral form of (7),

$$
\sigma^2 = \int_{x=-\infty}^{x=\infty} (x-m)^2 \, \mathrm{d}P_x \qquad \text{(Variance, normal} \qquad (6.10)
$$
$$
\text{distribution)}
$$
$$
= \frac{1}{\sigma_w \sqrt{2\pi}} \int_{-\infty}^{\infty} (x-m)^2 \exp\left[-\frac{(x-m)^2}{2\sigma_w^2}\right] \mathrm{d}x
$$

➡ $\qquad \sigma^2 = \sigma_w^2$

The standard deviation σ of the normal distribution is thus simply σ_w, the width parameter of the distribution (3). This is not accidental; the constant in the exponential is adjusted to make this so. In practice, the symbol σ is used in (3) without subscript, and it is called the standard deviation, which, of course, is what it is. Hereafter we do so also.

Measurement significance

As noted above, if one sets $\sigma = m^{1/2}$ and expects a large number of events, the Poisson distribution is approximated by the symmetric normal distribution. In this case, one can invoke the normal-distribution probabilities in Table 2, e.g., the probability of exceeding $\pm 3\sigma$ is 0.27%. Thus, if the pixel of a CCD is expected to record 100 photons during a given exposure time, the standard deviation will be $\sigma = 10$, and the probability of lying outside the 70–130 range will be only 0.0027.

If one measures 130 photons in this experiment, one might ask if the source really brightened or if this is simply a statistical fluctuation. There is one chance in $1/(0.0027) = 370$ that this excursion (or a greater one) would happen solely from statistical fluctuations. One might then be inclined to suspect the source had actually brightened, but it would be better to obtain more data to convert this into a 4σ or 5σ result, the latter with a probability of being a statistical fluctuation of only 6×10^{-7}. Additional data can improve one's confidence greatly that the effect is real, if it is. Otherwise, the statistical significance would most likely decrease.

It is sometimes tempting to report such a result as real if a repeat measurement is difficult or impossible, e.g., if it would require an entire week on a big telescope or if the source was in a rare flaring state. The temptation is even greater if the indicated result is of sufficient importance to win one great fame. Because of this, one must always try to repeat the measurement or at least to take great care to understand all the factors that went into the probability calculation that makes such an event seem real. One can also earn fame as a fool by over interpreting data.

Statistical traps

Statistical arguments can seem convincing even if they are wrong. Here we mention two rather common traps in such arguments. The first is to overlook the effect of repeated measurements, or *multiple trials*. Assume that one makes many measurements and finds a 5σ effect in one of them. There is a probability of a statistical fluctuation this large *in one given trial* of only 6×10^{-7}. But if the measurement consisted of an examination of each of 4 million pixels of a CCD chip exposed to the sky, one has 4 million trials, each of which could produce the 5σ effect.

In this case, the expected number of such fluctuations, the *expectation value*, among the 4 million pixels is $(6 \times 10^{-7}) \times (4 \times 10^6) = 2.4$. Thus one would not be surprised to find several statistical fluctuations of this magnitude on the CCD exposure. One must conclude therefore that the 5σ effect could easily be a statistical effect and is unlikely to be the representation of a real celestial source.

A related error is to calculate a probability after examining some data. This *a posteriori probability* (after the fact) calculation can easily overlook other unconscious trials. For example, if on my birthday, I happen to see a license plate with my birth-date numerals on it, I would be amazed. I could calculate that in my state there are $\sim 10^6$ plates and that I only see about 100 per day. Thus the probability of seeing it accidentally on this day is only 10^{-4}, one chance in 10 000. I might be tempted to conclude that there was a real psychic phenomenon at work here, namely that the owner of the car is drawn subconsciously to my locale.

My misconception here is that I did not take into account that there are lots of different plate numbers that are significant to me such as my birth-date numbers reversed or reordered, the birth dates of my wife or daughters or other relatives, my bank pin number, or important constants in physics, etc. etc. These amount to other experiments or trials that I should include in my calculation. If there are 100 such numbers, the probability would drop to 1 in 100, and that is not so unusual. Once about every 100 days, I might well see a number that amazes me, based only on the statistics of random events, not on psychic phenomena. Similar issues can arise in astronomy.

This error is particularly insidious because one can never know after the fact what events might have appealed to one as unusual. The proper thing to do is to define *a priori* (before the fact) the questions you will ask of your data, that is, before you take them. If a set of data show an interesting effect, a posteriori, one should test one's hypothesis that this effect is real by taking additional data. But sometimes one can not, because telescope time is unavailable, or because one may have already used all the examples the sky has to offer. Again great care is required in reporting results.

Background

Data are often contaminated by background events. For example, if one is counting x rays from a neutron-star binary system in a proportional counter with a simple tubular collimator, the background will consist of two major components: (*i*) counts due to cosmic ray particles that fail to trigger anticoincidence logic, and (*ii*) counts from the diffuse glow of x rays from the sky, known as the *diffuse x-ray background*. Commonly, two measurements will be made, one with the astrophysical source in the field of view and one with the telescope (or collimator) offset from the source so it measures only background. If the telescope/detector produces a sky image, one can measure intensities on and off the source in a single exposure. We start with a brief discussion of the propagation of errors.

Propagation of errors

The discussion above has pertained to the nature of errors on measured quantities. After the data are taken, one invariably manipulates then to obtain other quantities. For example one may divide the accumulated number of counts by accumulation time to find the rate of photon arrivals. Or, one might subtract the background to get the true source counts. What are the errors on the calculated quantities? The four basic arithmetical operations cover most practical cases.

A simple non-statistical argument can be used to give some insight into the process. Assume the measurement errors on a quantity never exceed some fixed value. Consider measurements of a length x and a length y, each accurate to 1 mm (maximum). The maximum possible error in the sum would thus be 2 mm. Restating this, let z be the sum, $z = x + y$. If dx and dy are the deviations (errors) in these quantities, the deviation is its differential, dz = dx + dy. The maximum possible deviation would be $|dz|_{max} = |dx|_{max} + |dy|_{max}$. The maximum error is thus the sum of the individual maximum errors. The subtraction $z = x - y$ would yield the same result, namely that the individual errors are summed.

A similar argument shows the *fractional* error of a product is the sum of the individual *fractional* errors. Let $z = xy$ to find that $dz = x \, dy + y \, dx$, or $|dz/z|_{max} = |dx/x|_{max} + |dy/y|_{max}$. The same result follows for the quotient of two variables.

The above argument assumes that excursion in one variable will be matched by excursions in the other in the direction to maximize the error. In fact the measurements, of x and y would most likely be uncorrelated. A maximum positive excursion in x would most likely not be associated with a maximum excursion in y. The errors in the calculated product or sum are thus, on average, less than the extreme values found above.

In this more realistic case, the individual variables, x and y, vary independently of one another about their respective means with normal distributions characterized by standard deviations σ_x and σ_y. It can be shown (not here) that in a summation or subtraction, the variance σ_z^2 of the sum or difference is the individual variances of x and y *added in quadrature*,

➡ $$\sigma_z{}^2 = \sigma_x{}^2 + \sigma_y{}^2 \qquad \text{(Error in a sum or difference)} \qquad (6.11)$$

For example, if $\sigma_x = \sigma_y$, the standard deviation in the sum (or difference) would be $\sigma_z = \sqrt{2}\,\sigma_x$ instead of $2\sigma_x$ given by the simplified argument. If $\sigma_x \gg \sigma_y$, the error in x more strongly dominates the overall error than under the former assumptions, giving $\sigma_z \approx \sigma_x$. Similarly, the fractional error squared of a product or quotient is the sum of the fractional standard deviations, *added in quadrature*,

➡ $$\left(\frac{\sigma_z}{z}\right)^2 = \left(\frac{\sigma_x}{x}\right)^2 + \left(\frac{\sigma_y}{y}\right)^2 \qquad \text{(Error in a product or quotient)} \qquad (6.12)$$

In each case, the "maximum value" derivation above gave the correct relation, but for the addition in quadrature. Remember that one adds *absolute* errors (in quadrature) when adding or subtracting measured quantities and adds *fractional* errors (in quadrature) when dividing or multiplying such quantities.

Background subtraction

The subtraction of background makes use of these tools. Let the expected number of source counts detected in a given time interval Δt be S, and let the number of expected background counts *in the same or equivalent time interval* be B. The on-source measurement will thus yield $S + B$ counts, and an off-source measurement of the same duration will yield B counts. The desired signal S is simply the difference of the two measured quantities,

$$S = (S + B) - (B) \qquad \text{(Signal counts; equal exposures)} \qquad (6.13)$$

The measured quantities $S + B$ and B will exhibit fluctuations which propagate through the subtraction process to yield a net error on the calculated S.

The two measurements of B and $S + B$ are quite independent of one another; different photons and different background counts are involved. Thus the fluctuations in the two rates will be uncorrelated. The variance on the difference S is thus, from (11),

$$\sigma_s^2 = \sigma_{s+b}^2 + \sigma_b^2 \tag{6.14}$$

where the two variances on the right refer to the two measured quantities, $(S + B)$ and (B), respectively. The two standard deviations σ_{s+b} and σ_b are obtained from the Poisson distribution which gives a standard deviation of $\sqrt{m}$. Thus, we have $\sigma_{s+b} = \sqrt{(S + B)}$ and $\sigma_b = \sqrt{B}$, where we approximate the mean values m with the measured values. Thus (14) becomes

$$\sigma_s^2 = S + B + B = S + 2B \tag{6.15}$$

If σ_s is much smaller than S, the measurement is of high quality and vice versa. If S were 3 times σ_s, the result would be called a "3σ result". If the result is less significant than 3σ, one should question whether the source was detected at all. This significance can be described with the ratio, S/σ_s.

➡ $\qquad$ Significance $\equiv \dfrac{S}{\sigma_s} = \dfrac{S}{\sqrt{S + 2B}}$ $\qquad$ (Number of standard $\qquad$ (6.16)
$\qquad\qquad\qquad\qquad\qquad\qquad$ deviations or S/N)

where S and B are the accumulated source and background counts in some fixed time. This expression is also known as the *signal-to-noise* (S/N) ratio.

The intensity of a source is best represented by the source event rate r_s (counts/s) which is independent of data accumulation time. In our case, with equal on-source and off-source accumulation times (Δt),

$$r_s = \frac{S}{\Delta t} \pm \frac{\sigma_s}{\Delta t} \qquad \text{(counts/s)} \tag{6.17}$$

According to (12), division of S by the precisely known Δt does not change the fractional error from its initial value σ_s/S. We therefore divide σ_s by Δt to maintain this ratio for the standard deviation of the derived source rate, $\sigma_{rs} \equiv \sigma_s/\Delta t$. If Δt is not precisely known; one would have to factor its uncertainty into σ_{rs}.

If the counts S and $S + B$ were accumulated over different time intervals, which is often the case, one could follow our logic leading to (16) to find an expression for S/N in terms of the rates r_s and r_b (counts s^{-1}) and accumulation times t_{s+b} and t_b.

Low and high background limits

Two limiting cases of (16) may be given, one wherein the background counts are much less than the source counts, and vice versa. We assume again that the accumulation time Δt is the same for both on-source and off-source measurements.

The low-background ($B \ll S$) case gives

$$\frac{S}{\sigma_s} \approx \frac{S}{\sqrt{S}} = \sqrt{S} = \sqrt{r_s \Delta t} \qquad \text{(Background negligible; } B \ll S) \quad (6.18)$$

which shows that the significance increases as the square root of the number of counts. We also express the counts in terms of the rate and the duration of the measurement and find that the significance increases as $(\Delta t)^{1/2}$. If one has a 2σ result that looks tantalizing, but hardly convincing at only 2σ, one could take more data. To increase the significance to 5σ, one would have to increase the duration Δt by a factor of $(5/2)^2 = 6.25$.

The significance for the high background case ($B \gg S$) is, again from (16),

$$\frac{S}{\sigma_s} \approx \frac{S}{\sqrt{2B}} = \frac{r_s \Delta t}{\sqrt{2r_b \Delta t}} = \frac{r_s}{\sqrt{2r_b}}\sqrt{\Delta t} \qquad \begin{array}{c}\text{(Background} \\ \text{dominates; } B \gg S)\end{array} \quad (6.19)$$

The background counts are also expressed as the background rate r_b and the duration of the measurement Δt. Again the significance increases as the square root of the duration of the observation. It can take a lot of observing time to increase the significance by a substantial amount.

Let us compare the S/N ratios of our two hypothetical detectors, one high B and the other of low B, but otherwise similar. Expose both to a source yielding the same signal rate r_s. As above, in (19), r_b is the background rate of the high-B detector; so that by definition $r_b \gg r_s$. The background rate of the low-B detector does not enter our approximations; see (18). From (18) and (19),

➡ $$\left(\frac{S}{\sigma_s}\right)_{B \gg S} = \sqrt{\frac{r_s}{2r_b}}\left(\frac{S}{\sigma_s}\right)_{B \ll S} \qquad \begin{array}{c}\text{(Comparison of} \\ \text{sensitivities)}\end{array} \quad (6.20)$$

Since $r_s \ll r_b$, the expression tells us that the significance is much less in the high-background case than for the low-background case, again for similar exposures. This agrees with one's common sense; a source should be detected with higher significance in the absence of background counts.

Bright and faint source observations

Focusing instruments are essentially low-background systems. The detection of only 3 x-ray photons during an observation, in one resolution element of the focal plane could be highly significant because the background in any given resolution element is so low. If the expected background in that element is only 0.1 counts during the observation, the probability of this background giving rise to the 3 x rays in that element is only 1.5×10^{-4} according to the Poisson formula (1). Thus the 3 x rays would be a highly significant detection. Such instruments are the only way to detect the distant faint sources in the cosmos. This discussion assumes that only this one resolution element is of interest, say because a known source in another frequency band is located there.

How does the significance of a detection in a given time depend on source intensity, i.e., on the rate r_s? When the source intensity dominates the background, as in focusing instruments, the statistical noise arises from the source itself. When source intensity increases, so does the statistical noise. Thus the significance in the weak-background case grows rather slowly with source intensity according to (18), specifically as $\sqrt{r_s}$. When the background dominates, the significance (19) scales as r_s. A source with twice the intensity (rate) yields a measurement with twice the significance (in the same time) because the statistical noise depends only on the background rate which does not change as r_s increases.

This comparison is illustrated in (20) which is valid as long as the source in the high-background detector is still weaker than the background in that detector ($r_s \ll r_b$). As r_s/r_b increases, but while it is still well below unity, the sensitivity of the high-background detector moves toward the sensitivity of the low-background detector; the advantage of the low-background detector decreases as the source brightens. When finally the source becomes so bright in the high-background detector that it exceeds its high background, the weak-background limit (18) applies to *both* detectors. Then the sensitivities of both detectors become identical, given that they are similar in all other respects.

In reality, such detectors are not similar. Other factors such as different effective areas at different energies control the relative sensitivities. For example, non-focusing (high-background) x-ray detectors, such as proportional counters with mechanical collimators, can be constructed to have a large effective area at energies up to 60 keV and higher. In contrast, focusing (low background) systems reach only to $\sim$8 keV, and the practical effective areas are typically much less, especially at the higher x-ray energies. Thus, for bright sources, a high-background large-area system can yield a higher significance (S/N) in a given time than can a low-background system.

The Rossi X-ray Timing Explorer makes use of large collecting areas to measure the temporal variability of bright x-ray sources on time intervals of a millisecond or less, thus probing the motions of matter in the near vicinity of neutron stars and black holes. This advantage in timing accuracy comes at the cost of not having the high angular and spectral resolution that are possible with focusing x-ray missions.

Comparison to theory

The result of the data processing above is that one ends up with a series of numbers, each of which has its own uncertainty, $y_i \pm \sigma_i$. This uncertainty σ_i is generally *not* the square root of y, which would be the case only if the process obeyed Poisson statistics and if the numbers were not further processed, for example by division by the exposure time to get a rate.

Finding parameters and checking hypotheses

Often one will want to compare the data to some theoretical expectation in order to (*i*) derive some parameter or parameters in the theory or (*ii*) to check whether the data are consistent with the theory (or hypothesis). For example one might measure how many cars come down the street every hour in order to find the average rate, but also to test whether the rate is indeed constant.

Suppose we make two measurements, and count 20 cars the first hour and 30 the second. We could conclude that the average rate was 25 cars per hour. What is the error on this result? Our raw unprocessed numbers obey Poisson statistics so the errors are the square roots, $N_1 = 20 \pm 4.47$ and 30 ± 5.48. To obtain the average we added the two numbers and divided by 2. Propagating the errors according to (11) and dividing by 2, we find the average rate to be 25 ± 3.53. This is our best estimate of the true value. The uncertainty is the standard deviation of the average. It is somewhat uncertain because it is derived from (11) which is based on a normal distribution. Note that there is a significant probability that the true value could be 2 standard deviations (or more) from 25, namely as much as 31.7 or even 35 at 3 standard deviations, or as low as 15 on the low side.

Can we also argue that the rate in this experiment increased from one hour to the next? It depends on the sizes of the error bars. To evaluate this, calculate the value of the difference of the two numbers which is expected to be zero if the underlying rate did not change. The difference and its calculated uncertainty, from (11), are $N_2 - N_1 = 10 \pm 7.07$. Is this result consistent with zero? Recalling that successive measurements can fluctuate by greater amounts than one standard deviation, it is actually quite consistent with zero. The measured value is only $10/7.07 = 1.4$ standard deviations from zero.

What is the probability that a true value of zero could fluctuate to a value as high as 10 or as low as -10? From Table 2, the probability of a 1.4 standard deviation fluctuation is 16%; one in six sets of measurements would show such a fluctuation. Our data thus can not exclude the constant-rate hypothesis. If we had found a probability significantly less than 1%, we would have seriously questioned the constant rate hypothesis and concluded with some confidence that the rate actually increased.

With more data points and with more complicated theoretical functions, one can ask the same questions: what are the parameters of the theory that best fit the data, and are the data consistent with the theory? There are formal ways to address these.

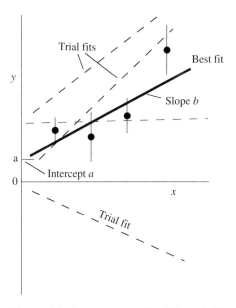

Figure 6.9. Least squares fits. The solid line is a by-eye fit to the data points in an effort to minimize χ^2 given in (21), namely the sum of the squares of the deviations, the latter being in units of the standard deviations. The dashed lines are fits that would have larger values of χ^2 and thus are less good, or terrible, fits.

For the former we introduce the *least squares fit* and for the latter the closely related *chi square test*.

Least squares fit

Comparison of data to theory can be carried out with a procedure known as the least squares fit. Consider the data points and theoretical curves in Fig. 9. Each data point is taken at some value x_i and has a value and uncertainty $y_i \pm \sigma_i$ indicated with vertical *error bars*. At each position x_i, calculate the deviation, $y_{ob,i} - y_{th,i}$, of the (observed) data point $y_{ob,i}$ from the value $y_{th,i}$ of a theoretical curve at that point. Then write this deviation in units of the standard deviation σ_i of that data point and square the result. This yields a value that is always positive regardless of whether the deviation is positive or negative. Then sum over all data points to obtain the quantity called *chi squared*, χ^2,

$$\chi^2 \equiv \sum_i \left[\frac{y_{ob,i} - y_{th,i}}{\sigma_i} \right]^2 \qquad \text{(chi square definition)} \qquad (6.21)$$

where the subscript i indicates the values of y and σ at the coordinate x_i. In practice one usually calculates the σ_i from the actual data values $y_{ob,i}$ because one does not know the true mean.

The theoretical values $y_{\mathrm{th},i}$ can be based on any function that might be appropriate for the data, for example a straight line, $y_{\mathrm{th},i} = a + bx_i$. In our car-counting example above, the independent variable x_i would be the time of a measurement. The linear expression reflects the hypothesis that the car numbers are increasing or decreasing linearly with time, or that they are constant at a if $b = 0$.

A number of straight lines can be drawn on the plot of Fig. 9, and a value for χ^2 can be calculated for each one. The solid line is an estimated, by eye, fit that attempts to minimize the deviations while giving the most weight to the points with the smallest error bars. This is the "least squares fit". Formally, one would attempt to minimize (21) which gives the most weight to points with small values of σ_i. The dashed lines in the figure clearly have larger values of χ^2 and hence would not be the "best-fit" curve.

The minimization of (21) for a given set of data can be carried out empirically with many trials of different straight lines. For well behaved functions, like our straight line, the minimization can be carried out analytically with calculus. In the minimization process, one is varying the trial theoretical function, which in our example is $y_{\mathrm{th},i} = a + bx_i$. The parameters of the theory, a and b, are being varied. When the minimum χ^2 is obtained, one knows the *best fit values* of a and b for the straight line hypothesis.

In astrophysics, one might measure the power from a star at a series of different frequencies and compare it to the theoretical blackbody spectrum which has the form,

$$I(v,T) = K \frac{v^3}{e^{hv/kT} - 1} \qquad \text{(Blackbody spectrum)} \qquad (6.22)$$

where v is the frequency, T the temperature, K the proportionality constant that is dependent in part on the distance to the star (greater distance yields less power), and finally h and k the Planck and Boltzmann constants respectively. The independent variable in this case would be v, and the spectral parameters to be determined are K and T. Minimization of χ^2 would yield best fit values of K and T.

Chi square test

A best-fit theoretical curve is not necessarily consistent with the data. When compared to the expected fluctuations in a set of data, the fit may be too bad (high χ^2) or too good (low χ^2).

One can compare a theoretical function that might underlie the data to the data points themselves to find if in fact the function is consistent with the data taking into account statistical fluctuations. The answer in general will be expressed in terms of probabilities, rather than a flat yes or no. The theoretical function being tested need not be the best fit function, but in most cases it would be.

A common and powerful test for this purpose is known as the *chi square test.* It makes use directly of the value of χ^2 calculated from the data and the trial function as given in (21) together with the number of *degrees of freedom f.* The latter is the number of data points n less the number of variable parameters p in the theoretical function. For our straight line function, the number of such parameters, a and b, is two, and for our blackbody function it is also two, for K and T. If for each case we had $n = 20$ data points, the number of degrees of freedom would be $f = n - p = 18$.

Suppose that the theoretical trial function is the true underlying curve that describes the data. In other words, if one made many very precise measurements, the data would faithfully track this curve. Now consider our less than precise set of measurements with an associated value of χ^2. One could then ask the question: *If the trial function were the true function, and I made another similar set of measurements, what is the probability I would find a greater χ^2 than this?* In other words, what are the chances that the data from the second set would deviate from the theoretical function more than do the set of measurements I already have in hand?

The answer lies in the *chi square statistic.* If each individual data point y_i in subsequent measures is distributed about the true (theoretical) value as a normal distribution with the standard deviation σ_i assigned to it, there is a theoretical basis for calculating this probability. If the probability turns out to lie between 0.1 and 0.9, it would indicate that the fluctuations in our data are comparable to those expected in another set of data, given the σ_i of the several data points. Thus one would consider the data (values and error bars) to be consistent with the trial function. It would indicate that more or less, on average, the points lie within one or at most two, standard deviations of the function, as shown in Fig. 10a.

If the probability turns out to be very low, say, less than ~0.02, it implies that our data have improbably large deviations (large χ^2) compared to the fluctuations expected from the individual error bars, as shown in Fig. 10b. In this case, we would question whether or not the trial function is the appropriate one; our hypothesis may be false. Alternatively, it could mean that the error bars on each point were underestimated. This would be the case if we had neglected to include, or had underestimated, systematic errors.

If the probability is greater than ~0.98, it would imply that our data clustered too tightly about the theoretical line as in Fig. 10c. This, too, is suspect because the data do not fluctuate as much as basic statistics would require. In our car-counting example, it would be as if we obtained the following counts for each of 5 successive hours: $24, 27, 25, 23, 26$. The root-mean-square deviation about the mean is only 1.4 when, from Poisson statistics or the normal distribution, we expect $\sigma = \sqrt{25} = 5$. The fit to our hypothesis of a constant rate is too good!

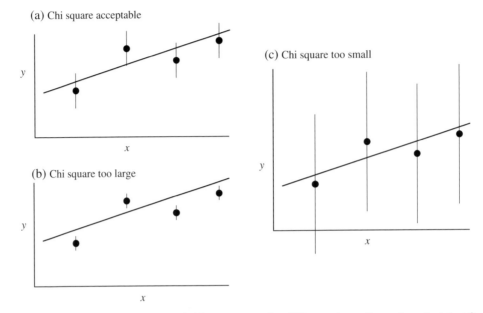

Figure 6.10. Results of chi square tests for different sizes of error bars (σ_i), but for the same data points x_i, y_i. (a) Moderate error bars. The chi square is acceptable because the average deviation is on the order of 1 standard deviation. (b) Small error bars. The deviations measured in units of σ_i are very large, leading to an unacceptably low χ^2 probability that fluctuations in another trial would exceed these. (c) Large error bars leading to an unacceptably high probability.

What can we conclude in this case? Most likely something is wrong with the error bars; our model of the fluctuations could be wrong. For example, the cars could be passing at regular intervals in some sort of procession such as a parade, rather than at random times. This would lead to smaller fluctuations in the number of car passages in successive time intervals; Poisson statistics would not apply. (The remaining small error bars might represent some small jitter in the regularity of car arrival times.) Smaller error bars would serve to raise the χ^2 value and lower the probability. This could then make the data consistent with the trial function.

We do not derive here the χ^2 statistic that yields the desired probabilities, but it can be tabulated or plotted as in Fig. 11. One enters the figure with the value of χ^2 and the number of degrees of freedom f as defined above. The curves give the probability that another measurement would yield a greater χ^2. Any probability between 0.1 and 0.9 can be considered to be consistent with the trial function curve, and values less than 0.02 and greater than 0.98 raise serious doubt about the appropriateness of the trial function.

Some insight into the behavior of χ^2 follows from its definition (21). For normal fluctuations about the expected values, the deviation in the numerator of one term

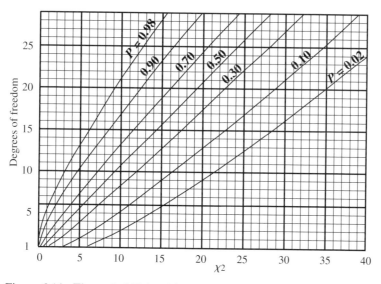

Figure 6.11. The probabilities P for the chi square test. The ordinate is the number of degrees of freedom (number of data points less number of variable parameters in trial function), and the abscissa is the value of χ^2. The curves give the probability that χ^2 would have a greater value in another set of measurements. [Adapted from Evans, *The Atomic Nucleus*, McGraw-Hill, 1955, p. 776, with permission.]

in the summation (21) should equal, on average, $\sim 1\sigma$, but note the σ in the denominator; the deviations are expressed in units of σ. One such term should, on average, thus be about unity. Summation over the n terms yields a χ^2 value of about n. Thus if there were 20 data points, we might expect χ^2 to lie somewhere in a broad band of values about $\chi^2 \approx 20$ if the model and our error bars are both correct. If, instead, we find $\chi^2 \approx 40$, the deviations of our data would be unreasonably large. If $\chi^2 \approx 5$, the deviations would be unreasonably small.

It is actually the number of degrees of freedom, $f = n - p$, that comes into play here. Suppose we are fitting a straight line, with its two parameters, a and b, and have only two data points. In this case the best-fit line would go exactly through both points, and we would have $\chi^2 = 0$. It is only a third data point that, with the first two, begins to yield a non-zero χ^2. Thus in our example, we would have $f = 20 - 2 = 18$, and we would expect χ^2 to lie in a broad band of values about 18, more or less. This is in accord with the probability limits 0.1 to 0.9 for satisfactory data in Fig. 11.

Astronomers often use the *reduced chi square* $\chi_\nu^2 \equiv \chi^2/f$, a value expected to be near unity for the correct model. How close to unity depends critically on f. For satisfactory fits, namely probabilities in the range 0.1–0.9, one requires 0.49 $<\chi_\nu^2 <1.6$ if $f = 10$, but if $f = 200$, one requires $0.87 <\chi_\nu^2 < 1.13$. If there are a lot of data points, one expects χ_ν^2 to be very close to unity. In either case, if your data do not fall within the prescribed probability limits, try another model or check your assumptions on systematic errors.

Problems

6.2 Position-insensitive detectors

Problem 6.21. (Requires material in Section 6.5.) The magnitude of the charge pulse from a proportional counter fluctuates in value from one incident x ray to another, even when the incident x rays all have the same energy, e.g., those obtained from an ^{55}Fe (iron 55) radioactive source. (a) Consider the detection of 6.0-keV x rays in an argon-filled proportional counter. What is the standard deviation in units of keV of these fluctuations if they arise mostly from Poisson fluctuations in the number of ion pairs created by the initial photoelectron? Assume that there are no escape photons, and consider only the first generation of ion pairs, those created by the several initial photoelectrons with a combined energy of 6.0 keV. What is the fractional energy resolution, defined as the FWHM of the response curve divided by the mean energy, at this x-ray energy? (b) What are the fractional energy resolutions at energies 2 keV and 30 keV? [Ans. ~0.4 keV, ~15%; ~25%, ~7%]

6.3 Position-sensitive detectors

Problem 6.31. Demonstrate that there is a maximum of the electric potential V that could trap a free electron within the n-type region of a CCD. Let the n-type region, of thickness d, contain only fixed positive charges with a charge density $+20\rho_0$. The depleted portion of the p-type region has thickness $5d$ and fixed charge density $-\rho_0$. The undepleted portion with its piled up free holes results in a zero electric field at $x = 0$ (see sketch).

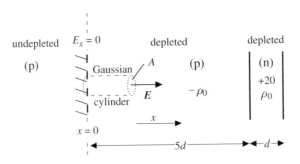

Neglect the effect of any residual free charges in the depleted regions or of the electrodes or insulator. Proceed by constructing a Gaussian pillbox with its base at $x = 0$ (see sketch). From Gauss's law, find the electric field $E_x(x)$ in both depleted regions and, from the results, find the potential $V(x)$ in both regions. Set

$V(0) = 0$ at $x = 0$. Find the position x of the maximum V. Make plots of $E_x(x)$ and $V(x)$. What is the significance of the position where $E_x = 0$? [Ans. $x_{max} = 5.25d$]

6.4 Gamma-ray instruments

Problem 6.41. Each of the 36 EGRET spark-chamber modules yields x and y coordinates of the track of an electron that passes through the module (see text). Two particle tracks are the typical consequence of the conversion of a gamma ray (Fig. 5a). In a given module, these lead to two x coordinates and two y coordinates which intersect at four locations, two that are valid and two that are spurious. (a) In comparing the spark locations from module to module, do the spurious locations form tracks that are discontinuous and thus easily seen to be non-physical? (b) Can you think of a simple modification to the experimental arrangement that would remove this ambiguity? In fact such a feature was incorporated into EGRET. (c) The efficiency for a given spark chamber to produce a spark is only 80–85%. Does this help remove this ambiguity? Consider the cases where a chamber produces one spark or no sparks.

6.5 Statistics of measurements

Problem 6.51. (a) Add three rows to Table 1 for values of the Poisson function for the expected means, $m = 0.01, 0.1$, and 0.3, each for $x = 0, 1, 2, 3, 4$. (b) Plot histograms (by hand or with computer) for these values and also those for $m = 1$ on a single plot. Use an expanded scale; truncate the plot after $x = 4$. Comment on the behavior of the four histograms.

Problem 6.52. (Makes use of probabilities from previous problem.) (a) How many x rays from a distant point-like star must be detected (in one resolution element or pixel) at the focus of an x-ray telescope for the x-ray detection to be considered real (with fair confidence)? Let the background (averaged over all pixels) yield exactly 1.00×10^{-2} counts per pixel during the exposure. Require that there be less than a 2% chance that a fluctuation in the background rate could give your number. The location of the star in the sky is known, as is the particular pixel where its x-ray image would appear. What kind of statistics are appropriate here? (b) If one is searching for a point-like x-ray image whose position is not known, one must consider all 10^5 pixels of the image. How many x rays from the point source must be detected in this case? Require that the probability for a background fluctuation up to your value in any one of the 10^5 pixels be less than 2%. [Ans. very small integer values]

Problem 6.53. (a) Demonstrate that the normal distribution is normalized to unity according to (4). (b) Calculate by numerical integration, say with 20 steps, the probability that a measurement obeying the normal distribution will deviate from the expected mean by more than 4 standard deviations in the positive direction. [Ans. (b) See Table 2]

Problem 6.54. The expression for the significance S/σ_s of a counting measurement derived in the text (16) is valid if the exposure times for the on-source and off-source measurements are equal. Here, let these times be different and designate them t_s ($\equiv t_{s+b}$) and t_b respectively. (a) Write the counting rate r_s (cts/s) of the source alone in terms of the measured rates r_{s+b} and r_b, and find the standard deviation σ_{rs} of the rate r_s in terms of these rates and the times t_s and t_b. Note that the number of background counts B_s detected "on source" may differ from the number B_b detected "off source". Be careful to distinguish total counts, $S + B_s$ and B_b from the counting rates r_{s+b}, r_b and r_s. (b) Use your answer to write directly the expression for the significance r_s/σ_{rs}, of the source rate. Find an expression for the time t_s that yields the maximum significance if the observing time $T = t_s + t_b$ is fixed. In other words, how should the time T be apportioned between on-source and background measurements? Hints: substitute $t_b = T - t_s$, define $g \equiv t_s/T$ and solve for the optimum value of g while noting that $r_{s+b} = r_s + r_b$. (c) Evaluate your expression for three cases: $r_s \ll r_b$; $r_s = r_b$; $r_s \gg r_b$. [Ans. (b) $1 - g = \left[1 + \left(1 + \frac{r_s}{r_b}\right)^{1/2}\right]^{-1}$; (c) $t_s = T/2, \sim 0.6T, T$]

7

Multiple telescope interferometry

What we learn in this chapter

Radio astronomers have learned to overcome the limitations of diffraction with **interferometry**, the use of two or more telescopes viewing the same source at the same time. The instantaneous beam of two telescopes is an interference **fringe pattern** on the sky. As the earth rotates, the pattern sweeps across a postulated point source yielding a **time varying interference** signal when the signals from the two telescopes are **summed** or **multiplied**. Simple examples show that two telescopes on the rotating earth can, in most cases, locate the **position of a point source**. Each brief two-telescope observation with a given **baseline** (telescope separation and relative orientation) can be described as a point on a two-dimensional plot (**Fourier plane**) of the x and y **spatial frequencies**. For each such point, the detected oscillatory signal yields a value of the complex **visibility function** $V(b)$ which is one **spatial Fourier component** of the sky brightness distribution. **Large arrays of telescopes** making repeated observations as the earth rotates provide additional points in the Fourier plane and thus additional Fourier components. With sufficient coverage of the Fourier plane, the **Fourier transform** of $V(b)$ yields a reasonable approximation of the **true sky brightness function**. This process is called **aperture synthesis**.

Interferometry dominates **radio astronomy**, e.g., the US VLA and the Australian AT arrays. The greatest antenna spacings yield the highest angular resolution. **Very Long Baseline Interferometry** (VLBI) with radio telescopes on different continents yields angular resolutions of <1 milliarcsecond (mas) and as low as 0.5 mas if one detector is in high earth orbit as is the Japanese HALCA satellite (**space VLBI**). **Optical interferometry** is likely to become important in the near future because new **8-m class optical observatories** are being built with one or more pairs of telescopes in close proximity with $\sim$100 m spacings, potentially yielding resolutions of $\sim$1 mas.

7.1 Introduction

Interferometry is an observational technique wherein the interference of electromagnetic waves is used to extract the highest possible angular resolution allowed by basic physical principles. We have seen in Section 5.5 how interference within a single telescope aperture due to atmospheric density fluctuations can lead to speckles from which one can construct, in principle, the image expected in the absence of such fluctuations. Speckle interferometry could also be called *single-aperture interferometry*.

Here we consider how interference between the signals received from two or more separate telescopes viewing the same object can yield extremely high angular resolution. This could be called *multi-aperture interferometry* but the usual name is *long baseline interferometry*. One can think of an array of telescopes viewing the same source as being several small segments of a huge hypothetical telescope of diameter comparable to the spacing between the outermost telescopes of the array. It turns out that one can attain, with some limitations, the higher angular resolution of the hypothetical single large telescope.

The simplest arrangement is two telescopes separated by the distance B known as the *baseline*. (We redefine the baseline as a vector $\boldsymbol{B}$ below.) In this case, the hypothetical large telescope would have a diameter B, and its angular resolution would be, from (5.8), $\theta_{min} = 1.22\lambda/B \approx \lambda/B$, where λ is the wavelength of the radiation. If the separation B is large compared to the individual telescope diameters d, the potential angular resolution will be much better (smaller θ_{min}) than that of one telescope alone with $\theta_{min} \approx \lambda/d$. Because the two telescopes would make up only a small portion of the hypothetical large telescope, the image quality would be terrible. It is much better if more telescopes are used to fill the hypothetical aperture. Nevertheless, even in the two-telescope case, one can obtain information about the angular structure on the fine scale represented by $\theta_{min} \approx \lambda/B$.

Interferometry is now carried out routinely at radio wavelengths. The long wavelengths of radio waves yield very poor angular resolutions for a single telescope because radio wavelengths are so large. The improvement of resolution obtained by interferometry is thus a necessity for radio astronomers.

Arrays of radio telescopes spreading over tens of kilometers yield angular resolutions less than $1''$. Telescopes on different continents working together routinely yield $\lesssim 0.001''$ resolution, and earth-bound telescopes working with a satellite can probe angular scales several times smaller than that. The latter two arrangements are called *Very Long Baseline Interferometry (VLBI)* and *Space VLBI* respectively. An array of telescopes will produce sky maps with imperfect point-source response functions because the telescopes do not form a complete large telescope. Nevertheless, high-quality maps are obtained through (*i*) a judicious spacing of the individual telescopes, (*ii*) the continuously changing orientation of the individual telescopes relative to the celestial source due to the rotation of the earth or the motion of a satellite, and (*iii*) sophisticated analysis techniques.

Long baseline interferometry at optical wavelengths is now coming into use with the construction of new modern telescopes in groups of two or four. Interferometry of x rays has just been demonstrated in the laboratory. In this chapter, we use radio interferometry as the basis of our discussion. The principles are the same at other (shorter) wavelengths, but the technical hurdles are greater.

7.2 Two-telescope interference

The underlying principle of interferometry is most simply demonstrated with a two-telescope system. Here we describe two such observations. The first is carried out from an observatory on the equator and the second from the North Pole. In each case an overhead point source is observed. First, we present an overview of the basic phenomenon.

Principle of interferometry

Radio interferometry is based on the interference of electromagnetic waves. Consider the reception of a plane radio wave from a distant point source that is directly above two radio telescopes, separated by a short distance B (*baseline*; Fig. 1a). The wavefronts in Fig. 1a arrive at the two telescopes in phase because the source is directly overhead. If the coaxial cables carrying the signals (electromagnetic waves) from the telescopes bring the signals together, the electric and magnetic fields are added to yield radio-frequency oscillations of large amplitude. This is constructive interference.

If earth rotation causes the source direction to be displaced from the zenith by the angle $\theta = \lambda/2d$ (Fig. 1b), the wavefronts arrive at the two telescopes exactly out of phase. The addition of the signals ideally yields a zero output (destructive interference). At increasing angular displacement $\Delta\theta$, this pattern repeats with constructive interference at $\Delta\theta \approx \lambda/B$, $2\lambda/B$, $3\lambda/B$, etc., for approximately normal incidence ($\theta \approx 90°$). The angular separation between adjacent constructive directions is, for $\Delta\theta \ll 1$,

$$\Delta\theta \approx \frac{\lambda}{B} \qquad \text{(Angle between directions of constructive} \qquad (7.1)$$
$$\text{interference; } \Delta\theta \ll 1; \theta \approx 90°)$$

If $\lambda = 1$ m and $B = 100$ m, then $\Delta\theta = 1/100 \approx 0.5°$; the earth rotates $0.5°$ in 2 minutes.

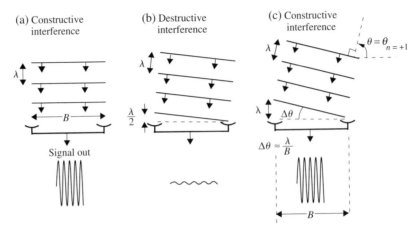

Figure 7.1. Two telescopes on the earth's equator, east–west of one another, receiving a plane wave from a distant source at different angles relative to the "baseline" connecting the telescopes. The signals are added (or multiplied) to yield a single output signal. (a) and (c) Constructive interference yielding a strong signal. (b) Destructive interference yielding, ideally, no signal. As the earth rotates, carrying the telescopes to different orientations, the output signal from a single source will alternately be strong and weak.

There are thus many directions of constructive interference from which the signal of a detected source could arrive and be detected. If the observer does not know, a priori, the location of the source, he or she can infer only that a source detected at maximum strength could lie in any one of these many directions. The intervening directions, e.g., at destructive interference, are excluded as possible source positions.

The rotation of the earth plays a key role here. It continuously reorients the telescopes and causes the directions of constructive/destructive interference to scan across a celestial radio source. This causes the amplitude of the summed radio-frequency oscillations to vary between a maximum value and zero as the angle θ of the source increases uniformly; see "signal out" in Figs. 1a,b,c. This oscillation, or modulation, due to interference is at a much lower frequency than the radio frequency ω. The modulation allows one to identify the times of maximum signal when the source is in a direction of constructive interference.

The observation of Fig. 1 yields position information only in the left–right direction. The two telescopes can not locate a source in the other direction (in and out of the paper) because a source displaced a modest amount in this direction would exhibit maxima at the same times as the overhead source.

The determination of celestial source positions requires knowledge of the precise location of the telescopes relative to the stars at the times of the maxima. This is provided by knowledge of the geographic locations of the telescopes and the precise times of the occurrence of the maxima. The precise times specify the orientation of the earth.

This modulation of the signal between maxima and minima as the earth rotates also provides information about the *angular size* of the source. If the source is larger than the angle between the directions of strong reception, $\Delta\theta = \lambda/B$, different parts of the source will simultaneously span the directions of constructive and destructive interference. Thus the signal will modulate very little (if at all) as the earth rotates. In other words, the presence of strong modulation of the signal indicates that the angular size $\Delta\theta_s$ of the source is less than about $\Delta\theta$, i.e., $\Delta\theta_s \lesssim \lambda/B$.

Further positional and size information is obtained with telescope pairs of different spacings B_i and with different orientations, e.g., if the baseline runs in and out of the paper in Fig. 1. The baseline of a pair of telescopes is usually defined as a vector *B* with magnitude equal to the telescope spacing and with direction of the line joining the two telescopes. It is defined in inertial space, and it changes orientation as the earth rotation carries the telescopes to different relative positions in inertial space. Thus, *B* is continually changing. An array of telescopes provides many two-telescope vector baselines *B* simultaneously, and the rotation of the earth reorients them providing even more baselines during the course of a day. The information can yield high quality maps of the sky.

Equatorial observation

We now examine in detail a specific case, namely observations of an overhead source by two telescopes that lie relatively close together and east–west of one another on the earth's equator (Fig. 2a). We first consider that the two antennas are transmitting, rather than receiving in order to understand the *telescope beam*. Then we return to the (astronomical) receiving case and find the nature of the positional information provided by the observations.

Transmission of radiation

Let the two equatorial telescopes in Fig. 2a each transmit purely sinusoidal waves with the same frequency in the direction approximately normal to their baseline. The two electromagnetic waves will interfere with one another as do the waves in a *ripple tank* experiment wherein two adjacent probes disturb the water surface with an oscillatory motion at a given frequency. An observer at the probes or telescopes finds that there are view directions of constructive interference (solid lines with arrows in Fig. 2a) alternating with directions of destructive interference. In the constructive directions, the phases of the two waves are always in phase. The

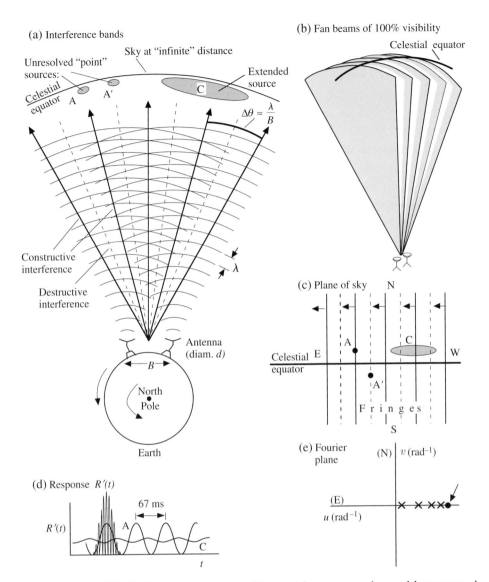

Figure 7.2. Radio interferometry with two telescopes on the earth's equator observing a region of the sky directly overhead on the celestial equator. (a) Directions of constructive (solid lines) and destructive (dashed lines) interference. At the time shown, the radiating point source A would be visible whereas point source A′ would not. Extended source C would be only partially visible. (b) Fan beams of constructive interference. (c) Fringe pattern on sky showing (solid) lines of 100% visibility and dashed lines of invisibility. As the earth rotates the pattern translates toward the east (left). (d) Power received from sources A and C as the earth rotates, known as the *response* $R'(t)$. The rapid (radio-frequency) electromagnetic oscillations of the radiation are illustrated (at greatly reduced frequencies) for the first peak. The fringe period of 67 ms is for a baseline of 10 km; see text. (e) Two-dimensional Fourier (u,v) plane showing (dot) the "spatial" frequency component measured in this experiment. If the observation were to be repeated 4 times during the next 6 hours, the points marked "×" at decreasing spatial frequencies would be sampled.

system thereby broadcasts a large power in these directions. No power is radiated in the directions midway between them; dashed lines in Fig. 2a.

In three dimensions the directions of constructive interference are fan beams (Fig. 2b). Each "fan" represents a plane of directions of constructive interference. These intercept the celestial sphere to give *visibility lines*, the solid vertical lines in Fig. 2c, also known as a *fringe pattern*. The visibility lines are circles on the celestial sphere that approximate great circles for directions nearly normal to the baseline. (A *great circle* on a sphere lies in the plane containing the center of the sphere, like the earth's equator or the meridian lines of longitude. The plane of a *small circle* does not intersect the center, e.g., the earth's latitude parallels.) As we learn below, the fan or fringe pattern actually is not a set of discrete lines, but rather is sinusoidal in the direction normal to the lines of maxima. This sinusoidal multiple-fan pattern is in effect the *beam* of the two-telescope system.

For directions approximately overhead, the angular spacing $\Delta\theta$ between the directions of constructive interference is, from (1), $\Delta\theta = \lambda/B$. If the telescopes are separated by $B = 10$ km and are broadcasting at frequency $v = 6$ GHz, the wavelength is $\lambda = c/v = 50$ mm, and the angular spacing is $\Delta\theta = 5 \times 10^{-6}$ rad $= 1''$. If the telescopes are on different continents, $B \approx 10^4$ km, the spacing becomes $\Delta\theta = 0.001''$

These spacings are very small compared to the beam size θ_{beam} of one of the individual telescopes. If the dish diameter is $d = 25$ m,

$$\theta_{\text{beam}} \approx \frac{\lambda}{d} = \frac{0.050}{25} = \frac{1}{500} \text{ rad} \longrightarrow 6.9'. \qquad \text{(Beam size)} \qquad (7.2)$$

The two telescopes would broadcast into this relatively wide $7'$ beam of directions, but this beam would contain ~ 420 cycles of the $1''$ interference pattern (for $B = 10$ km) or 420 000 cycles of the $0.001''$ pattern (for $B = 10^4$ km). The times for one fringe cycle to pass over a given celestial position due to the earth's rotation are 67 ms and 67 μs respectively. The period of the 6-GHz electromagnetic wave is much shorter; $P = 1/v = 1.6 \times 10^{-10}$ s $= 0.16$ ns. The 67 ms fringe period for $B = 10$ km is shown in Fig. 2d.

Reception

Now, consider the reception of radiation by the E–W equatorial telescopes. Let point-like source A (Fig. 2a) emit a broad band of radio frequencies, and let the two telescope receivers be tuned to the same single frequency. Since source A happens to lie in a zone of constructive interference in the figure, a wavefront from it will arrive at the two telescopes in phase. If the electric vectors of the electromagnetic waves from the two telescopes are summed, the result will oscillate at the radio frequency *with a large amplitude*. This is just the situation shown in Fig. 1a.

If, on the other hand, the point-like source of the radiation is in a direction of destructive interference (e.g. A′), the summed telescope signals will add to zero as in Fig. 1b. In intermediate regions, an intermediate signal would be obtained. An extended source C in general would have a greatly reduced, or zero, response depending upon the exact distribution of source brightness on the sky.

The lines of visibility (constructive interference) on the sky are again those shown in Fig. 2c; they run north–south (Fig. 2b). As the earth rotates, the pattern moves along the sky in the eastward direction (arrows), and source A passes from visibility to invisibility and back again with a sinusoidal response of period 67 ms or 67 μs for the cases given above. The corresponding frequencies are 15 Hz and 15 kHz. The 6 GHz radio frequency is greater by factors of more than 10^8 and 10^5 respectively.

Earth rotation

The response $R'(t)$ in Fig. 2d is a plot of the power (amplitude squared) of the summed electric vectors of the electromagnetic signals received from source A at the two detectors as a function of time as the earth rotates. A low-pass filter averages out the very rapid radio-frequency oscillations shown in the first peak. (The prime in R' is used here to reserve R for the case where the average power is subtracted from the response curve.)

Here we have considered only a brief observation near the zenith. If another observation had occurred several hours later, the telescopes would have been carried by the rotating earth to a new orientation relative to the source. This results in a reduced effective baseline (baseline projected normal to the source direction). In turn, this yields an increased spacing of the visibility lines according to (1). As we demonstrate below, this additional set of lines helps refine our knowledge of the positions and angular sizes of the sources in the field.

The earth's rotation is thus doubly helpful: it causes the fringe pattern to pass over the source to yield the oscillatory detection that indicates the presence of a source and helps locate its position (Fig. 2d), and it reorients the telescopes so the spacing of the fringe pattern changes, which refines the results. (Figure 2e will be discussed later.)

Position of source

How is the response $R'(t)$ used to provide information about the location of a point source of unknown celestial position? Knowledge of the physical locations of the telescopes in inertial space (derived from their location on the earth and the angle of the earth's rotation) allows one to locate precisely the positions of the visibility lines on the sky at any given time. One can plot these lines on the sky at an instant when the response of a point source is at one of its maxima.

Figure 2c is such a plot. The solid *lines of position* represent the positions upon which the source must lie plus or minus the uncertainty. If the signal is quite strong, the time of the visibility maximum could be quite precisely known, and the uncertainty in position quite small, say $\pm 2\%$ of the line spacing. In this case, the position of the source would be restricted to about 4% of the fringe period. Nevertheless, it could still lie on, or close to, any one of the ~ 420 visibility lines (for $B = 10$ km) within the overall $7'$ extent of our hypothetical beam.

If the observation of our point source were repeated after several hours, as suggested above, the new visibility lines with their greater spacing would be another set of lines of position upon one of which the source must lie. Clearly the source can lie only at positions that agree with both sets of lines, i.e., where the lines overlap within their uncertainties. This greatly restricts the possible locations of the source. Similarly, the telescopes could be relocated so that they lie, say, northeast/southwest of one another. The lines of position on the sky would then run normal to this direction (northwest/southeast). The intersections of these lines with the N–S lines would further restrict the possible positions of the source.

North-Pole observation

The fringe patterns on the sky can rotate as well as translate due to the earth's rotation. The rotation is best illustrated with two telescopes placed near the North Pole and viewing a region encompassing the north celestial pole (NCP; Fig. 3a). In our sketch, the baseline and telescope beam are not centered directly on the NCP. The fringes will rotate on the celestial sphere as the earth rotates. In the frame of reference of the earth, the source follows a circular path centered on the NCP and completes a cycle in 24 h (Fig. 3b). As the source passes through the fringe pattern, the response $R'(t)$ will oscillate rapidly for part of the circle and slowly at others (Fig. 3c). In a typical arrangement, the fringe angular spacing would be $\lesssim 1''$ so that a source more than $1°$ from the NCP would exhibit many thousands of fringe transits per day.

The time profile of these oscillations is unique to each possible source location: the number of oscillations per day gives the radius of the circular path, and the phase of the slow–fast modulation gives the azimuth. The rotation of the fringes on the sky thus yields two-dimensional positions whereas the equatorial observation of Fig. 2 yielded only one-dimensional lines of positions. If a visibility line (or an anti-visibility line) happens to lie directly on the NCP, sources at two opposing positions on the circle will yield identical responses. In these special cases, a given response pattern will be consistent with two opposing positions as candidates for the real source.

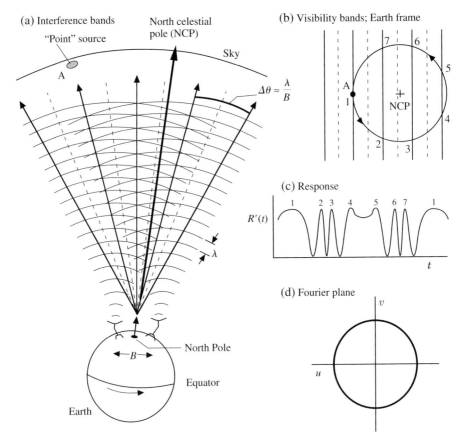

Figure 7.3. Radio interferometry with two telescopes at the North Pole observing point source A which is near, but not at, the celestial pole. (a) Interfering waves and directions of 100% visibility (solid lines) and invisibility (dashed lines). Point source A is on one of the former at the time shown. (b) Movement of point source A in frame of reference of the earth. The track is a circle about the north celestial pole (NCP). (c) Response $R'(t)$ of the source A. It oscillates alternately slowly and rapidly; numbers indicate source positions given in (b). (d) Fourier plane showing circular track sampled by a continuous 24-h measurement with fixed projected baseline.

A pseudo image of the sky can be constructed graphically for a North-Pole observation such as that of Fig. 3. Again, we assume that a single point source is in the field of view. We further assume realistically that the visibility lines are closely spaced (e.g., 1″) and that the NCP is sufficiently distant so that many fringes (say, ~700) cross the source during a 12-h period, or ~1 per minute on average. The procedure is the same as before, namely to make a short observation, say of 10 min, and to plot lines of visibility on the sky at a moment when the response is maximum. These are the lines of position; the source must lie on one of them.

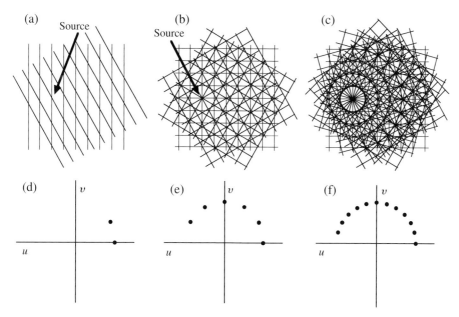

Figure 7.4. Lines of 100% visibility on the sky for the situation of Fig. 3 with a single point source in the field of view. The lines are drawn for each orientation at a time of maximum response; they are *lines of position* upon which the source could lie. (a) Lines obtained from two brief observations separated by 2 h (earth rotation angle of 30°). (b) Lines obtained from four additional observations at 2-h intervals. The location of the point source (arrows) is becoming apparent. (c) Twelve sets of lines with 15° angular intervals (1 h) between observations. (d,e,f). Fourier u,v planes showing as dots the angles and spacings sampled in (a,b,c).

Two such observations separated by ∼2 h would yield two such sets of lines of position (Fig. 4a). The two sets are rotated from one another by 30°, the rotation angle of the earth in a 2-hour period. The source must lie somewhere on each set, so the true position must be at one of the many intersections. If the observations are repeated at 2-h intervals for 10 h, one can plot six sets of fringes at 30° intervals extending over 150° of rotation (Fig. 4b). In this case, the intersections begin to indicate the true source position (see arrows). If observations are made every 1 h for 11 h, the source position stands out markedly (Fig. 4c). This shows unambiguously that sufficient information lies in the data to locate uniquely a source position, except for a possible false image on the other side of the NCP.

All-sky fringe pattern

Heretofore, we considered only sources nearly overhead, i.e., at $\theta \approx 90°$; this angle is defined in Fig. 1c. The location of the great and small circles of constructive

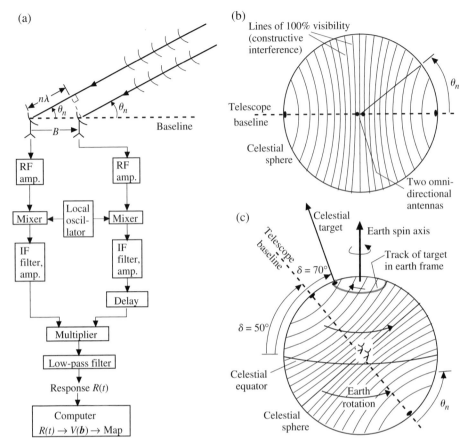

Figure 7.5. (a) Geometry which defines the directions of 100% visibility for two telescopes separated by distance B. The angle θ_n measured from the baseline is the angle of the nth line of 100% visibility where the $n = 0$ line is normal to the baseline. The electronic processing logic flow is shown. (b) Nineteen lines of 100% visibility on the celestial sphere spaced at equal intervals Δn of line number n. (c) Celestial sphere showing visibility lines and the earth's rotation axis for the case of the telescopes being placed such that their baseline intercepts the celestial sphere at declination $\delta = \pm 50°$. The earth's rotation causes the visibility pattern on the celestial sphere to rotate about the north celestial pole, i.e., about the earth's spin axis. The telescopes are continuously repointed toward the source region as the earth rotates. The track of a celestial source at $\delta = +70°$ is shown in the earth frame of reference. Note that the view is from outside the celestial sphere.

interference at all angles $0 < \theta < 180°$ follow from the geometry of Fig. 5a. Let θ_n be the angle between the baseline and the direction of constructive interference such that the path to one telescope is exactly $n\lambda$ longer than that to the other. Here n is an integer and λ is the wavelength of the electromagnetic wave. The geometry yields

$$\cos\theta_n = \frac{n\lambda}{B} \quad \text{(Angle of constructive interference; } n = \text{integer;} \qquad (7.3)$$
$$-(B/\lambda) \le n \le +(B/\lambda)$$

where B is the magnitude of the length of the baseline (in meters), n is an integer that ranges from $+B/\lambda$ to $-B/\lambda$ as θ_n ranges from $0°$ to $180°$. Thus the $n = 0$ line is at $\theta_0 = 90°$. This gives the angles of all the visibility lines. On the celestial sphere (Fig. 5b), the lines form great circles ($n = 0$) or small circles ($n \ne 0$) about the baseline. Although only 19 lines are drawn in the figure, the actual number can be very large, $2B/\lambda$, or twice the baseline given in number of wavelengths. In the figure, the lines shown are drawn for equal intervals of Δn, at $n\lambda/B = +0.9$, $+0.8 \ldots 0 \ldots -0.8, -0.9$, in order of increasing θ_n and decreasing n.

The separation between adjacent lines follows from (3),

$$d(\cos\theta_n) = \frac{\lambda}{B}\,dn \qquad (7.4)$$

➡ $$d\theta_n = -\frac{\lambda}{B}\frac{dn}{\sin\theta_n} \qquad \text{(Angular spacing of visibility lines)} \qquad (7.5)$$

Here $dn = 1$ yields the separation of adjacent bands. The bands normal to the baseline ($\theta_n \approx 90°$) are separated by $|d\theta_n| = \lambda/B$ in agreement with (1), but the angular spacing increases toward the poles of the baseline as $\sin\theta_n$ decreases. In our equatorial observation (Fig. 2), this gives the increase in line spacing noted above when the source moves away from the zenith.

The effect of earth rotation is shown in Fig. 5c for the two telescopes placed at an arbitrary orientation, in this case with the baseline projecting to declination $\delta = \pm 50°$. The telescopes (at the center of the celestial sphere) are carried with the earth's rotation; hence the entire fringe pattern rotates about the earth's spin axis. As the pattern undergoes a $360°$ rotation, a source, being fixed on the celestial sphere, is scanned by a range of line spacings $\Delta\theta$ and line directions. This is best seen in the frame of reference of the earth in which the fringes are fixed and the stars move. The shaded track in Fig. 5c shows the circular track of a source (at $\delta = 70°$) in this rotating frame of reference.

During the earth rotation, the pointing directions of the telescopes must continuously change to keep the source in their individual fields of view. Sources not at high declinations will set and rise each day; they can not be observed when they are below the horizon. This *earth occultation* eliminates some of the possible line crossings and adversely affects the quality of the resultant images.

Point-source response

Here we derive formally the response $R_{PS}'(t)$ of a two-telescope system to a point source.

Wavefront samples

The two telescopes sample the incoming electromagnetic wave at two places in the plane wavefront. The two telescopes are configured to select the same component of the incoming (vector) $\boldsymbol{E}$ field (the same *linear polarization*); thus the measured scalar components E_1 and E_2 can be used to describe the electric fields at telescopes 1 and 2. These fields may be written as

$$E_1 = E_0 \cos \omega t \qquad \text{(Electric field; tel. 1)} \qquad (7.6)$$

$$E_2 = E_0 \cos(\omega t - \phi) \qquad \text{(Electric field; tel. 2)} \qquad (7.7)$$

where ω is the radio angular frequency and ϕ is the phase delay in radians that corresponds to the path-length difference $n\lambda$ in Fig. 5a, where n, the equivalent number of wavelengths, need not be an integer. The phase delay may be written with the aid of (3) which is generalized to allow non-integer n so that $\theta_n \rightarrow \theta(t)$,

$$\phi(t) = 2\pi \frac{n\lambda}{\lambda} = 2\pi n = 2\pi \frac{B}{\lambda} \cos \theta(t) \qquad \text{(Phase delay; radians)} \qquad (7.8)$$

The earth's rotation causes the angle $\theta(t)$ to vary with time, thereby continually shifting the phase $\phi(t)$ of E_1 relative to E_2. This leads to *beating* of the two waves as they move into and out of phase with one another. The *beat frequency* is the frequency with which the visibility lines pass over the point source, or the frequency of the slow modulation of the response function $R'(t)$ shown in Fig. 2d.

The interference can be viewed in another way: the earth's rotation causes one telescope to steadily move closer to (or farther from) the source compared to the other. The telescope getting closer (relative to the other) detects the wavefront at shorter time intervals, i.e., at a slightly higher frequency; this is the Doppler effect. The signals from the two telescopes thus have slightly different frequencies, and this leads to the observed beats.

Summed waves

Electric fields are additive; two fields at a given position and time yield a net field obtained by vector addition. After detection of the two wave samples from our two telescopes, the fields could be added by bringing them together with coaxial cables. In our case (same polarizations), one may add the scalar components algebraically to obtain the net E field, from (6) and (7),

$$E = E_0 \cos \omega t + E_0 \cos(\omega t - \phi) \qquad \text{(Summed EM waves)} \qquad (7.9)$$

The sum of these two cosine functions may be rewritten, with the aid of the appropriate trigonometric relations, as a product of cosines that separate out the phase ϕ dependence,

$$E = 2\,E_0 \cos\left(\omega t - \frac{\phi(t)}{2}\right) \cos \frac{\phi(t)}{2} \tag{7.10}$$

This function describes a high frequency oscillation at radio frequency ω, which is modulated by a slowly varying cosine function, $\cos(\phi/2)$, which forces the oscillations to zero amplitude at $\phi = \pi, 3\pi, 5\pi, \ldots$ radians (Fig. 6a). The slower oscillations are due to the waves going into and out of phase with one another as the earth rotation continuously repositions the telescopes. As noted above, the radio and beat frequencies will actually differ by a large factor, 10^6 or more.

The power in an electromagnetic wave is proportional to the square of the summed electric field,

$$E^2 = 4\,E_0^2 \cos^2\left(\omega t - \frac{\phi(t)}{2}\right) \cos^2 \frac{\phi(t)}{2} \qquad \text{(Proportional to} \tag{7.11}$$
$$\text{power in wave)}$$

which is illustrated in Fig. 6b. Since the $\cos^2(\phi/2)$ term may be written as $(1 + \cos\phi)/2$, this modulation has a sinusoidal shape as the phase delay ϕ changes, but it is offset from zero so it never goes negative.

The large difference in the radio and fringe frequencies allows one to carry out a series of time averages on the function (11), each over a time long compared to the period of the rapid radio-frequency oscillations but short compared to the period of the oscillations of $\phi(t)$. This will remove the rapid radio-frequency oscillations from the signal. This type of averaging is called a *low-pass filter* because it lets low frequencies pass through but blocks the higher frequencies. The output of the low-pass filter follows from the fact that the average of a $\cos^2$ function over an integral number of cycles is 1/2. Invoking this average for the rapid $\cos^2[\omega t - (\phi/2)]$ term, we have

$$E_{\text{LP}}^2 = 4E_0^2 \frac{1}{2} \frac{1 + \cos\phi(t)}{2} = E_0^2[1 + \cos\phi(t)] \tag{7.12}$$
$$\text{(After low-pass filter)}$$

This function, plotted in Fig. 6c, has the form shown previously in Fig. 2d.

The energy flux density (W/m^2) in the wave is the magnitude of the *Poynting vector* $\mathscr{F} = E \times \underline{B}/\mu_0$, where $\underline{B}$ is the magnetic field, underlined here so as not to confuse it with baseline B used in this chapter. Since Maxwell's equations tell

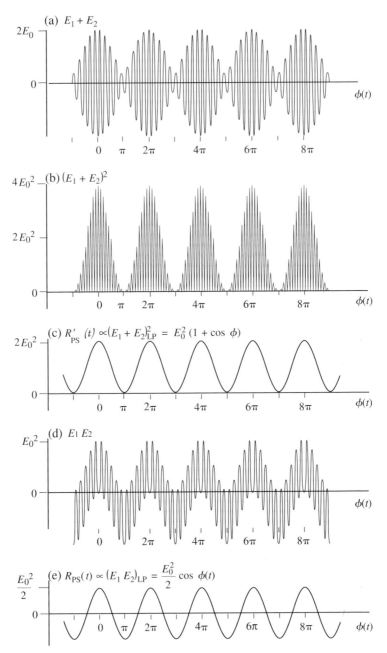

Figure 7.6. Interference of electromagnetic waves $E_1 = E_0 \cos \omega t$ and $E_2 = E_0 \cos(\omega t - \phi)$. The rapid modulations represent the high frequency ω of the electromagnetic waves, or sometimes 2ω, while the slow oscillations are due to the changing relative phase ϕ between the two interfering signals as the earth rotates. (a)–(c) The waves are added, squared, and then passed through a low-pass filter. (d)–(e) The waves are multiplied and then sent through the low-pass filter. These plots are qualitative sketches.

us that E is perpendicular to $\underline{B}$ and $|\underline{B}| = |E|/c$ (in SI units),[1] the magnitude of the Poynting vector becomes[2]

$$|\mathscr{F}(t)| = \frac{E_{\mathrm{LP}}^2}{\mu_0 c} = \frac{E_0^2[1 + \cos\phi(t)]}{\mu_0 c} \qquad \text{(W/m}^2; \text{ energy flux} \atop \text{density)} \qquad (7.13)$$

This expression is also known as the *response* $R_{\mathrm{PS}}'(t)$ to a point source,

➡ $$R_{\mathrm{PS}}'(t) = |\mathscr{F}| \propto E_0^2[1 + \cos\phi(t)] \qquad \text{(Summed point-source} \atop \text{response)} \qquad (7.14)$$

Note that we define R' to be the response *after* low-pass filtering. Each peak in the function (13) or (14) indicates a change of the path-length difference to the source by one wavelength λ.

Multiplied waves

The same variation can be obtained if the waves are multiplied rather than added. To motivate this, take the square of the summed waves (9) directly,

$$E^2 = [E_0 \cos\omega t + E_0 \cos(\omega t - \phi)]^2 \qquad (7.15)$$
$$E^2 = E_0^2[\cos^2\omega t + 2\cos\omega t \cos(\omega t - \phi) + \cos^2(\omega t - \phi)] \qquad (7.16)$$

The first and last terms on the right side are separately the fluxes of the two signals; the time-average value of each is the constant $E_0^2/2$. Thus the second term contains all the interference information; it is known as the *interference term*.

The interference term is simply twice the product of the two waves (6) and (7). The product

$$E_1 E_2 = E_0^2 \cos\omega t \cos(\omega t - \phi) \qquad \text{(Product of waves)} \qquad (7.17)$$

may be rewritten to isolate the variation of $\phi(t)$, again with the aid of trigonometry relations,

$$E_1 E_2 = \frac{E_0^2}{2}[\cos\phi(t) + \cos(2\omega t - \phi)] \qquad (7.18)$$

Here the rapid radio-frequency term $\cos(2\omega t - \phi)$ is summed with the slow interference term $\cos\phi$ (Fig. 6d). A low-pass filter averages the rapid oscillations to zero while the beat term is not affected appreciably (Fig. 6e). The point-source response function thus becomes

---

[1] We use $|V|$ for the magnitude of the vector V, that is, the root of the vector dot product $(V \cdot V)^{1/2}$.

[2] The symbol S is usually used for the Poynting vector, but we reserve it for a related quantity, spectral flux density (W m^{-2} Hz^{-1}); see Section 8.2.

$$R_{\mathrm{PS}}(t) \propto (E_1 E_2)_{\mathrm{LP}} = \frac{E_0^2}{2} \cos \phi(t) \qquad \text{(Multiplicative} \qquad (7.19)$$
$$\text{point-source response function)}$$

This expression is the multiplicative response to a point source after the two signals have passed through a multiplier followed by a low-pass filter. Only the beat-oscillation term remains. The basic temporal variation is essentially identical to the summation case except for a constant offset and different additive and multiplicative factors; see (14) and Fig. 6c.

The function $R_{\mathrm{PS}}(t)$ oscillates about zero rather than always being positive; we use R without the prime to signify that the average is zero. The amplitude of the response is again proportional to E_0^2 and hence to the energy flux in the wave $(\mathrm{W/m^2})$. The phase shift $\phi(t)$ may again be expressed in terms of the observatory parameters (8), to yield

$$R_{\mathrm{PS}}(t) \propto E_0^2 \cos \phi(t) = E_0^2 \cos \left[2\pi \frac{B}{\lambda} \cos \theta(t) \right] \quad \text{(Multiplicative} \qquad (7.20)$$
$$\text{response)}$$

The function (20) can be written in a more general form. Let the baseline vector $\boldsymbol{B}$ be directed toward the right-hand telescope in Fig. 7a with magnitude $B(\mathrm{m})$. Introduce s as a unit vector in the direction of the source and θ the angle between s and $\boldsymbol{B}$, consistent with our earlier definition. The term $B \cos \theta(t)$ in (20) can then be expressed in terms of a dot product, and (20) becomes

➡️ $$R_{\mathrm{PS}}(t) \propto E_0^2 \cos \left[\frac{2\pi}{\lambda} \boldsymbol{B}(t) \cdot \boldsymbol{s} \right] \qquad \text{(Point-source visibility} \qquad (7.21)$$
$$\text{function)}$$

The time dependence in (21) arises from the variation of $\cos \theta(t)$ which in turn arises from the changing orientations $\boldsymbol{B}(t)$ of the two telescopes in inertial space. The expression (21) is a fundamental equation of interferometry.

Radio astronomers invariably use the multiplication scheme these days. The summation scheme is quite sensitive to variations in amplifier gains. It was a major step in radio interferometry when multiplication was first implemented. The multiplication can be carried out quite simply with modern digital logic as the signals arrive. The radio frequencies ω of the incoming signals are first reduced to an intermediate frequency (IF; see below). The two signals are then each sampled several times during each cycle of the IF, and the amplitude and sign of the wave are converted to a digital value (e.g., 3 bits) for each sample. The two values are then multiplied digitally in real time and the answer recorded.

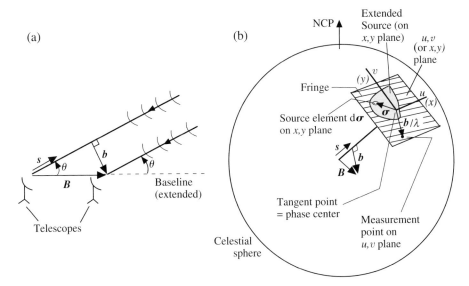

Figure 7.7. (a) Definitions of the baseline vector B which has units of meters, the projection of the baseline b (also meters), and the unit vector s directed to a point near the source, called the phase center. (b) Tangent plane to the celestial sphere with tangent point at the phase center. The origin of the x, y coordinates (rad) is at the phase center. The coincident Fourier plane has coordinates u, v (rad^{-1}) which are the components of b/λ, where λ is the wavelength of the radio frequency. Solid angle elements $d\sigma$ of an extended source (shaded) are located on the tangent plane with the vector σ (components x, y). The fringes which run normal to b are shown as lines on the tangent plane.

The response R_{PS} derived here, (21), is for the observation of a point source with two telescopes. For a given short observation, the amplitude of the modulation gives the source intensity, and its phase gives position information in one direction, as discussed above.

Signal processing

The basic elements of the signal processing are illustrated in Fig. 5a. It is rather difficult to work directly with radio frequencies, so, as in ordinary radio receivers, the frequency from each telescope is converted to a lower *intermediate frequency* (IF) provided by a local oscillator which beats against the amplified radio frequency in a *mixer*. The IF signals are then amplified. The multiplication is carried out on the amplified IF signals.

A time delay is also included in the line from the telescope closest to the source, so the two samples of a given wavefront (from a hypothetical point source within the field of view) arrive approximately simultaneously at the multiplier. This is done with a network of delay cables which are switched in and out periodically as the path length $n\lambda$ changes due to the earth's rotation. A given fixed delay adds a constant phase to the fringe pattern on the sky, or to the response $R_{PS}(t)$, and so does not alter the rate at which the fringes pass over the source region. The effect of the delay is to place the *zero-phase position* (as in Fig. 1a) near the source being observed. For a fixed delay, the zero-phase position moves through the sky as the earth rotates.

The added delay is necessary to prevent the washing out of the fringes due to the finite bandwidth of the telescope receivers. With the delay, the two path lengths from the source to the multiplier are nearly equal; for a source at the zero-phase position, one is comparing two portions of the *very same* wavefront. The wave consists, in fact, of a sum of wavelets of slightly different frequencies that are included in the bandwidth $\Delta \nu$ of the receiver. Equivalently, a range of wavelengths is present, and there is thus a spread in the number of wavelengths that comprise the path length difference to the two telescopes $\boldsymbol{B} \cdot \boldsymbol{s}$ (Fig. 7a). If this spread approaches one wavelength, there is no longer a well-defined phase difference between the two signals, and the fringes would be washed out. A small path difference minimizes the spread.

After the multiplier, the signal is fed through a low-pass filter to yield the response function $R(t)$ for the region of the sky under observation. Multiple sources in the field of view would yield a function more complex than the point-source response derived here.

Fourier plane

The particular spacing and orientation of the two telescopes at a given time can be described by a single point in the *Fourier plane*, also known as the *u,v plane*. This turns out to be a useful way to look at the accumulation of many antenna spacings and orientations used in the study of a given source. Here we simply acquaint the reader with the concept of the plane. There are two ways to describe it, one as a plot of *spatial frequencies* and the other as a projection of the baseline vector $\boldsymbol{B}$ onto a plane tangent to the celestial sphere at the source.

Spatial frequencies

At a given time, the angular spacing $\Delta\theta$ (rad or rad/cycle) between adjacent visibility lines (on the sky) of a single antenna pair is known as the *spatial wavelength*. The inverse wavelength $(\Delta\theta)^{-1}$ is the *spatial frequency* (cycles/radian or rad^{-1}), that is, the number of visibility lines per radian on the sky measured normal to the lines. The quantities u and v are components of the spatial frequency (rad^{-1}) in the two orthogonal directions, E–W and N–S respectively.

On a Fourier-plane plot (Fig. 2e), the spatial frequency components, u and v, of a given antenna pair for a brief observation are plotted as a point where the abscissa and ordinate represent u and v respectively. The spatial frequency is the radial distance $(u^2 + v^2)^{1/2}$. The azimuth of a plotted point represents the antenna baseline orientation at the time of the observation, specifically the direction perpendicular to the visibility lines. Each plotted observation must be sufficiently short that the spatial frequency and antenna orientation may be represented as a point. One long continuous observation during which these quantities change would be plotted as a line, that is, a track of points.

The distribution of brightness on the sky can best be reconstructed if signals from a large number of telescope-pair spacings and orientations are used. Many pairs can be plotted on the Fourier plane as a large number of points for a brief observation or as a series of tracks for an extended observation.

The position of the dark dot (arrow) on the u (E–W) axis of Fig. 2e indicates that the N–S visibility lines of our overhead equatorial observation (Fig. 2c) measure position and structure in the E–W direction. Lines that lie in the E–W direction would measure N–S position and structure; they would be plotted on the vertical v axis. Lines at any other angle would be plotted at the appropriate azimuth.

If additional observations are made over the next several hours, the projected baseline is reduced, and the visibility line spacings (spatial wavelengths) become greater; see (5) and Fig. 5b. These correspond to smaller spatial frequencies which are plotted on the u axis closer to the origin (crosses in Fig. 2e). They would improve the final image. Better maps require the measurement of even more *Fourier components* along the horizontal axis and also components off the axis.

In other words, the u,v plane must be filled in with many measurements. Generally, the more components measured, the higher will be the quality of the final sky map. According to Fourier theory, the number for a complete reconstruction down to some small angular scale is finite. In practice, one rarely reaches this number. Points symmetrically opposite each other in the plot provide redundant information because rotation of the fringe pattern by $180°$ yields fringes with the same spacing and orientation. Thus it is generally not necessary to fill more than half the plane.

The Fourier plane plots for the North-Pole observations of Fig. 4 are given in Figs. 4d,e,f for the three cases illustrated. Each brief observation yields a dot on the plot. In each case, the fringe spacing, and hence the spatial frequency, is the same so all points are at the same radius $(u^2 + v^2)^{1/2}$. If the observation had carried on continuously for 12 hours, the plot would have been a continuous half circle. The u,v plot for a 24-h North-Pole observation would be a complete circle; Fig. 3d.

In the case of the 12 discrete observations (Fig. 4c,f), the source stands out dramatically, but it is hardly a perfect image because of the limited number of observations. The rings surrounding the source in Fig. 4c are known as *artifacts*.

Additional measurements that add points at different radii in the Fourier plane would tend to suppress the rings surrounding the source, thus improving the contrast of the source relative to the surrounding region. Such data could be obtained with different telescope spacings.

Projected baseline

One can visualize the Fourier (or u,v) plane as the plane tangent to (and fixed to) the celestial sphere near the center of the source region (Fig. 7b). As noted, this tangent point is called the phase center. The phase center is in the direction s (Fig. 7a) and is fixed on the celestial sphere (unlike the zero phase position described above). The vector s is fixed in inertial space as the earth rotates.

An observer viewing the earth from the phase center would observe the baseline B to be foreshortened unless it were exactly perpendicular to the line of sight. The foreshortened distance is the baseline projected normal to the line of sight, $b = B \sin \theta$. The projection may also be considered a vector, $b = s \times (B \times s)$, as shown in Fig. 7a.

The quantity plotted in the Fourier plane, according to our discussion above, is the inverse of the visibility-line spacing (rad^{-1}) given in (5) for d$n = 1$. It has length r in the u,v plane,

$$r = [u^2 + v^2]^{1/2} \equiv \frac{1}{\mathrm{d}\theta} = \frac{B}{\lambda} \sin \theta = \frac{b}{\lambda} \quad \text{(rad}^{-1}\text{; inverse} \quad \text{(7.22)}$$
$$\text{fringe spacing)}$$

where we suppress the subscript n.

Equation (22) shows us that the inverse angular spacing (rad^{-1}) of the visibility lines, i.e., the *spatial frequency*, is equal to the length of the projected vector b in units of λ. The azimuthal direction of the vector b according to the distant observer is normal to the visibility lines and is the direction along which positional or structural information is obtained. This is the azimuth to be plotted on the u,v plane. The vector b/λ thus defines the position in the Fourier plane that represents any given brief observation. Since the vector b is always normal to s, it may be considered to lie in the plane of the u,v surface, as shown in Fig. 7b. It follows that the u,v coordinates are the components of b/λ for any given observation.

As the earth rotates, the distant observer at the phase center looking back at the telescopes sees the baseline B continually changing orientation, and hence the projection b continually changes length and azimuth. The associated track of the head of the vector b/λ defines a track on the u,v plane that characterizes the sustained observation. At any given time, a large projected baseline implies a measurement with closely spaced visibility lines or a high spatial frequency, and a short projected baseline implies a measurement with widely spaced visibility lines or a low spatial frequency.

7.3 Mapping the sky

The construction of a proper map of the sky requires a more sophisticated method than the simple drawing of 100% visibility lines as in Fig. 4. These methods are a necessity if multiple point sources or extended structure are in the field of view. The process of constructing an image is called *aperture synthesis*, and the methods based on Fourier transforms are known as *Fourier optics*. Before introducing the Fourier method, we will describe a more intuitive "shading" or *cross-correlation* method that is an extension of the line-drawing method and also is closely related to the Fourier method. We continue to consider a two-telescope system.

Cross-correlation or "shading" method

If there are multiple sources in the field of view, the *observed* response $R'(t)$ would include the contributions of the several sources. (Here we choose the response function R' that is always positive, Fig. 2d). In the case of a rotating fringe pattern (Fig. 3), the responses of the different sources cancel or reinforce each other depending on the rotation angle of the earth. The modulation would go to zero if two sources (of equal intensities) were perchance modulating exactly out of phase with about the same frequency, and it could become very large a bit later when they are in phase. In general, the combined response for multiple sources would seem quite random with peaks of varying heights at irregular spacings.

Bins on the sky

The shading method is a simple extension of the line-drawing method of Fig. 4. Instead of drawing lines at times of maximum response as in Fig. 4, the entire sinusoidal two-telescope point-source fringe pattern (i.e., the two-telescope beam) is binned (shaded) onto a map of the sky as shown in Fig. 8a. This is done for each (small) time interval of the observation. The time intervals are of constant duration and are taken to be much shorter than the times between fringe maxima at a given bin.

Furthermore, the amplitude of each such sinusoidal shading is proportional to the measured antenna response in the time interval. This gives the most weight to the times when a source is at maximum visibility. The amplitude ("darkness") at a given sky bin is recorded as a number proportional to the value of this adjusted sinusoid at the bin. If the response is at peak, the sinusoidal shading would be pronounced, the analog of drawing one set of lines in Fig. 4. If the response is zero, no shading would occur.

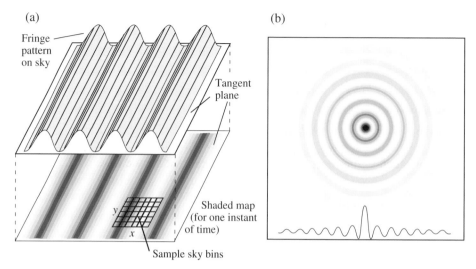

Figure 7.8. Shading method of creating a sky map from interferometry data. For each time interval, each bin on the sky is assigned a value which is the product of the instantaneous value of the fringe pattern at the bin position and the instantaneous value of the actual response. (a) Sinusoidal shading pattern for one instant of time. Its amplitude will be large if a celestial source is on a visibility line. (b) Shading pattern for a single point source near the north celestial pole for a continuous 12-h observation as in Fig. 3. The central peak is located at the source position; cf. Fig. 4c. The random noise in a real observation would add irregularities to the map. The Bessel-function profile is sketched qualitatively.

The contributions to each bin from all time intervals of the observation are then summed. The sinusoids from the many time intervals may have different orientations, amplitudes and phases. The resultant map of summed values is the desired representation of the sky.

In other words, at each instant of time, one paints onto the sky map the regions the telescopes can "see" proportionally to how much signal $R'(t)$ is detected by the telescopes. When a source (or combination of sources) is most visible, the sinusoid peaks run across the source positions and they are binned with large amplitudes because R' is large. The sinusoids with maxima at the locations of the sources are thus given the most weight. In addition, stronger sources yield higher R' when they are visible, so their locations will consistently receive higher contributions. In this way, a sky map can be built up from the data. After superposition of the sinusoids for all the time intervals, the maximum values will lie at the source positions, and stronger and weaker sources will appear as such.

A 12-h observation of a single point source in the vicinity of the North Pole would, with such an analysis, yield circular shaded rings such as those of Fig. 8b. They are similar to a damped cosine function and are clearly a representation of the rings of our simple line-drawing method (Fig. 4c). This pattern is an example of a *Bessel function*. The rings are, as before, an artifact due to the imperfect filling of the Fourier plane.

For larger numbers of telescopes, each pair can be treated similarly. The shadings from all telescope pairs can be superimposed in a single set of bins. The final result of this process is a sky map that represents the data in an unbiased manner. It will nevertheless include signal, artifacts, and noise. Multiple bright sources in the sky would be evident, but each would appear with the artifacts characteristic of the observation, such as the rings in our two-telescope North-Pole case.

Cross-correlation

Formally, in the shading method, the quantity that is summed for a given sky position α,δ at time t is the product of: (*i*) the fringe pattern on the sky (e.g., Fig. 8a), which is the interferometric response at time t to a unit point source at α,δ, or from (14), $R_{PS}'(\alpha,\delta,t) = 1 + \cos\phi(\alpha,\delta,t)$, and (*ii*) the detected power $R'(t)$. The latter could contain information about several different sources in the sky, including their strengths. This product is summed over all time intervals of the observation to obtain the final shading of the sky bin at α,δ. The summation in integral form is

$$C'(\alpha,\delta) = \int_{t_1}^{t_2} R'(t)R_{PS}'(\alpha,\delta,t)\,dt \quad \text{(Cross-correlation function} \quad (7.23)$$
$$\text{for creating map)}$$

The integral (23) has the form of a cross-correlation function, seen before in (5.5). The product $R'R_{PS}'\,dt$ is the intensity one applies to the sky bin at α,δ at a given time, and the correlation $C'(\alpha,\delta)$ is the shading value at α,δ after summing all time intervals. One repeats the integral for each sky position α,δ and thus creates a sky map that shows the multiple sources in the beam with their respective strengths.

The function (23) can be understood to be a correlation as follows. The integral tests the extent to which the data function $R'(t)$ correlates with, or matches, the expected response $R_{PS}'(\alpha,\delta,t)$ for a source at position α,δ. (The latter function is a "trial function". The correlation integral has a large value if the two functions R' and R_{PS}' vary similarly in time and lesser values if the variations are uncorrelated.

If there is actually a source at the trial position α,δ and no other in the field, the data function $R'(t)$ will vary exactly as the expected function R_{PS}' for that position, and the correlation value will be larger than elsewhere in the field. Even if there are several sources in the field causing R' to be relatively complex, the correlation function will still recognize the similarities between the data and the trial functions at the appropriate positions α,δ, and thus will produce large values at the several source positions.

One can see more clearly that similar functions in a correlation yield large values as follows. Subtract the time average values from the two functions to obtain new functions, $R \equiv R' - (R')_{av}$ and $R_{PS} \equiv R_{PS}' - (R_{PS}')_{av}$. These new functions will therefore have time averages of zero and the latter will have the form, from (14),

$$R_{PS}(t) \propto E_0^2 \cos \phi(t) \tag{7.24}$$

Substitute the above definitions into (23) and use the definition of a time average $\int R \, dt = R_{av} T$, where T is the total integration interval, to obtain

$$C'(\alpha,\delta) = \int_{t_1}^{t_2} R(t) R_{PS}(\alpha,\delta,t) \, dt + \text{constants} \tag{7.25}$$

We thus find that

$$C'(\alpha,\delta) = C(\alpha,\delta) + \text{constants} \tag{7.26}$$

where $C(\alpha,\delta)$ is defined to be the unprimed correlation function based on the functions with zero averages,

$$C(\alpha,\delta) \equiv \int_{t_1}^{t_2} R(t) R_{PS}(\alpha,\delta,t) \, dt \tag{7.27}$$

Expression (26) shows that maps based on the two correlation functions will be essentially identical except for a constant offset.

We can thus examine the behavior of (27) in lieu of (23). Suppose the two functions within the integral are identical; they will both be positive at the same times and negative at the same times because the average of each is zero. As noted regarding (5.5), their product will thus be positive at all times, and the integration (summation) over the entire observation will thus be large. If, instead, the functions are different in a random way, the products will vary randomly from positive to negative, and the integration will yield a value near zero.

The cross-correlation thus reveals the extent to which the observed function R' contains a signal R_{PS}' indicative of a point source at sky bin α,δ. The cross-correlation process thereby picks out the locations of the sources in the field of view, with amplitudes proportional to the source strength. This is the more conventional way to describe the equivalent shading approach. It has its own rationale, independent of the shading approach, even though the two are equivalent. If there are more than two telescopes, one would carry out a global cross-correlation operation that includes the contributions from each telescope pair.

Equal weighting of time intervals

Finally, let us note a problem with this shading method as described here; it weights all time intervals equally. This can lead to uneven shading if some Fourier components are sampled more than others. For example, consider the 12-hour north-pole observation of Figs. 3 and 4c. Suppose that during half of this time, the earth turned only 1°. The fringes would continue to cross the source but the orientation (rotation angle) of the pattern would change negligibly. Then suppose that the earth rotated rapidly, so that it covered the remaining 179° in the next 6 hours. During the slow period, the shading method would pile up lots of wave-like patterns, all at nearly the same rotation angle, and during the fast period the waves would be spread out over a large range of angles.

Since one-half of the shadings derive from the slow period, the resultant map would show large-amplitude parallel sinusoidal ridges running through the concentric circles; the pattern of Fig. 8a would be superposed on that of Fig. 8b with equal weighting. This is indeed a fair representation of the data, but the large ridges would not represent the real sky very well. Rather, one might prefer to give equal weight to each wave orientation, even if one received more exposure than the others. The Fourier method described below does just this. One could also modify the cross-correlation method to take into account such an imbalance of exposures.

Fourier analysis of sky brightness

The goal of our mapping has been to determine the intensity of the sky at every point (within, say, the 7′ beam of our sample individual telescopes) where the brightness can vary arbitrarily from sky bin to sky bin. Several point sources in an otherwise empty sky are a limiting special case that is more easily solved. A continuous distribution can be described as a sum of *Fourier sinusoidal (spatial) waveforms*; the description of the brightness in these terms is known as *Fourier analysis*. We make use of complex numbers in this section. The reader may choose to read only the overview in the following subsection.

Principle of aperture synthesis

Imagine that the brightness of the sky actually varies as a one-dimensional sinusoid that varies along the equator as a cosine wave, e.g., $(1 + \cos \phi)/2$, in the E–W direction, and with no variation in the N–S direction. It would resemble ocean waves frozen in time, like the waves of Fig. 8a. Let the spatial period of the brightness be $1.0''$. In this case, the sky brightness would consist of only one Fourier spatial wavelength, $\Theta_1 = 1.0''$.

Let the fringe pattern of our antenna system have a similar spatial wavelength $d\Theta = \Theta_1 = 1.0''$. This would be ideal for the detection of our artificial $1.0''$ sinusoidal sky distribution. Consider the equatorial observation, Fig. 2. As the earth rotates, the sinusoidal fringe pattern scans eastward over the sinusoidal sky distribution, alternately yielding large and small responses as it comes into and out of phase with the sky distribution. The variation of the response $R(t)$ would have large amplitude and would be sinusoidal, the same as for a point source.

Now let the brightness of the sky vary with a different period, say, $0.9''$. At any single instant of time, it would be out of phase with the $1.0''$ fringe pattern at some sky positions and in phase at others. As the earth rotates, the situation would remain the same in that, at any fixed time, some positions on the sky would still be in phase and others out of phase. Thus the response $R(t)$ would not modulate as the earth rotates. Our $1.0''$ fringe pattern would not detect this $0.9''$ distribution.

Thus, we find that our $1.0''$ fringe pattern will preferentially detect a $1.0''$ sky distribution and not other spatial wavelengths. In fact, our pattern will select the $1.0''$ component of the sky distribution even if the sky brightness contains many other spatial wavelengths. Knowledge of the other spatial wavelengths of the sky brightness can be obtained only if other antenna spacings are used to yield the matching fringe periodicities. For example, a larger projected antenna spacing b could yield a $0.9''$ fringe pattern. This would detect a $0.9''$ spatial component of sky brightness. The amplitude of the response $R(t)$ at each wavelength reveals the strength, or amplitude, of the corresponding component of sky brightness.

According to Fourier theory, any arbitrary function can be synthesized from a sum of sinusoidal waves with the appropriate amplitudes and phases. If these *Fourier components* are known, the arbitrary function can be constructed from the component sinusoids. In our case, a measurement with a given b yields one of the needed Fourier components. A complete set of b to some maximum resolution, with appropriate periodicities and orientations, can be used to construct a map of the surface brightness distribution of the sky.

The case of a single point-like source is an interesting special case. As long as the antenna fringe pattern has wavelength larger than the angular size of the source, a large response will be obtained, as in Fig. 2. This tells us that a true point source (a *delta function* of sky brightness) consists of *all* spatial Fourier wavelengths. A real source with (small) angular extent would be represented by all spatial wavelengths that are longer than the angular size of the source.

It follows that a single equatorial measurement (Fig. 2) would not distinguish between a point source and an extended wave of sky brightness. To be assured that the sky brightness truly consists only of a single small source rather than an extended sinusoid, one would have to measure all these spatial wavelengths with different baselines and would expect to detect fringes (modulation) at each of them. If modulation were found at only one of them, as in our equatorial case, one would be forced to conclude that the sky brightness distribution is a sinusoidal brightness wave as just described!

Similarly, a single polar (rotating) observation (Fig. 3) would not distinguish between a single point source and a bright spot surrounded by sinusoidal-like

(Bessel-function) rings of decreasing brightness (Fig. 8b). Measurements with different baselines would be required.

Arbitrary sky brightness distribution

The response functions illustrated heretofore were for point sources only. Here we present the response to an extended source of arbitrary brightness distribution. Let the unit source vector s refer to the direction of the phase center (tangent point of the Fourier plane, Fig. 7b). The expression (21) gives the response $R_{PS}(t)$ to a point source at that position. Now define any other nearby sky position with the vector sum $s + \sigma$. The vector σ gives the offset from the phase center in radians. It is plotted on the tangent plane of Fig. 7b with position components x, y measured in angle (rad) in the E–W (right ascension) and N–S (declination) directions, respectively. We have also described the u, v Fourier plane as a similar tangent plane, but it is a different mathematical space where the u, v coordinates of b/λ are plotted in units of rad^{-1} (actually cycles/rad).

Let the specific intensity or brightness of the source as a function of position on the sky be designated $I(\sigma)$ (W m^{-2} sr^{-1}). This quantity is power received onto 1 m^2 per unit solid angle from element dσ by a telescope centered on position σ of the source. (We introduce specific intensity in Section 8.4.) The response of the interferometer to a two-dimensional solid-angle element dσ of the source may be represented as the product of the unit point-source response, from (21), and $I(\sigma)$ dσ.

Integration of the product over the entire source region then yields the total response R to the extended source. One must carry out the integration over the entire beam of the individual telescopes, to $> 7'$ in our earlier example. Thus,

$$R(t) \equiv \int_{\text{beam}} I(\sigma) \cos\left\{\frac{2\pi}{\lambda} B(t) \cdot [s + \sigma]\right\} d\sigma \qquad (7.28)$$

(Response, extended

source; W/m^2)

where B is the telescope separation (baseline) in meters and λ is the wavelength of the radio-frequency radiation. (Radio astronomers often define B as the dimensionless B/λ, but we choose not to do so.) This defines the magnitude of $R(t)$ in terms of the actual sky brightness I. Note that dσ is the solid angle of the element (rad^2 = sr) which we have indicated with dΩ in other contexts (Sections 3.3 and 8.4).

Distortion due to the tangent-plane approximations is small for small beam sizes (usually $< 1°$), and σ is nearly perpendicular to s in this approximation. Since b is by definition the component of B perpendicular to s and since σ is nearly perpendicular to s, we have the approximation

$$B \cdot (s + \sigma) = B \cdot s + B \cdot \sigma \approx B \cdot s + b \cdot \sigma \qquad (7.29)$$

Substitute into (28),

$$R(t) = \int_{\text{beam}} I(\sigma) \cos\left\{ \frac{2\pi}{\lambda} B(t) \cdot s + \frac{2\pi}{\lambda} b(t) \cdot \sigma \right\} d\sigma \qquad (7.30)$$

In inertial space, the baseline B and its projection b vary with time as the earth carries the two telescopes to different orientations. (In the earth frame of reference, it is the direction to the phase center $s(t)$ that varies with time.) Expand the cosine function and suppress (but do not forget) the time dependence of B and b,

$$R(t) = \cos\left[\frac{2\pi}{\lambda} B \cdot s \right] \int_{\text{beam}} I(\sigma) \cos\left[\frac{2\pi}{\lambda} b \cdot \sigma \right] d\sigma$$

$$- \sin\left[\frac{2\pi}{\lambda} B \cdot s \right] \int_{\text{beam}} I(\sigma) \sin\left[\frac{2\pi}{\lambda} b \cdot \sigma \right] d\sigma \qquad (7.31)$$

This is the desired response function. Carrying out the integration over the source region yields the response from the two-antenna system.

Visibility

The expression (31) can be written more compactly as a complex expression of which $R(t)$ is the *real part*,

$$\Rightarrow \quad R(t) = \text{Re}\left\{ \exp\left[i\frac{2\pi}{\lambda} B \cdot s \right] \int_{\text{beam}} I(\sigma) \exp\left[i\frac{2\pi}{\lambda} b \cdot \sigma \right] d\sigma \right\} \qquad (7.32)$$

(Extended source)

which can readily be confirmed by expanding the exponentials according to the definition, $e^{+i\zeta} \equiv \cos \zeta + i \sin \zeta$ and retaining only the real part. (We drop the "Re" label from here on.) The term outside the integral is the sinusoidal response expected for a hypothetical point source at phase center; see (21) and Fig. 6e. The integral modifies this function to give the correct response for the extended source.

The structure information is contained completely in the integral which is called the *visibility* $V(b)$,

$$\Rightarrow \quad V(b) \equiv \int_{\text{beam}} I(\sigma) \exp\left[i\frac{2\pi}{\lambda} b(t) \cdot \sigma \right] d\sigma \qquad \text{(Visibility)} \qquad (7.33)$$

where b will usually vary with time as the earth rotates. The units of λ, b and σ are, respectively, m, m, and the dimensionless "rad". Thus, introducing $k \equiv 2\pi/\lambda$, the observational response (32) becomes

$$\Rightarrow \quad R(t) = V(b) \exp[ik B \cdot s] \qquad (7.34)$$

Again, the exponential is the unit point source response for a point source at s. The visibility $V(b)$ is seen to be the amplitude of $R(t)$, the measured oscillatory response to $I(\sigma)$ at the projected fringe spacing b. It also contains the phase adjustment required to give the correct response $R(t)$ for the actual sky distribution $I(\sigma)$. It is thus a complex number.

For a sufficiently brief observation, the fringe spacing and direction (or equivalently b) changes very little while the changing orientation θ of B in inertial space (Fig. 7a) still leads to fringes translating across the source. The visibility function $V(b)$ thus is usually a slowly varying function. A brief measurement of $R(t)$ might detect the passage of thousands of fringes at some (nearly) fixed b. An exception is an observation very, *very* close to the celestial pole, where the changing orientation of the fringes, and hence of b and $V(b)$, can dominate the effect of fringe translation.

As stated, the exponential $\exp[ik B(t) \cdot s]$ in (34) describes the frequency and phase of the fringe oscillations of a hypothetical point source of unit intensity at phase center in terms of the changing path-length difference $B(t) \cdot s$ (Fig. 7a), while $V(b)$ is a complex multiplier that contains the phase offset and amplitude that derives from the source structure. Since the vectors B and s are known (in principle) at any given instant, a measurement of $R(t)$, both phase and amplitude, determines $V(b)$ for that particular time, or equivalently for that particular b. Knowledge of $V(b)$ for many different b allows one to construct an image of the source region. They are the Fourier components of the source structure.

In the Cartesian coordinates of the tangent plane (Fig. 7b), the components of σ are x, y, and, in the associated Fourier plane, the components of b/λ are u, v. Thus, one can expand the dot product in (33),

$$V(u,v) = \iint I(x,y) \exp[i2\pi(ux + vy)]\, dx\ dy \qquad \text{(Visibility} \qquad (7.35)$$
$$\text{function)}$$

The components u and v are expressed as inverse radians while x and y are given in radians; the argument of the exponential is thus appropriately dimensionless. The function $V(u,v)$ is another form of the *visibility function*.

Phase of visibility function

Let us examine the phase information contained in the visibility function $V(b)$. Consider the exponential in (33). Recall that the projected vector b is directed normal to the fringes and that b/λ is the number of fringe cycles per radian (spatial frequency). Thus, $b \cdot \sigma/\lambda$ in (33) is the number of fringe cycles separating the phase center and the position σ (see Fig. 7b). Multiply by 2π to obtain $kb \cdot \sigma$, the phase offset in radians between the two positions; this is the argument of the exponential in (33). The real part of this exponential (the only part of concern to us) is $\cos(k\boldsymbol{b} \cdot \boldsymbol{\sigma})$.

As an example, let the argument $k\boldsymbol{b} \cdot \boldsymbol{\sigma}$ be a multiple of 2π radians at some instant of time for some element $d\boldsymbol{\sigma}$ at $\boldsymbol{\sigma}$. The cosine function is thus unity, and $V(\boldsymbol{b})$ simply contributes the amplitude $I(\boldsymbol{\sigma}) \, d\boldsymbol{\sigma}$ to the phase-center point response in (34) for this particular $\boldsymbol{\sigma}$; the visibility function contributes no phase shift. This is expected because the fringes cross the two positions (phase center and $d\boldsymbol{\sigma}$) exactly in phase with each other (compare to Fig. 7b). The actual response $R(t)$ due to this element, in this case, is the same (in phase) as that of the phase center. A unique case of this is when both positions are simultaneously viewed by the same fringe, namely when $k\boldsymbol{b} \cdot \boldsymbol{\sigma} = 0$, or $\boldsymbol{b} \perp \boldsymbol{\sigma}$.

If, on the other hand, a source element is such that the phase center and source element give out of phase responses, we have $k\boldsymbol{b} \cdot \boldsymbol{\sigma} = \pi$; the cosine becomes -1. In this case, $V(\boldsymbol{b})$ would shift the phase-center response by $180°$ to give the actual response $R(t)$ to this element.

For a (fixed) value of $\boldsymbol{b}$, the integration (33) will yield a visibility of the form $V(\boldsymbol{b}) = I_0 \exp[ik\boldsymbol{b} \cdot \boldsymbol{\sigma}_{\mathrm{eff}}]$ where $\boldsymbol{\sigma}_{\mathrm{eff}}$ is an effective coordinate produced by the integration over the extended source. Thus $V(\boldsymbol{b})$ is seen to contain an amplitude I_0 and a phase angle, $k\boldsymbol{b} \cdot \boldsymbol{\sigma}_{\mathrm{eff}}$, for that value of $\boldsymbol{b}$. Multiplication by the phase-center response, $\exp[ik\boldsymbol{B}(t) \cdot \boldsymbol{s}]$, yields, according to (34), the actual response $R(t)$ for the extended source,

$$R(t) = I_0 \exp[ik(\boldsymbol{B}(t) \cdot \boldsymbol{s} + \boldsymbol{b} \cdot \boldsymbol{\sigma}_{\mathrm{eff}})] \qquad (7.36)$$

Here we see explicitly that the argument of the exponential in $V(\boldsymbol{b})$ becomes a phase shift in the actual response; it is added to the phase angle at phase center in the argument of the exponential. In our examples above, the phase shifts $k\boldsymbol{b} \cdot \boldsymbol{\sigma}_{\mathrm{eff}}$ were 0 radians and π radians respectively. This discussion illustrates how $V(\boldsymbol{b})$ provides the phase and amplitude needed to correct the phase-centered response to the actual response for each value of $\boldsymbol{b}$.

Sky brightness

The measured visibility $V(\boldsymbol{b})$ is thus a Fourier amplitude/phase that allows us to reconstruct the sky brightness $I(\boldsymbol{\sigma})$. Those familiar with Fourier theory will recognize that (33) is the Fourier transform of the brightness distribution $I(\boldsymbol{\sigma})$. It can be shown (not here) that the expression can be inverted to obtain the sky brightness distribution,

$$\Rightarrow \quad I(\boldsymbol{\sigma}) \equiv \int_{\substack{u,v \\ \text{plane}}} V(\boldsymbol{b}) \exp\left[-i\frac{2\pi}{\lambda}\boldsymbol{b} \cdot \boldsymbol{\sigma}\right] d\boldsymbol{b} \qquad \begin{array}{l}\text{(Sky brightness} \\ \text{distribution)}\end{array} \qquad (7.37)$$

This is the *inverse Fourier transform*. It is an integral over the various values of *b* provided by the antennas during the course of the entire observation. Since b/λ describes a point in the u, v plane, it is also an integral over the u, v plane. One such integration yields the brightness I at the single chosen sky point σ, or equivalently α, δ. Subsequent integrations for other σ yield the brightness at other points.

The integral (37) is in effect a sum of all the Fourier sine waves (exponentials), each with the appropriate amplitude and phase adjustment $V(b)$ obtained from the measured response of the telescope pair for a given *b*. The reconstruction integral (37) yields a map of sky intensities, as did the cross-correlation function (27). Note that (37) weights all *b* equally while (27) weights all time intervals equally.

We thus see that the measurement of the complex visibilities $V(b)$ for many points in the Fourier (u, v) plane allows one to integrate over the Fourier plane to obtain the sky brightness distribution $I(\sigma)$, a reconstructed image or map of the sky. If half of the Fourier plane is sampled with a "complete set" of samplings, the sky brightness $I(\sigma)$ could be completely recovered to some specified resolution. Since the Fourier plane is never completely filled in practice, the resultant map will miss some Fourier components and therefore will show artifacts.

Cleaning algorithms

The maps obtained with these techniques represent the content of the entire data set. Unfortunately, the artifacts invariably associated with incomplete u, v sampling cause the map to be less than perfect. For a given observation, one knows the artifact associated with a point source. Thus radio astronomers are able to use "clean" algorithms to eliminate the artifacts to some extent and hence to produce maps that more closely approximate the real sky. There are several related methods for accomplishing this. One general approach is called *Clean* and the other the *Maximum Entropy Method*. We describe the former here.

Consider our two-telescope North-Pole experiment of Fig. 3. One knows, a priori, that the image of a point source in the Fourier reconstructed map is the ring pattern of Fig 8b. This is actually the beam of the interferometer for the particular observation; it is an average beam derived from the entire observation. This response to a point source is also called the point spread function (psf, introduced in Section 5.4). For any observation, the psf can be obtained directly from the sampling of *b* in the u, v plane by means of a Fourier transform.

It turns out that every point in the map from our polar observation has the ring sidelobes of the point source function. Even the noise bumps have the ring pattern, though they would be lost in the noise. The whole map is a sum of the Bessel-function ring point-source responses. A more complete array at a moderate latitude would yield a net beam pattern for the entire observation that would be more concentrated, i.e., with smaller sidelobes. The ring pattern of our hypothetical two-telescope polar observation is an extreme example of sidelobes.

The Clean method may be roughly described as follows, using our hypothetical polar observation as an example. First the brightest point in the reconstructed sky map is located. This most likely would represent the dominant source in the field of view. Its best-fit position and intensity are obtained through comparison to the psf. A portion of this best-fit source, say 10% to 50% of its intensity, complete *with its rings*, is subtracted from the $V(\boldsymbol{b})$ data. The result is used to create another map where, again, 10–50% of the amplitude of the brightest point with its rings is subtracted. This could well be the same dominant source.

As this process is iterated, other sources will stand out as the brightest points and 10% to 50% of their amplitudes will be subtracted one by one, always with the ring-pattern sidelobes. At some point the highest points will be background noise ("sources"). When the background has been well sampled, the subtractions can cease. In this way, all sources in the data have been found and their intensities determined, down to and including the noise.

A *cleaned map* is then constructed from the subtracted data, in which the subtracted sources are plotted with a suitable Gaussian width, but without their sidelobes, or rings in our case. The background noise in this map is due to background and source noise in the original map. It is important to realize that the subtraction of a given source does not remove the noise due to its intrinsic fluctuations.

This procedure is attractive because it shows the sky as we believe it to be, but it can be misleading as the final map presents more information than the data actually contain. In the present case, we tacitly assumed that there were only point sources in the field of view. If there are more sources and/or faint extended features, the reconstructed map may still contain spurious features. Thus cleaning of sky maps must be carried out with caution.

One also sees cleaning carried out in other branches of astronomy, e.g., to remove the fuzziness of an image taken with the Hubble Space Telescope when its focus was impaired. It is legitimate to do this if one realizes the assumptions made and the limitations of the process.

7.4 Arrays of telescopes

Multiple baselines

The utilization of multiple telescope spacings and orientations will improve image quality by removing artifacts. For our observation near the North Pole, the addition of different spacings obtained with additional telescopes would yield additional rings at different radii in the image, in effect flattening the region surrounding the point source. Observations with shorter spatial wavelengths (larger b) would also serve to restrict the angular size of the point source. As the Fourier plane is filled with more projected baselines b, one fills in more and more pieces of a hypothetical large single telescope, and the Fourier synthesized map becomes closer and closer to the actual sky.

In general, the baseline b of a single telescope pair tracks out an elliptical path on the u, v plane as the earth rotates. Our two examples in Figs. 2 and 3 were special limiting cases: a line and a circle respectively if the observations had been continuous in time. An array of n telescopes will contain $n(n-1)/2$ pairs of telescopes and hence that number of baselines. The optimum arrangement is one wherein none of these pairs have the same spacing and orientation, i.e., so that none have the same projected baseline b. Also, the Fourier plane should be more or less uniformly sampled by the tracks of these vectors as the earth rotates.

Radio arrays

There are currently a number of radio interferometric arrays in operation. There are several arrays containing at least eight telescopes extending over a few kilometers or more. Arrays of this size yield angular resolutions of down to $0.1'' - 1''$. Examples are the Westerbork array in the Netherlands, the Very Large Array (VLA) in the US (New Mexico), the MERLIN array in England, and the Australian Telescope (AT). These operate at centimeter and longer wavelengths.

The VLA in New Mexico has 27 telescopes, each of 25-m diameter. It can operate on frequencies ranging from 74 MHz to 50 GHz (4 m to 7 mm). The antennas are arranged on three arms, each of length 21 km (Fig. 9a). The telescopes are on tracks so the arm length can be reduced to 600 m for lower-resolution studies. The number of telescope pairs in the VLA is $27(27-1)/2 = 351$. A similar array with larger dishes and optimized for long wavelengths ($\gtrsim 1$ m), a Giant Meter-wavelength Radio Telescope (GMRT), has more recently begun operations in Pune, India.

Figure 10 is a radio image taken with the VLA of the famous *twin quasars* (labeled A and B). The discovery of these two identical objects at optical wavelengths was the first demonstration of gravitational lensing. The light rays from a single distant quasar are bent by the gravitational pull of an intervening galaxy. They arrive along two different paths, so two apparent quasars (A, B) are seen. The radio image seen here also shows lobe structure (C, D, E; matter ejected from the quasar) as well as the closer lensing galaxy (G). Note the high resolution of this image, $\sim 0.4''$. The two images A and B are separated by only $6''$.

Figure 11 shows a VLA "image" of the famous radio galaxy Cygnus A. This image was constructed from an unusually wide range of projected baselines that allowed the observers to extract amazing detail in this extended source. The galaxy

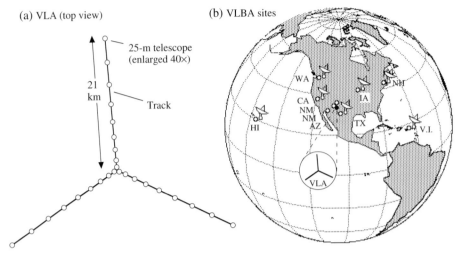

(a) VLA (top view)

25-m telescope
(enlarged 40×)

21
km

Track

(b) VLBA sites

Figure 7.9. (a) The Very Large Array (VLA) in New Mexico with three arms, each of 21 km length. The 27 individual telescopes are each 25 m in diameter. (b) Locations of the 12 telescopes of the Very Long Baseline Array (VLBA). They span 8000 km and operate up to 43 GHz (7 mm). [(b) Courtesy S. Olbert]

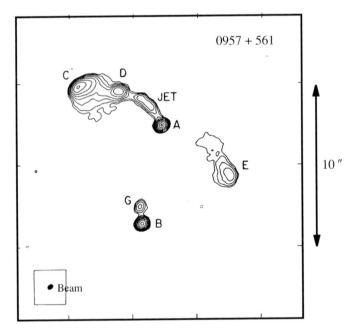

Figure 7.10. The twin quasar 0957+561 viewed at radio wavelengths with the VLA. It shows the two quasars (A and B), lobes (C, D, E), and the intervening lensing galaxy (G) with a resolution (FWHM beam size) of 0.4″. [From Roberts, *et al., Astrophys. J.* **293**, 356 (1985)]

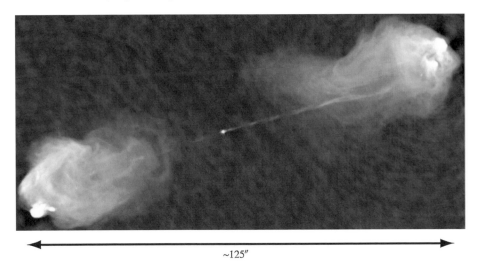

~125"

Figure 7.11. Radio map of the galaxy Cyg A at 5 GHz from VLA showing the active core, the jets, the extended lobes, and hot spots. The large range of structure scales was obtained from a wide variety of baselines. The resolution is about 0.5"; the angular extent of the radio emission is 2.1' or about 500 000 LY. [From Carilli and Barthel, *A&A Reviews* **7**, 1 (1996). See also Perley, Dreher, and Cowan, *Astrophys. J. Lett.* **285**, L35 (1984)]

is distant about 850 MLY based on its redshift parameter $z = 0.056$ and a Hubble constant of 20 km s^{-1} MLY^{-1}; see Ch. 9.5. The central source is an elliptical galaxy with an active nucleus which is probably a massive black hole ($\sim 10^8$ $M_\odot$). The large lobes extend up to about 1.2' from the central source, or about 300 000 LY. They arise from energetic particles being ejected into jets that feed the lobes. The central source and the jets are clearly seen as is extensive filamentary detail in the lobes. Bright spots near the outer edges of the lobes indicate where the material in the jets runs into the relatively dense intergalactic material, creates a shock, and dissipates its energy. The sharp boundaries at the outer edges of the lobes also indicate the interface with the intergalactic medium.

Millimeter observations were traditionally carried out with single dishes, because interferometry is more difficult at these high frequencies. Now, arrays have been constructed and are in use at Nobeyama in Japan, at Hat Creek CA in the US, and in the mountains of southern France (IRAM interferometer). The Submillimeter Array of telescopes in Hawaii reaches 0.3 mm wavelengths and resolutions of 0.1". An international consortium is developing an array of 64 12-m telescopes to be installed at high altitude in Chile, the Atacama Large Millimeter Array (ALMA).

Very long baseline interferometry (VLBI)

It is possible to place telescopes on the earth's surface with spacings approaching the earth's diameter. Telescopes on different continents attain fringe spacings (and resolutions) less than 1 milliarcsecond (mas). While they provide exceptional angular resolution, only limited numbers of baselines can be attained, and extremely good timing is required at the several sites if the taped signals are to be brought together for correlation (multiplication) at a later time at a central site. In the past, atomic clocks were carried from site to site to synchronize clocks, but now global timing systems make that unnecessary.

Observations have long been made among observatories on the several continents. Certain times are reserved by all for coordinated observations. The principal arrays are the European Very Long Baseline Interferometry Network (EVN), the US-based VLBI Network Consortium, the more recently completed Very Long Baseline Array (VLBA) in the US, and the AT array expanded to include other more distant Australian telescopes.

The VLBA consists of ten 25-m telescopes (45 baselines), eight in the continental US with one in Hawaii and one in St Croix, Virgin Islands (Fig. 9b). The telescopes can operate with the VLA and span 8000 km, which yields angular resolution (i.e., fringe spacings) $\Delta\theta = \lambda/D \lesssim 1$ mas, for a fair number of combinations of the available baselines and operating frequencies (330 MHz to 43 GHz, or wavelengths 0.9 m to 7 mm).

The orbiting Japanese satellite HALCA, launched in 1997, carried out radio interferometry in conjunction with ground-based telescopes. Its elliptical orbit to 3.3 earth radii yielded baselines several times longer than could be obtained on earth. At $\lambda = 60$ mm, resolutions of $\sim$0.5 mas were obtained. Its orbital motion with a 7-h period continuously and rapidly changed the baseline vectors between it and ground-based telescopes. This resulted in a more rapid sampling of the various Fourier components and better filling of the Fourier plane. Scheduling of such observations is complicated because the satellite's orbit leads to frequent occultations of a given source by the earth, from the perspective of a satellite observer.

Optical and x-ray interferometry

Optical "long baseline" interferometry is on the threshold of becoming a major tool in optical astronomy. A number of new generation telescopes (8-m class) are or have been built with multiple telescopes separated typically by $\sim$100 m. Examples are the twin 10-m Keck telescopes in Hawaii and the four 8-m Very Large Telescopes (VLT) of the European Southern Observatory in Chile. The much shorter wavelength of optical light compared to radio, a factor of $\sim$10^{-5}, yields angular resolutions of

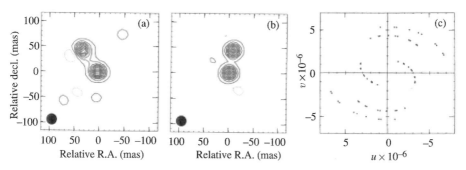

Figure 7.12. Long baseline *optical* interferometry of binary star Capella with an effective beam size of ∼20 mas (dark circle) on (a) 1995 Sept. 13 and (b) 1995 Sept. 28. The change in orientation and spacing of the two stars due to the 105-d binary period is clearly apparent. The u,v plane coverage for observation (a) is shown in (c); the value of "5" on the axis represents a projected baseline $b = 5 \times 10^6 \lambda$. [From Baldwin *et al.*, *A&A* **306**, L13 (1996)].

∼1 mas for telescope spacings of only ∼100 m. Space-borne optical interferometry missions are now being planned.

A notable result from ground-based optical interferometry is the observation of the binary star Capella obtained with the British COAST interferometer (Fig. 12). These observations were made with three telescopes stopped to 140-mm diameter with a maximum baseline of 6.1 m. The effective beam size is only 20 mas and the binary separations are 55 mas and 45 mas respectively. The two components are clearly resolved in exposures taken 15 days apart. The changed orientation due to the 105 d binary period is clearly evident.

At even shorter wavelengths, the x-ray band, laboratory experiments are demonstrating the possibility of x-ray interferometry, which must be carried out in space. In general the shorter the wavelength, the greater are the technical difficulties of carrying out long baseline interferometry. For example, the positions of the several telescopes must be maintained or known to less than a single wavelength!

Problems

7.2 *Two-telescope interference*

Problem 7.21. Verify the trigonometric conversions from (9) to (10) and from (17) to (18).

Problem 7.22. (a) The two telescopes of Fig. 2a on the earth's equator are separated E–W by 100 km and are tuned to a radio frequency of 1 GHz. They are observing a point source A on the celestial equator. What is the spatial wavelength of the visibility fringes for a source near the zenith, in radians and arcsec? What is the temporal period of the response function, i.e., what is the time between maxima? How many cycles of the radio frequency can occur in this time? (b) If the telescopes are at the North Pole with the same spacing and operating at the same frequency, what is the shortest temporal fringe period if the source is 15' from the north celestial pole (NCP)? [Ans. ~40 ms; ~10 s]

Problem 7.23. Consider a point-like radio source approximately at the zenith (at $\theta \approx 90°$, measured from the baseline) of two equatorial radio telescopes at time $t = 0$ (e.g., Fig. 2a). The exact position of the source is not known, and you wish to reduce the position uncertainty. The source is observed for a brief period at $t = 0$ when it is approximately overhead, again briefly at $t = 1$ h, and similarly at $t = 2$ h. The uncertainty in the position of each measured line of position is $\pm 2\%$ of the line spacing. Consider only regions relatively close to the source so the angular fringe spacing (spatial wavelength) may be taken to be constant in the source region for a given observation. (a) Describe the regions on the sky to which the $t = 0$ data alone would restrict the possible source positions? What is the fringe spacing in terms of λ / B? What fraction of the sky in this region is excluded? (b) How are the possible regions further restricted if the 1-h data are taken into account? Use equation (5) for the fringe spacing and determine where the lines of position with their error regions from the two observations overlap common positions. Assign line numbers to the lines from each observation with the zeroth lines from the two observations coincident on the sky. Consider line numbers out to about 100. (c) How do the 2-h data further restrict the positions (or do they)? What are the (0-h data) line numbers less than ~100 that remain viable locations for the source? [Ans. (c) 0, ~ ±30, . . .]

Problem 7.24. Consider the situation illustrated in Fig. 5c. The line connecting two radio telescopes (the *baseline*) lies 40° from the spin axis of the earth; it intercepts the sky at declination $\delta = \pm 50°$. An observation of a celestial source at $\delta = 70°$ carries on for 24 h (there are no earth occultations). The telescopes are separated by 1 km and are operated at 10 GHz. (a) What are the smallest and largest spacings between adjacent 100% visibility lines (in arcseconds) at the source position during this 24-h period? (b) Make a sketch as viewed from above the NCP of the situation shown in Fig. 5c; show 100% visibility lines and the circular track of the source (in the earth frame). From your plot construct the *approximate* track of the source on the Fourier plane. Is there value in making measurements over the full 360° or will 180° suffice? Hint: consider the source to be at several different positions along its circular path. (c) At what latitudes λ on the earth could one set up the two telescopes to have the baseline intercept the celestial sphere at $\delta = +50°$, and how would they be oriented? Assume the telescope separation is much less than the earth radius, and that the telescopes are both at the same altitude, i.e., the local terrain is flat. Show your answers on sketches similar to Fig. 5c. How does the requirement that a source at $\delta = 70°$ never be occulted by the earth further restrict the telescope locations? As the earth rotates, does the baseline remain fixed at $\delta = 50°$? Explain. [Ans. (a) $\sim 7''$, $\sim 20''$]

7.3 Mapping the sky

Problem 7.31. (a) Sketch qualitatively a few cycles of the response function $R(t)$ (14) for a solitary source with no background for the following two cases. (*i*) a solitary point source and (*ii*) a source extended enough to modestly demodulate the response function (which remains sinusoidal; see next problem). Assume $\phi(t) \propto t$. Label the ordinates of the two plots with R_{max}, R_{min}, and R_{av} ("av" = average). Find an expression for the flux density $\mathscr{F}$ (W/m^2) from the source, *independent of its angular extent*, in terms of R_{max}, R_{min} and an unknown multiplicative (constant) coefficient. (b) Repeat (a) for the case where there is a constant (in time) background R_B, in the telescope response (from instrumental noise and/or diffuse sky emission) in addition to a solitary source. Make plots for (*i*) point and (*ii*) slightly extended sources. Label the ordinates with R_{max}, R_{min}, R_{av}, and R_B. Express $\mathscr{F}$ in terms of R_{max}, R_{min}, R_B and a constant coefficient. Could one distinguish cases (a)(*ii*) and (b)(*i*)? If not, what other interferometric measurements would help distinguish them. Hint: how could one determine R_B in the two cases? [Ans. (b) $\mathscr{F} \propto [(R_{max} + R_{min})/2] - R_B]$

Problem 7.32. Show that a modestly extended source that partially demodulates the response function gives rise to a response function that is sinusoidal in form with the *same* frequency as the point source response. Consider an extended source of arbitrary brightness distribution to be a sum of point sources with different intensities. The net response will then simply be the sum of sinusoidal point source response functions with different phases α_i and amplitudes A_i, but with each having the frequency Ω(rad/s) that is characteristic of the fringe passage over the source; $\phi(t) = \Omega t$ in (24). Evaluate the simple case where there are only two such elements, #1 and #2 that give rise to a response of the form $R(t) = A_1 \cos \Omega t + A_2 \cos(\Omega t - \alpha)$ where the time average has been set to zero. Show that this can be written in the form $R = C \sin(\Omega t - \beta)$, and argue that this demonstrates the desired point. What are the values of C and β in terms of A_1, A_2 and α?

$$\left[\text{Ans. } \tan \beta = \frac{A_2 \sin \alpha}{A_1 + A_2 \cos \alpha} ; C = \left(A_1^2 + 2A_1 A_2 \cos \alpha + A_2^2 \right)^{1/2} \right]$$

Problem 7.33. Find the relation between the two versions of the cross-correlation functions, $C'(\alpha, \delta)$ (23) and $C(\alpha, \delta)$ (27). In other words, what are the "constants" in (25)? [Ans. Two are zero; the third is not]

Problem 7.34. Here we find the response (28) of a two-telescope system to a point source and to two one-dimensional sinusoidal sky brightness distributions. The purpose is to illustrate that the system selects only one spatial (angular) frequency component of the sky intensity distribution. First we rewrite (28) in one dimension.

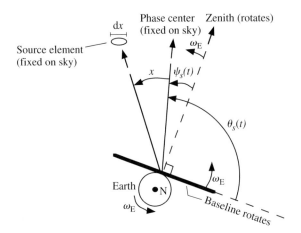

(a) Consider the two-equatorial-telescope observation of Fig. 2 with baseline B and operating at wavelength λ. The antenna fringe pattern translates normal to the fringe planes at a constant rate. Let the intensity (brightness) distribution $I(x)$ of the sky vary in only one direction, the variation direction of the fringe pattern (left–right in Fig. 2 and in the sketch above). The angular position x (rad) of a sky position is measured from the phase center at θ_s which is fixed on the sky and chosen to be near the center of the region being observed. The earth rotates at angular velocity ω_E (rad^{-1}), such that $\psi_s = -\omega_E t$ where ψ_s is the position of the phase center relative to the rotating zenith (see sketch). Demonstrate that (28), the two-dimensional expression for $R(t)$ may be rewritten for this one-dimensional case as

$$R(t) = \int_{\text{beam}} I(x) \cos\left[\frac{2\pi}{\lambda} B(\omega_E t - x)\right] dx$$

for a source region nearly overhead ($x \ll 1$ rad and $\psi_s \ll 1$ rad).

(b) What is the response function $R(t)$ for the following three sky distributions: (i) The sky brightness $I(x)$ is sinusoidal over the entire scanned region with $I(x) = A\cos[2\pi u(x - x_0)]$ where u is in rad^{-1} with magnitude B/λ and x_0 is a reference position on the sky. The spatial frequency of the sky distribution thus matches that of the interferometer fringes. Set $m = 2\pi u$ and $n = 2\pi B/\lambda$ to simplify your work. The ratio B/λ is the "cycles per radian" of the fringe pattern on the sky at a fixed time. (ii) The sky brightness is sinusoidal with frequency twice that of the fringes, $u = 2B/\lambda$; (iii) the sky distribution is a one-dimensional "point source" (actually, a line running parallel to the fringes): $I(x) = A\delta(x - x_0)$ where $\delta(x - x_0)$ is the delta function. Comment on the significance of the several answers you get. Be careful to distinguish variation over sky coordinates and over time.

$$\left[\text{Ans.(b)}(i)\ \frac{A\lambda}{2B} \cos\left[\frac{2\pi}{\lambda} B(\omega_E t - x_0)\right]\right]$$

7.4 Arrays of telescopes

Problem 7.41. (a) The VLA array has 27 telescopes as shown in Fig. 9a. Nine are spaced along each of three arms of length 21 km arranged like three spokes of a wheel with equal angles (120°) between the spokes. Consider an arrangement of the telescopes wherein the distances from the center on each arm scale as 1, 4, 9, 16, etc. with the outermost telescopes at the end of the arms. There is no telescope at the center. (a) A *brief* observation is carried out of a source at the zenith. Make a Fourier-plane u, v diagram showing the sampled points due to the inner 6 antennas only, i.e., those at radii 1 and 4, for the time of the observation. Align one spoke in the N–S direction. Hint: how many points should be in the plot? (b) If the VLA were at the North Pole observing an overhead source, how long an observation is needed to get the maximum u, v coverage with this 6-antenna configuration? Will the time differ for the 27-antenna configuration? (c) The array can operate at a frequency of 5 GHz. In this case, what is the approximate source angular size that could just barely be resolved, i.e., the smallest angular size that would appear as extended? Consider both the 6-antenna and 27-antenna cases. [Ans. 15; 4 h; $\sim$5″, $\sim$0.3″]

8

Point-like and extended sources

What we learn in this chapter

The flux of radiation arriving from a distant **point (unresolved) source** may be described with the **spectral flux density** $S(\nu, t)$ (W m^{-2} Hz^{-1}) which gives the flux as a function of frequency ν and time t. Integration of S over the frequency interval of the detector yields the **flux density** $\mathscr{F}$ (W/m^2). In turn, integration of $\mathscr{F}$ over the antenna area yields the detected **power** $\mathscr{P}$ (W), and similarly, integration of $\mathscr{P}$ over the time interval of the observation yields the **fluence** $\mathscr{E}$ (J). If the source is assumed to radiate isotropically with flux $\mathscr{F}(r)$ at distance r, its **luminosity** L(W) is simply $4\pi r^2 \mathscr{F}$.

Optical astronomers traditionally describe flux densities with a historical logarithmic **magnitude** scale where the brightest stars have magnitude zero and the faintest the human eye can see is 6. Magnitudes are defined for different **spectral bands**. **Bolometric magnitude** describes the flux over the entire optical band (extending into the IR and UV). **Absolute magnitude** is a measure of luminosity; it is magnitude adjusted for distance to the source.

Celestial objects with measurable angular sizes are called **resolved** or **diffuse** sources. The flux is described completely with **specific intensity** $I(\nu, \theta, \phi, t)$ (W m^{-2} Hz^{-1} sr^{-1}) which describes the variation of flux with position θ, ϕ on the sky. Integration of I over the solid angle of a source yields the above-mentioned spectral flux density S. **Surface brightness** $B(\nu, \theta, \phi, t)$ (W m^{-2} Hz^{-1} sr^{-1}) describes the radiation leaving the surface of a celestial body. It can be shown to be identically equal to the detected specific intensity, $B = I$, a general relation that follows from **Liouville's theorem**. Looking into the source itself, the power generated per cubic meter is the **volume emissivity** j (W m^{-3} Hz^{-1}). For an **optically thin source** of known thickness along the line of sight, it is quite simply related to the specific intensity a distant observer would measure.

8.1 Introduction

The quantities used to describe incoming radiation are presented in this chapter. These quantities should describe the rate of energy flow (power), its variation with time, its variation with frequency of the electromagnetic wave, and the direction(s) from which it flows. Note that "frequency" can also refer to oscillations of intensity as seen in pulsars.

The most general description of the radiation would allow for every point on the sky to have an arbitrary brightness and polarization that varies arbitrarily in time at every frequency of the radiation. The real sky brightness is significantly simpler; the flux from many sources of radiation does not vary substantially in time, and many of the sources are point-like objects. Nevertheless the sky does exhibit a wide range of objects, from point-like stars to diffuse or extended nebulae (e.g., H II regions, supernova remnants, and galaxies), and finally variable-intensity objects (e.g., variable stars, pulsars, supernovae, gamma and x-ray bursts, and quasars).

It is convenient to divide celestial sources into two major groups, *point-like* and *diffuse*. A celestial object at a great distance from the earth can be considered *point-like* if its angular size measured from the earth is much less than the beam width or angular resolution of the antenna or telescope in question. Point-like sources are often called *unresolved* objects because the telescope can not resolve their small angular sizes. If the source is comparable to, or larger than, the beam in angular extent, the telescope can detect the finite size. In this case, the object is called a *diffuse* or *resolved* source.

The beam width (Ch. 5) is due to the intrinsic parameters of the telescope/detector system such as diffraction and also, in the case of optical light, the earth's atmosphere. An optical object of angular size much less than $1''$ may be considered to be point-like for ordinary (non-interferometric) observations with ground-based optical telescopes because of the atmosphere. A sun-like star of size $\sim 10^9$ m at the relatively close distance of 10^{17} m (10 LY) subtends an angle of only 10^{-8} radians, or $0.002''$, and easily qualifies as a point-like object for ground-based observers. The human eye has a resolving power of only about $1'$; anything smaller in angular size appears as a point-like object even though it might appear quite large in angular size when viewed through binoculars or a telescope.

In this chapter, we present quantities used for the measurement of point sources, including the *spectral flux density S* (W m^{-2} Hz^{-1}) and the astronomical unit of (logarithmic) flux density called the *magnitude*. We then address diffuse sources and introduce the quantity that gives their brightness per unit solid angle on the sky (e.g., per steradian or per square arcsecond), namely the *specific intensity I* (W m^{-2} Hz^{-1} sr^{-1}). This latter quantity is amazing. It is independent of the distance to the source and, remarkably, is equal to the intensity emitted from the surface of the celestial object, the *brightness B* (W m^{-2} Hz^{-1} sr^{-1}).

8.2 Unresolved point-like sources

Spectral flux density

The flux from an unresolved (point-like) object may be considered to be a parallel beam of light or a plane wave originating at "infinity". It impinges on the telescope with a given amount of energy deposited per second, per square meter, and per unit frequency interval at frequency v (e.g., in the interval $v - 0.5$ Hz to $v + 0.5$ Hz). This is known as the *spectral flux density*,

$$S(v) \equiv \text{Spectral flux density} \qquad (\text{W m}^{-2}\,\text{Hz}^{-1}) \qquad (8.1)$$

but it could be more properly called the *spectral energy flux density* to distinguish it from a photon flux density (photons m^{-2} Hz^{-1}). In fact it is often called simply the "flux density". Here we reserve that term for another related use.

The definition serves to standardize the values of spectral flux density reported in the literature from telescopes with different areas and frequency bands (*bandwidths*). The standard quantity is simply the observed power divided by the area of the telescope and by the bandwidth in Hz. Thus observations with different equipment of the same source should yield approximately the same value for S.

The motivation for this normalization is similar to that for reporting an automobile speed as distance *per unit time*, e.g., 100 km/h, rather than in terms of the actual elapsed time of the journey, 20 km/12 min. The standard reference time makes it much easier to compare the average speeds from journeys of different durations.

The actual energy received by the telescope per second (power $\mathscr{P}$ in watts), in a narrow frequency band Δv, is the product of the spectral flux density (averaged over the band Δv), the *effective area* A_{eff} of the telescope, and the *bandwidth* Δv,

$$\mathscr{P} = S(v)_{\text{av}} A_{\text{eff}} \Delta v \qquad (\text{W}) \qquad (8.2)$$

For example, a relatively bright celestial radio source might yield a spectral flux density $S(v)$ at the earth of

$$S(v) = 1.0 \times 10^{-26}\ \text{W m}^{-2}\,\text{Hz}^{-1} = 1.0\ \text{Jy} \qquad (\text{jansky}) \qquad (8.3)$$

at frequency $v = 100$ MHz. This particular spectral flux density is known as 1.0 *jansky*; Carl Jansky was the discoverer of radio radiation from the (MW) Galaxy.

Now, suppose that this signal is detected with a perfect antenna of diameter 16 m ($A_{eff} \approx 200$ m^2) that is tuned to $\nu = 100$ MHz with a narrow bandwidth $\Delta\nu = 10^4$ Hz. This means that the detection system accepts radiation in the frequency band 10^4 Hz wide at $\nu = 100$ MHz, i.e., 99 995 000 Hz to 100 005 000 Hz. Further assume that $S(\nu)$ is constant, or nearly so, across the band $\Delta\nu$ so that $S(\nu) \approx S(\nu)_{av}$. The actual power received by the antenna would then be, substituting into (2),

$$\mathscr{P} = 1 \times 10^{-26} \times 200 \times 10^4 = 2 \times 10^{-20} \text{ W} \tag{8.4}$$

This is a very small amount of energy; it would take 0.5×10^{20} s to accumulate the energy of one joule, enough to light a 1 W bulb for 1 s. This time corresponds to about 10^{13} years or ~ 1000 times the age of the universe! The detection of such small (and even smaller) amounts of power requires high-sensitivity detectors. Do not be misled by such tiny power levels; they are associated with tremendous total power outputs from the source, as we demonstrate below.

The calculation above was made under the assumption that the spectral flux density $S(\nu)$ is constant over the bandwidth of frequencies accepted by the antenna. For a bandwidth narrow compared to the measured frequency, this is often a reasonable assumption. However, it is possible that much of the energy lies in a spectral line, an enhancement of flux in a very narrow, e.g. 10 Hz, frequency band. In this case, the spectral flux density could be close to zero for a large part of the 10^4 Hz band. A proper calculation of the power received by the antenna for a variable $S(\nu)$ requires that the product $S(\nu)\,d\nu\,A$ be integrated (summed) over the frequency band,

$$\mathscr{P} = A_{eff} \int_{\nu_1}^{\nu_2} S(\nu)\,d\nu \qquad \text{(W; power)} \tag{8.5}$$

where A_{eff} is the effective area of the telescope, taking into account inefficiencies that dissipate some of the incident energy, and where the integration is over the bandwidth of detected frequencies. In practical cases, the effective area is a function of the frequency $A_{eff}(\nu)$, so it too would go inside the integral.

The spectral flux density is properly a vector $\boldsymbol{S}$, where $S \equiv |\boldsymbol{S}|$, because the flux at any point in space must have a direction; it is a "flow". The direction of the flux may be determined by immersing a test surface of fixed size into the flow and then rotating the surface to various orientations. When the flux through the surface reaches its maximum value, the surface normal lies along the flow lines. This is analogous to the vector current flux density $\boldsymbol{J}$ (A m^{-2}) in electromagnetic theory. Do not confuse the quantity $\boldsymbol{S}$ (W m^{-2} Hz^{-1}) used in this text with that used in electromagnetic theory for the Poynting vector (W/m^2).

Flux density

The total power flowing across unit area is called the *flux density* $\mathscr{F}$ (W/m^2). It is obtained by integrating the spectral flux density $S(\nu)$ over the frequency band of interest,

➡️
$$\mathscr{F} \equiv \int_{\nu_1}^{\nu_2} S(\nu)\, d\nu \qquad \text{(W/m}^2\text{; flux density)} \qquad (8.6)$$

Again, this quantity is properly a vector since it is a flow that has direction; a surface immersed in the flow can be oriented to give the direction of flow as above. For our purposes, we usually use the scalar quantity $\mathscr{F} \equiv |\mathscr{F}|$. In electromagnetic theory, this vector is the quantity known as the Poynting vector; see (7.13).

The flux $\mathscr{F}$ can be written in terms of the average over frequency of the spectral flux density S by multiplying and dividing the right side of (6) by $\Delta\nu = \nu_2 - \nu_1$ and recalling the definition of an average, $S_{\text{av}} \equiv \int S\, d\nu / \Delta\nu$. Thus, $\mathscr{F} = S_{\text{av}}\Delta\nu$.

Luminosity

The luminosity L of a source is, in its usual meaning, the total power output (W) summed over *all* frequencies. In practice, one usually must specify the band of frequencies (of radiation) that are being measured, e.g., the visual V band, the entire optical band, or the 1–10 keV x-ray band. Normal stars like the sun emit only a tiny fraction of their power in the radio and x ray, and these emissions were long unknown, so these bands were ignored in traditional astronomy. The luminosity over the entire optical band including the spillover into the adjacent infrared and ultraviolet bands is called the *bolometric luminosity*; it is this which is usually given the symbol L where $L \equiv L_{\text{bol}}$.

If the distance r to a source is known, an estimate of the luminosity in a specified band $\Delta\nu$ is obtained by multiplying the flux density $\mathscr{F}$ (W/m^2 in the band $\Delta\nu$) by the area of a sphere centered on the source with its surface passing through the earth (antenna) as shown in Fig. 1,

$$L_{\Delta\nu} = 4\pi r^2 \mathscr{F}_{\Delta\nu} \qquad \text{(W; isotropic emission)} \qquad (8.7)$$
$$= 4\pi r^2 S_{\text{av}} \Delta\nu$$

where the (unconventional) subscript $\Delta\nu$ reminds one that the luminosity is restricted to the chosen band $\Delta\nu$.

An assumption implicitly adopted in (7) is that the emission from the source is isotropic, that is, the energy is radiated equally into all directions. Only in this case does our antenna get its expected share of the total radiation that leads to the factor $4\pi r^2$ in (7). The assumption of isotropy is a common one because an antenna on

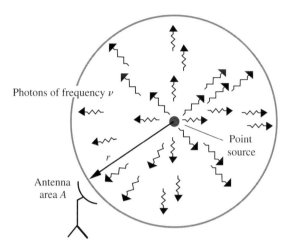

Figure 8.1. A point-like source at distance r from an antenna radiates equally in all directions. The luminosity of the source in a given frequency band is the flux density $\mathscr{F}$ (W/m^2) detected in that band multiplied by $4\pi r^2$.

the earth can sample only one emission direction, and isotropy is often the most reasonable guess. Some objects, e.g., pulsars, active galactic nuclei and gamma-ray bursts, emit beams of radiation that are demonstrably non-isotropic; in these cases, (7) is clearly incorrect.

Consider the luminosity of a hypothetical radio source radiating isotropically with a constant spectral flux density at the earth of $S = 1.0$ Jy over the 50–150 MHz band. If it is at the center of the Galaxy, at a distance of $\sim$25 000 LY (2.4 $\times$ 10^{20} m), its luminosity from 50 to 150 MHz ($\Delta\nu = 10^8$ Hz) would be

$$L = 1.0 \times 10^{-26} \times 10^8 \times 4\pi \times (2.4 \times 10^{20})^2 = 7 \times 10^{23} \text{ W} \qquad (8.8)$$

This is a lot of watts, equivalent to almost 10 billion trillion 100-watt light bulbs! It is about 1/600 of the sun's luminosity of 4×10^{26} W. The quasar 3C273 at a distance of 2.1 MLY with spectral flux densities ranging from 100 to 50 Jy over the range 100 to 1000 MHz, has a radio luminosity of $\sim 10^{36}$ W. Inclusion of optical, x-ray and gamma-ray radiation raises this to more than 10^{38} W, or $\sim 10^{12} L_\odot$ (solar luminosities). This would be in error, probably an overestimate, if the radiation is beamed.

Over a wide bandwidth, such as a factor of two change in frequency, the spectral flux density $S(\nu)$ is likely to change substantially; in this case one must integrate over frequency to obtain the luminosity,

➡ $$L = 4\pi r^2 \mathscr{F}_{\Delta\nu} = 4\pi r^2 \int_{\nu_1}^{\nu_2} S(\nu)\, d\nu \qquad \text{(W; isotropic emission)} \qquad (8.9)$$

One can often estimate roughly the luminosity without integrating by substituting for the integral the product of the bandwidth $\Delta \nu$ and a typical or average value of the spectral flux density S_{av} in that band. If the functional form of $S(\nu)$ is simple, formal integration is quite straightforward.

Fluence

Some astronomical sources emit occasional isolated bursts of radiation which might last for 10–100 seconds. It is convenient to define a quantity that gives the flux of energy integrated over the duration of the burst. This is called the fluence $\mathscr{E}$ (J m^{-2}), or more precisely, the energy fluence, which is defined as the time integral of the flux density,

$$\mathscr{E} = \int_{t_1}^{t_2} \mathscr{F} \, dt \qquad\qquad \text{(J m}^{-2}\text{; fluence)} \qquad (8.10)$$

Again this quantity is properly a vector $\mathscr{E} \equiv |\,\mathscr{E}\,|$. A spectral fluence (J m^{-2} Hz^{-1}) could also be defined if desired.

Note that the quantities above are all derived from an integrations of $S(\nu)$ over one or more of the variables: area, frequency, and time.

8.3 Astronomical magnitudes

The traditional quantity used by optical astronomers to describe the intensity or brightness of a star has been the *visual magnitude m*. Hipparchus, the Greek astronomer and mathematician ($\sim$135 BCE), with refinements by Ptolemy ($\sim$150 CE), classified stars as having magnitudes 1, 2, 3, 4, 5, 6 where "1" was the brightest and "6" the faintest. The current system maintains this rough calibration. A sixth magnitude star ($m = 6$) is about the faintest one can see with the unaided, dark-adapted eye on a clear, dark night. Thus the largest magnitudes represent the faintest stars. The brightest star is Sirius with a visual magnitude that is less than zero, namely $m_V = -1.5$, according to the modern definition discussed below. Actually, the brightest star in the sky is the sun with $m_V = -26.7$. Astronomers with large telescopes can now obtain images of stars as faint as, or fainter than, $m_V \approx 25$ or even reaching $m_V \approx 29$ with the Hubble Space Telescope.

Astronomers now distinguish between *apparent magnitude m* and *absolute magnitude M*. The former is the quantity used by Hipparchus; it is simply a measure of the apparent flux density of the star as measured from the earth. In contrast, the absolute magnitude is a measure of the luminosity L, the total rate of energy emission by the object; it is independent of distance to the object. Two physically identical stars at different distances from the earth would have the same absolute magnitudes but different apparent magnitudes.

A star's magnitude is typically measured over a fairly broad band of frequencies and these are specified as ultraviolet U, blue B, visual V, etc. Precise calibrations in terms of physical units (W/m^2) is not simple because a given detection system has an efficiency for photon detection that is a function of frequency, and this function differs for different types of detectors. Since the incoming frequency distribution of photons (spectral shape) differs from one astronomical body to the next, the ratio of responses of two dissimilar detector systems will vary from source to source. Astronomers have thus gone to great pains to adopt standard measurement systems and calibrations for the purpose of accurately measuring broadband fluxes (magnitudes).

Apparent magnitude

Here we present the definition of the *apparent magnitude* in terms of the response of a detection system to incoming radiation. Scales of *color magnitudes* are also defined.

Magnitudes and fluxes

A typical detector in the IR–optical–UV region has a response that is proportional to the *number* of photons impinging upon it in a given time. A quantitative measure of this is the *photon spectral flux density* $S_p(\nu)$ which is the number of photons (rather than their energy) at frequency ν per unit frequency, area, and time (m^{-2} Hz^{-1} s^{-1}). It is given simply by dividing our usual energy spectral flux density $S(\nu)$ (J m^{-2} Hz^{-1} s^{-1}) by the energy of each photon, $S_p(\nu) = S/(h\nu)$. The number of photons recorded by a detector in unit time is obtained from multiplication of S_p by an efficiency factor $\epsilon(\nu)$ and integration over frequency. This yields the detected *photon flux density* $\mathscr{F}_p$,

$$\mathscr{F}_p = \int_0^\infty S_p(\nu)\,\epsilon(\nu)\,d\nu \quad \text{(Detected photon flux density;} \quad (8.11)$$
$$\text{photons m}^{-2}\text{ s}^{-1})$$

where $\epsilon(\nu)$ is a function of frequency and varies from zero to one. It is the probability that a photon of frequency ν will register in the detector; detectors are not perfect.

The eye tends to classify stars according to logarithmic intervals of flux; thus Hipparchus's stars of magnitude 1 and 2 had approximately the same ratio of fluxes as did stars of magnitudes 5 and 6. The apparent magnitude m_2 of a star relative to that of a comparison star m_1 is now defined in terms of the logarithm (base 10) of the *ratio* of the photon flux densities $\mathscr{F}_p$ from the two stars,

$$\Rightarrow \qquad m_2 - m_1 \equiv -\frac{5}{2} \log \frac{\mathscr{F}_{p,2}}{\mathscr{F}_{p,1}} \qquad \text{(Definition of magnitude} \qquad (8.12)$$
$$\text{difference)}$$

where the subscripts refer to the two stars, numbered 1 and 2, and the coefficient is exactly 2.5.

The *zero of apparent magnitude* is set, in the optical band, to match that of a star of type A0V at distance 26.4 LY with surface temperature $T \approx 10\,000$ K. *Stellar type* may be thought of as a measure of the mass and elemental composition of the star. For a normal hydrogen-burning star, these properties determine the star's surface temperature and radius, which in turn determine the spectral distribution we measure.

Note from (12) that a ratio of 100 in fluxes represents a difference of 5 magnitudes,

$$m_2 - m_1 = -5.0 \qquad\qquad (\mathscr{F}_{p,2}/\mathscr{F}_{p,1} = 100) \qquad (8.13)$$

If star 2 provides 100 times more flux than star 1, its magnitude is *less* by 5. It is said to be "5 magnitudes brighter" rather than the correct, but confusing "5 magnitudes less". Equation (12) can be solved for the ratio of fluxes to yield

$$\frac{\mathscr{F}_{p,2}}{\mathscr{F}_{p,1}} = 10^{-0.4(m_2-m_1)} \qquad\qquad (8.14)$$

which becomes, for $m_2 - m_1 = 1$,

$$\mathscr{F}_{p,2}/\mathscr{F}_{p,1} = 1/(2.512 \ldots) \qquad\qquad (m_2 - m_1 = 1) \qquad (8.15)$$

Thus a difference of one magnitude represents a ratio of 2.51 in flux. The logarithmic scale of astronomical magnitudes is actually a base 2.51 system.

This magnitude system is somewhat annoying. Fainter stars have larger numbers, and the logarithmic interval for unit magnitude (a factor of 2.51) is hardly conventional. A factor of 10, or 2, or $e = 2.718$, would be more natural. But one of the beauties of astronomy is its roots in human history; the oldest written records include astronomical references. The magnitude system is a daily tangible reminder of the observations of Hipparchus and astronomical history. But we could likewise perpetuate furlongs, rods, and pecks, and this definitely would not be convenient.

Spectral color bands

The magnitude of a star can be measured in any of a number of wavelength bands. Thus when specifying a magnitude, the spectral band that was measured is usually specified. The "3-color" system mentioned above is the traditional system often used, namely the *UBV* system: ultraviolet, blue, visual (yellow). Each spectral

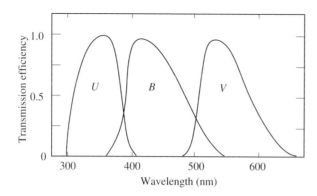

Figure 8.2. Transmission curves for the standard U, B, and V filters used for the measurement of flux densities of celestial objects. The taking of such measurements is called photometry. The ordinate is efficiency of transmission (1.0 is perfect transmission) and the abscissa the incident wavelength.

band is defined by a standard astronomical filter that can be placed in the beam of starlight near the focus of a telescope. Thus, when the flux is measured, only the desired portion of the spectrum reaches the detector (often a photomultiplier tube or CCD detector). The response functions (Fig. 2) do not have sharp frequency cutoffs because they are the measured responses of real, not ideal, filters and detectors.

Define the transmission efficiency of the blue filter combined with the detector efficiency as a function of frequency to be ϵ_B. The blue-band flux density is thus, after (11),

$$\mathscr{F}_{\mathrm{p},B} = \int_0^\infty S_{\mathrm{p}}(\nu)\,\epsilon_B(\nu)\,\mathrm{d}\nu \qquad \begin{array}{l} \text{(Photons m}^{-2}\text{ s}^{-1};\\ B\text{-band photon flux)} \end{array} \qquad (8.16)$$

The ratio of two such quantities, for two different stars, gives, through (12), the difference in their blue magnitudes, $m_{B,2} - m_{B,1}$. Observers can thus obtain accurate blue magnitudes of an uncataloged star or object by comparing it to a star with a well calibrated star blue magnitude.

The magnitudes in the several frequency bands may be denoted in two ways: for the ultraviolet band, both m_U and U are used (Table 1). The *bolometric magnitude* m_{bol} includes the flux over all relevant frequencies in the IR, optical, and UV regions (see below). Values of these apparent magnitudes are given for the sun in Table 1.

Spectral bands to lower frequencies than the V band are now in wide use (Table 2), thanks to detectors and films with extended responses, e.g., R (red), and I (infrared). These bands are also normalized so zero magnitude approximates the fluxes from a star of type A0V adjusted to the distance of Vega. Vega (α Lyrae) was traditionally used as this reference, but it is not easily accessible to southern hemisphere observers and it has dust clouds which give anomalous IR fluxes. The effective (average)

Table 8.1. *Solar apparent magnitudes*

m_U or U	-25.91
m_B or B	-26.10
m_V or V	-26.75
m_{bol}	-26.83

Table 8.2. *Photometry bands: Johnson–Cousins–Glass system[a]*

Filter	Effective wavelength λ_{eff} (nm)[b]	Passband $\Delta\lambda$(nm)[b] (FWHM)	Effective frequency ν_{eff}(THz)[c]	Passband $\Delta\nu$ (THz)[c] (FWHM)	Spectral flux density at $m = 0$ $S_0(\nu_{eff})$ (Jy)[d]
U	367	66	817	147	1780
B	436	94	688	148	4000
V	545	88	550	89	3600
R	638	138	470	102	3060
I	797	149	376	71	2420
J	1220	213	246	43	1570
H	1630	307	184	35	1020
K	2190	390	137	24.4	636
L	3450	472	87	11.9	281
M	4750	460	63	6.1	154

[a] Bessell, M. S. in *Astronomy & Astrophysics Encyclopedia*, ed. S. Maran, Van Nostrand Rienhold, 1992, p. 406. Glass, I. S., *Handbook of Infrared Astronomy*, Cambridge University Press, 1999.
[b] 1.0 nm (nanometer) $= 1.0 \times 10^{-9}$ m. The effective wavelength is the average wavelength in the band for the spectrum of a star of type A0V.
[c] 1.0 THz (terahertz) $= 1.0 \times 10^{12}$ Hz. The effective frequencies ν_{eff} and passbands $\Delta\nu$ were obtained from columns 2 and 3 by the present author, using $\nu_{eff} = c/\lambda_{eff}$ and $\Delta\nu = c(\Delta\lambda)/\lambda^2$. The former could give slightly different results than (18) depending on the spectral shape.
[d] 1.0 Jy (jansky) $= 1.0 \times 10^{-26}$ W m^{-2} Hz^{-1}. Approximate spectral flux density given for magnitude zero at the "effective frequency". At other magnitudes, m, the spectral flux density is: $S(\nu) = S_0(\nu)\,10^{-0.4m}$ (Jy).

wavelengths (see below) λ_{eff} and passbands $\Delta\lambda$ are given in Table 2 along with the corresponding frequencies ν and passbands $\Delta\nu$. The bands in the table are one of several such systems of standard filters.

Conversion from magnitudes to SI units

Since magnitudes are not physical (SI) units, it is often necessary to convert magnitudes to (spectral) flux densities S (W m^{-2} Hz^{-1}). The flux densities S_0 that correspond to the magnitude zero for any color are given in the rightmost column of Table 2. Each entry $S_0(v)$ is the spectral flux density at $m = 0$, with units 10^{-26} W m^{-2} Hz^{-1} (i.e., the jansky) at the *effective frequency* $v = v_{eff}$, which is roughly the central frequency of the band.

The flux densities S corresponding to magnitudes other than $m = 0$ can be calculated easily from S_0. The (energy) spectral flux density S decreases with increasing magnitude in the same manner as the photon flux density. Comparison to (14), for a reference magnitude $m_1 = 0$, yields

$$S(v) = S_0(v)\, 10^{-0.4(m-0)} \qquad\qquad \text{(W m}^{-2}\text{ Hz}^{-1}) \qquad (8.17)$$

where $S_0(v_{eff})$ is obtained from Table 2. This enables one to convert any magnitude in a given color band to a spectral flux density in SI units.

A value of $S_0(v_{eff})$ in Table 2 is nominally the spectral flux density at the indicated *effective frequency* of the filter passband for a Vega-type spectrum. The effective frequency is the average frequency of the recorded photons,

$$v_{eff} = \int_0^\infty v S_p(v)\, \epsilon(v)\, dv \left/ \int_0^\infty S_p(v)\, \epsilon(v)\, dv \right. \qquad \begin{array}{c}\text{(Effective} \\ \text{frequency)}\end{array} \qquad (8.18)$$

where ϵ is the efficiency function of the filter in question and S_p is the spectral distribution of the source being observed. For a Vega-like spectrum, the result should yield the frequencies quoted in Table 2, but see the caveat in footnote c to the Table.

One can multiply both sides of (18) by the Planck constant h to obtain the photon energy $h v_{eff}$ at the effective frequency on the left. The integral in the numerator is then the photon energy hv times the detected photon flux summed over all frequencies, i.e., the detected *energy* flux $\mathscr{F}$ (W m^{-2}). Since the denominator is the measured *photon* flux density $\mathscr{F}_p$ (s^{-1} m^{-2}), the energy flux $\mathscr{F}$ is simply the product of $h v_{eff}$ and the measured photon flux $\mathscr{F}_p$.

Unfortunately the effective frequencies depend on the spectral shape of the incident photon flux, i.e., on the relative numbers of low and high frequency photons arriving within the passband of the filter. Consider the extreme case where *all* the photons in the passband are concentrated in a narrow spectral line at the low-frequency end of the band. Clearly, the average frequency will occur at the low end of the band. In contrast, the average value will be near the midpoint of the band for a source with a flat, continuous spectrum, or one that linearly increases or decreases with frequency, if it dominates the spectral lines.

The effective wavelengths in Table 2, as noted, were calculated for the spectrum of an A0V star, like Vega, with an expression similar to (18) where λ replaces v and the weighting factor is $S_{\lambda p}$ (photon flux per unit wavelength) replaces S_p. We calculated the values of ν_{eff} in Table 2 from the λ_{eff} rather than from (18), but the differences should be minor.

The widths $\Delta\nu$ of the transmission function of the filters (Fig. 2) are the full width at half maximum, FWHM. The approximate photon spectral flux density S_p at the effective frequency may be found by dividing the width into the measured flux density $\mathcal{F}_p$

$$S_p(\nu_{eff}) \approx \frac{\mathcal{F}_p}{\Delta\nu} = \frac{1}{\Delta\nu} \int_0^\infty S_p(\nu)\, \epsilon(\nu)\, d\nu \quad \text{(Incident photons;} \atop \text{m}^{-2}\, \text{s}^{-1}\, \text{Hz}^{-1}) \qquad (8.19)$$

where we made use of (16).

The magnitude system is very much an instrument-driven system. As noted above, the efficiency functions of different types of color (filter) systems will generally be different. Thus, two different color systems will yield different ratios of blue responses $\mathcal{F}_{B,2}/\mathcal{F}_{B,1}$ for two stars that have different spectral shapes within the blue passband. That is, one filter system might produce a ratio of 1.5 indicating that star #2 is 50% brighter than #1, while the other system might yield 1.7.

The calibration of different broadband detection systems against one another is thus dependent on the distribution in frequency of the incoming radiation. Measurement of the distribution of energy from a star thus requires knowledge of that distribution! How does one do this? One either uses exactly the same detection system as other colleagues, or, if one uses a non-standard system, the measured response can sometimes be corrected iteratively to a standard color system. The latter is feasible if the spectrum is continuous and well behaved, but not so if unknown spectral lines are involved.

Measurements of color magnitudes of stars, known as *photometry*, is always accompanied by measures of *standard stars* with well determined magnitudes and spectral shapes. These allow one to correct for instrument and atmospheric effects. These days, most photometry is carried out with CCD detectors.

Color indices

The ratio of fluxes in two color bands for the same star, e.g. the ratio $\mathcal{F}_{p,B}/\mathcal{F}_{p,V}$ for the B and V bands, yields through (12) the difference of magnitudes, $m_B - m_V$. This quantity is called the *color index $B - V$*. This is a color as we would perceive it. We see an object as blue if it emits more blue flux than yellow flux. The blue magnitude alone does not indicate a color.

Colors can be defined for any two adjacent spectral bands, with the redder magnitude subtracted from the bluer, e.g., $U - B$, $R - I$, etc. Since these are given in magnitudes, larger values of a color indicate a redder color. The $B - V$ color is often used as an indicator of temperature for the quasi blackbody radiation from stars (Section 9.4).

Absolute magnitudes – luminosity

The luminosity of a celestial body can be cast into the system of magnitudes. The *absolute magnitude M* is a direct measure of luminosity, the power (W) emitted by the star in the specified spectral band. It is defined as the magnitude of a star *if* it were at the *standard distance of 10 parsecs.* (1 pc $\approx$ 3.26 LY $= 3.09 \times 10^{16}$ m; see (4.9) and associated discussion.) The use of a standard 10-pc distance removes the distance dependence and thus yields a quantity equivalent to a luminosity. The upper-case "*M*" is invariably adopted for absolute magnitude. Again, a subscript can specify the frequency band, e.g., M_V for the absolute visual magnitude. To distinguish this quantity from the ordinary magnitude m_V, the latter is sometimes called *apparent magnitude* because it is based on the distance-dependent appearance or faintness of a star.

For example, what is the absolute magnitude of a star at a distance of 10 000 pc with an *apparent* visual magnitude of $m_V = 20$? The star would be 1000 times closer if it were at 10 pc and would thus appear to be 10^6 times brighter ($\mathscr{F} \propto r^{-2}$). The magnitude difference for this flux change, from (12), is $-2.5 \times 6 = -15$. The *absolute* magnitude therefore is $M_V = 20 - 15 = 5$. A shortcut method makes use of the fact that a change of flux by a factor of 100 corresponds to 5 magnitudes (13). The change in flux by a factor of 10^6 thus corresponds to three factors of 100 brighter, or $3 \times 5 = 15$ magnitude brighter.

The absolute (visual) magnitude of the sun is $M_{\odot,V} = 4.82$, about the same value as in our example. Thus the sun would appear to be about $V = 20$ at the distance of 10 000 pc (32 600 LY) which is somewhat farther than the galactic center at 25 000 LY distance. The sun lies much closer than the standard 10 pc ($= 32.6$ LY) distance; it is only 499 light seconds ($= 1.58 \times 10^{-5}$ LY) distant (semimajor axis). The *apparent* magnitude of the sun turns out to be $m_V = -26.75$; the sun is "apparently" very bright indeed! Thus stars similar to the sun could in principle have apparent magnitudes m ranging from -26.75 to more than $+20$ due solely to their distances from the earth.

The expression that relates the apparent and absolute magnitudes is

$$\blacktriangleright \qquad m - M = -2.5 \log \frac{\mathscr{F}_p}{\mathscr{F}_{p,10}} = +5 \log \left(\frac{r(\text{pc})}{10(\text{pc})} \right) \qquad \text{(distance} \qquad (8.20)$$
$$\text{modulus)}$$

where r is the distance to the star in parsecs. This first equality in (20) follows from the definition (12) where the two stars being compared are identical but at different distances, in this case the actual distance r and the hypothetical distance of 10 pc. The inverse square relation between the flux and distance $\mathscr{F}_p \propto r^{-2}$ gives $\mathscr{F}_p / \mathscr{F}_{p,10} = (r/10\,\text{pc})^{-2}$. Substitution of this into the second term yields the third. The fraction in the latter term may also be expressed in terms of light years, i.e., $[r(\text{LY})/32.6\,\text{LY}]$. The apparent magnitude of the sun given just above was derived from (20).

Equation (20) is an important relation that one encounters often. It shows that the quantity $m - M$ is a direct measure of the distance of the star from the earth; this quantity is thus called the *distance modulus*. If the luminosity of a star is known from spectroscopic studies (which can provide a stellar *type* of known luminosity), the value of M is known. Measurement of the apparent magnitude m then yields the distance r to the star according to (20),

$$r = 10^{0.2(m-M)}\,10\,\text{pc} \tag{8.21}$$

Tables of stars may well give both their apparent and absolute magnitudes and leave it to you to find the distance from (21). Note that a distance modulus $m - M = 5$ magnitudes leads to a distance $10^1 \times 10\,\text{pc} = 100\,\text{pc}$, that 10 magnitudes gives 1000 pc, and that 15 gives 10 000 pc.

Bolometric magnitude

The *bolometric magnitude* m_{bol} was defined above (see Table 1). It is based upon the total *energy* flux in and near the optical band (IR, optical, UV). The distribution of radiation from a star approximates a blackbody spectrum with superimposed absorption lines; it peaks in the visible range for many stars and falls toward zero at shorter and longer wavelengths (See Chapter 11). The bolometric magnitude is a measure of the *energy* spectral flux density (W m^{-2} Hz^{-1}) integrated over the entire appropriate frequency band; that is, it is a measure of the bolometric flux density $\mathscr{F}$ (W/m^2).

The bolometric magnitude is often not measured directly because photometric detectors usually count photons in wide energy bands; they do not keep track of the exact energy of every photon. Furthermore, a large part of the flux may not be measurable because it is in a region of the spectrum that does not penetrate the atmosphere, for example, in the ultraviolet beyond 1.0×0^{15} Hz ($\lambda = 300$ nm) or in the near-infrared below $\sim 4 \times 10^{14}$ Hz ($\lambda = 750$ nm) (Fig. 2.1). However, the missing flux can often be estimated from the observable part of the spectrum.

The bolometric magnitude can be used in both the *apparent* and *absolute* senses. The absolute bolometric magnitude M_{bol} is the bolometric magnitude that would be measured if the emitting body were at a distance of 10 pc. Radiation at extreme wavelengths (radio or x-ray) is small for most stars, as noted above, and is traditionally not included in the bolometric magnitude.

Bolometric correction

The *bolometric correction* (BC) is the amount in magnitudes by which the visual magnitude, apparent or absolute, must be corrected to obtain the bolometric magnitude,

➡ $$m_{\mathrm{bol}} = m_V + \mathrm{BC} \qquad \text{(Apparent magnitudes)} \qquad (8.22)$$

$$M_{\mathrm{bol}} = M_V + \mathrm{BC} \qquad \text{(Absolute magnitudes)} \qquad (8.23)$$

The BC is a useful quantity because one can not directly measure the entire IR–optical–UV spectrum of a star (except from space). The BC is the same value in (22) and (23) because an additive correction in magnitudes represents a fractional change in flux which is the same for the apparent and absolute cases.

How does one obtain the bolometric correction? For many stars, the distribution of light with frequency will, as noted, approximate a blackbody spectrum. The temperature of a blackbody completely determines the distribution of light emitted by it. A hot body is blue-white and a cool body is reddish. The amount of light that falls outside the optical band is well known for a perfect blackbody. The spectra of actual stars deviate from the blackbody curve in ways that are now quite well known from direct measurements and theoretical studies, even for stars that have a large portion of their spectrum outside the optical range.

These studies provide a bolometric correction for each temperature and/or type of star so that one can look up a BC for any star whose type is known from spectroscopic studies. The BC is set to zero for a yellowish star (type "F0V") which is a bit hotter than the sun and a bit cooler than Vega. For much cooler or hotter stars, the visual filter misses large portions of the light; the correction is increasingly negative in both directions (Table 3). The table also lists sample stellar types.

Absolute bolometric magnitude and luminosity

The *absolute* bolometric magnitude is a direct measure of the *bolometric* luminosity of a star, i.e., the energy output integrated over the IR, visual, and UV, as appropriate for the temperature of the star. The bolometric magnitude M_{bol} of an arbitrary star is related, by definition, to its luminosity L as

$$M_{\mathrm{bol}} \equiv -2.5 \, \log(L/L_\odot) + 4.74 \qquad (8.24)$$

which has similarities to (12) and uses solar quantities, $L_\odot = 3.845 \times 10^{26}$ W and $M_{\mathrm{bol},\odot} = 4.74$, as a reference. Solve for the luminosity L to obtain the numerical relation between M_{bol} and L,

Table 8.3. *Star types: absolute magnitudes and bolometric corrections[a]*

Spectral type	Absolute visual mag. M_v	Effective temperature $T_{eff}(K)$	Bolometric correction BC	Absolute bolometric mag. M_{bol}
O5V	−5.7	42 000	−4.4	−10.1
B0V	−4.0	30 000	−3.16	−7.2
A0V	+0.65	9 790	−0.30	+0.35
F0V	+2.7	7 300	−0.09	+2.6
G0V	+4.4	5 940	−0.18	+4.2
G2V (Sun)	+4.82	5 777	−0.08	+4.74
K0V	+5.9	5 150	−0.31	+5.6
M0V	+8.8	3 840	−1.38	+7.4
M5V	+12.3	3 170	−2.73	+9.6

[a] For main sequence stars (class V) from J. Drilling and A. Landolt in *Allen's Astrophysical Quantities*, 4th Ed., ed. A. N. Cox, AIP Press, 2000, p. 388.

$$L = 10^{-0.4\,M_{bol}} \times 3.0 \times 10^{28}\ \text{W} \qquad (8.25)$$

This expression is valid for any normal gaseous star regardless of its spectral distribution (color).

8.4 Resolved "diffuse" sources

In most frequency bands, from radio to gamma-ray, the sky exhibits extended or diffuse emission in addition to point sources. This emission can be relatively localized like the emission from a supernova remnant such as the Crab nebula (a few arcmin in angular size) to emission that extends over the entire sky like the microwave background radiation which originated in the early universe. As stated previously, a diffuse source is one that can be resolved by a particular telescope because the source is larger in angle than the telescope beam or resolution. Recall (Chapter 5) that one can consider each pixel of an imaging detector at the focal plane of a telescope to have its own beam. The several quantities that allow one to measure and discuss diffuse emission are presented here.

Specific intensity

Here we introduce the quantity that takes into account the spread of directions from which the radiation arrives from an extended source. It could be called *directional spectral energy flux density*, but we will use the simpler, but less informative, name *specific intensity*.

Concept of specific intensity

A hypothetical antenna with a narrow beam and narrow bandwidth could survey
the entire sky and produce a map of the brightness of the sky. Your eye does this
when it surveys the sky and so does the radio antenna that produces radio sky
maps. Since different antennae have different areas, beam sizes, and bandwidths,
it is again convenient to normalize the measured power to unit area, etc. Here we
have the additional factor of beam size. Thus, we divide the power by the area,
bandwidth *and* solid angle of the antenna beam. Recall that the unit of solid angle
is the steradian (sr), and that in spherical coordinates the element of solid angle is
$d\Omega = \sin\theta \, d\theta \, d\phi$; see Fig. 3.7.

This normalization defines the *specific intensity* $I(\theta,\phi,\nu)$ with dimensions
W m^{-2} Hz^{-1} sr^{-1},

$$I(\theta,\phi,\nu) \equiv \text{Specific intensity} \qquad (\text{W m}^{-2}\text{ Hz}^{-1}\text{ sr}^{-1}) \qquad (8.26)$$

This quantity is a function of the angular location (θ,ϕ) on the sky and of the
frequency ν of the radiation. If one desires the actual power detected by a given
telescope, one simply multiplies $I(\theta,\phi,\nu)$ by the actual area, bandwidth and beam
solid angle of the telescope. If I is not constant over angle or frequency, this may
require integration as demonstrated below.

The specific intensity function can thus be thought of as the energy received
from the angular position (θ_t,ϕ_t) by the antenna per second, per square meter of
antenna aperture, per unit bandwidth at frequency ν, and per unit solid angle at θ_t,ϕ_t.
That is, it represents the energy per second that would be received each second by
a hypothetical antenna with area 1 m^2 facing each portion of the source, with a
bandwidth of 1 Hz, and with a beam of solid angle 1 sr, *if* the source fills the entire
beam with constant specific intensity. In fact, the specific intensity is, in almost all
instances, not constant over such large angles (57° × 57°), and most telescopes
have much smaller beams than 1 sr. Thus one must divide the measured power by
the actual solid angle of the beam just as one must divide by the actual area and
bandwidth of the telescope to estimate I.

One can develop this concept differently. Take an ideal telescope or antenna of
area A and point it to some celestial position, θ_t and ϕ_t (Fig. 3) where there is
a uniform background radiation. Take the solid angle of the beam $\Delta\Omega$ and the
antenna passband $\Delta\nu$ to be small with the former centered on the direction (θ_t, ϕ_t)
and the latter centered at frequency ν. The power, $\Delta\mathscr{P}$ (in watts) received by the
antenna must be proportional to the magnitude of each of the quantities A, $\Delta\nu$,
and $\Delta\Omega$. (We assume here the efficiency ϵ does not vary with frequency or with
angular position over the small beam and passband.) That is, if the bandwidth $\Delta\nu$
of the antenna is doubled, the power received is doubled; similarly doubling the

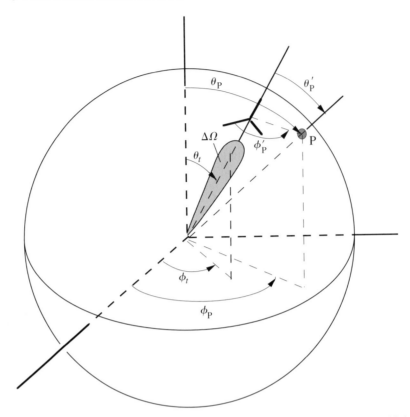

Figure 8.3. Antenna beam observing a diffuse source in the sky of which one element P is shown. The celestial coordinates of the antenna pointing direction are θ_t, ϕ_t, the coordinates of P are θ_P, ϕ_P. The coordinates of P relative to the telescope axis and to some azimuthal reference fixed to the telescope are θ'_P, ϕ'_P.

area A or the solid angle $\Delta\Omega$ doubles the power received. Thus $\Delta\mathscr{P} \propto A\Delta\nu\Delta\Omega$, or in differential form (e. g., $\mathrm{d}\Omega \ll \Delta\Omega_{\text{beam}}$),

$$\mathrm{d}\mathscr{P} \propto A \, \mathrm{d}\nu \, \mathrm{d}\Omega \qquad\qquad\qquad (\text{W}) \qquad (8.27)$$

where the element of solid angle may also be written as (3.14),

$$\mathrm{d}\Omega = \sin\theta_t \, \mathrm{d}\theta \, \mathrm{d}\phi \qquad\qquad\qquad\qquad (8.28)$$

The missing proportionality constant in (27) is simply the specific intensity $I(\theta_t, \phi_t, \nu)$ already defined,

$$\Rightarrow \qquad \mathrm{d}\mathscr{P} = I(\theta_t, \phi_t, \nu) \, A \, \mathrm{d}\nu \, \mathrm{d}\Omega \qquad (\text{W; Power; defines } I) \qquad (8.29)$$

In fact, this expression serves to define the specific intensity. The units of I (W m^{-2} Hz^{-1} sr^{-1}), as previously specified, balance the equation.

The specific intensity is properly a vector, $I\,(\theta,\phi,\nu)$ because the flow of energy at any one position in space has a well defined direction, as do the spectral flux density S and the flux density $\mathscr{F}$. We will typically use its magnitude, $I \equiv |I|$, which is a scalar.

Power received by antenna

If the specific intensity of the sky is *not* uniform over the received band of frequencies and beam angles, one must sum the contributions by each element of the sky to obtain the received power. In other words, one must integrate the specific intensity $I(\theta,\phi,\nu)$ over the observed angles and frequencies, where θ,ϕ are coordinates on the celestial sphere.

In addition, one takes into account the variation with angle of the efficiency, or response function, of the antenna, $\epsilon(\theta,\phi,\nu)$. This is again a dimensionless function with value ranging from 0.0 to 1.0. It will usually be at maximum on the antenna axis, and will be near zero outside the beam. It includes the effect of reduced off-axis projected area and the frequency-dependent diffraction limit on the beam size.

An antenna pointed at a fixed position on the sky will thus receive the following power,

$$\mathscr{P} = \int_\nu \int_\theta \int_\phi I(\theta,\phi,\nu)\, A(\nu)\, \epsilon(\theta,\phi,\nu) \sin\theta\; d\theta\; d\phi\; d\nu \qquad (8.30)$$

$$(\text{W}) \qquad \left(\frac{\text{W}}{\text{m}^2\,\text{Hz}\,\text{sr}}\right)(\text{m}^2) \quad (-) \qquad (\text{sr}) \quad (\text{Hz})$$

where the integration may now be taken over the entire sky because the efficiency function insures that the contribution from each part of the sky is properly weighted. The solid angle is written in terms of θ and ϕ, from (28), and the units of the several terms are shown.

One must be careful with the meaning of area A of the telescope in (30). The reflecting surfaces will not be perfectly efficient, transmission lines will exhibit *ohmic losses*, and detector efficiencies will not be perfect. Thus the total power detected, even on the beam axis, is less than that impinging on the geometric collecting area of the telescope. One often uses a hypothetical lesser *effective area*, $A = A_{\text{eff}} < A_{\text{geom}}$, which collects all the energy impinging on it along the telescope axis. In this case, the term $A(\nu)$ carries these efficiencies and $\epsilon = 1$ for radiation impinging along the telescope axis. If one uses $A = A_{\text{geom}}$, then ϵ would carry the on axis efficiency; in this case, we would have $\epsilon < 1.0$ for radiation impinging on axis.

The antenna efficiency $\epsilon(\theta,\phi)$ is expressed here in terms of the angles of a sky coordinate system θ,ϕ. Alternatively, it could be defined as a function of the angles in a coordinate system centered on the antenna itself (θ',ϕ') as illustrated in

Fig. 3 for the source element P. The efficiency ϵ of the telescope will generally be more easily expressed in terms of the these latter coordinates because it is usually symmetric about the forward direction at $\theta' = 0$. For the purpose of the integration (30), one must settle on one coordinate system for all the variables. Since $I(\theta,\phi,\nu)$ is expressed in terms of the sky-based system, the sky system is used in (30). It is possible to convert from $\epsilon(\theta',\phi')$ to $\epsilon(\theta,\phi)$ with appropriate transformation expressions.

We have shown in (30) how the power $\mathscr{P}$ observed with a real antenna is derived from the intrinsic true specific intensity $I(\theta,\phi,\nu)$ of the sky. We next discuss the converse operation wherein a real and imperfect antenna detects a power $\mathscr{P}$, but it is the specific intensity $I(\theta,\phi,\nu)$ we wish to know.

Average specific intensity

The specific intensity can be measured precisely only with an *ideal* narrow-beam and narrow-bandwidth antenna. The ideal measure of $I(\theta,\phi,\nu)$ with perfect angular and frequency resolution can not be attained from measurements with a real antenna. Our knowledge of the sky can be no better than the resolution of our telescopes.

A real measurement yields the *average* specific intensity I_{av} over the real antenna angles and frequencies. From (30) and the definition of an average,

$$\Rightarrow \quad I_{av} = \frac{\int_\nu \int_\theta \int_\phi I(\theta,\phi,\nu)\, A\, \epsilon(\theta,\phi,\nu)\, \sin\theta\, d\theta\, d\phi\, d\nu}{\int_\nu \int_\theta \int_\phi A\, \epsilon(\theta,\phi,\nu)\, d\Omega\, d\nu} \quad \begin{array}{c}\text{(Av. specific}\\ \text{intensity)}\end{array} \quad (8.31)$$

$$= \frac{\mathscr{P}}{\int_\nu \int_\theta \int_\phi A\, \epsilon(\theta,\phi,\nu)\, d\Omega\, d\nu}$$

where again the integration is over the entire sky, but since ϵ is small outside the antenna beam, the integration is effectively over the antenna beam. Also, we invoked $d\Omega = \sin\theta\, d\theta\, d\phi$.

The denominator of (31) represents the energy-gathering power of the antenna without reference to the specific intensity $I(\theta,\phi,\nu)$ of the sky. It is a quantity that can be calculated precisely for a given antenna and receiver system. An approximate value of I_{av} may be obtained without integration from an approximation of this quantity. Adopt an average efficiency ϵ_{av}, within a beam of effective solid angle $\Delta\Omega_{eff}$ steradians over an effective frequency bandwidth $\Delta\nu_{eff}$. Then (31) becomes

$$I_{av} \approx \frac{\mathscr{P}}{A\, \epsilon_{av}\, \Delta\Omega_{eff}\, \Delta\nu_{eff}} \quad (8.32)$$

One can estimate the average specific intensity I_{av} simply by dividing the observed power by the product, $A\, \epsilon_{av}\, \Delta\Omega_{eff}\, \Delta\nu_{eff}$.

Antennas with high angular resolution enable us to learn about the smallest features of the specific intensity of the sky. Unfortunately, all positions on the sky can not be observed with high resolution because it could take too long. The number of photons from a small region of sky is small and it takes a long time to accumulate enough for a significant measurement. In addition, there are many more positions on the sky that must be sampled if the same overall solid angle is to be measured.

In contrast, antennas with broad beams will view large regions of the sky and hence collect large numbers of photons, but the angular resolution will be poor. Such measurements provide the broad features of the specific intensity function. Similar statements may be made with regard to frequency resolution. In practice both high and low resolution measurements have their place. These several types of measurements tell us what the sky really looks like as a function of angular position and frequency.

Spectral flux density revisited

Relation to specific intensity

The spectral flux density $S(\nu)$ used for the study of point sources can be expressed in terms of the specific intensity. Consider the flux from an extended celestial source impinging upon a flat surface of unit area (a very simple antenna) and calculate the amount of flux passing through the surface area (Fig. 4),

$$\rightarrow \quad S(\nu) = \int_\theta \int_\phi I(\theta, \phi, \nu) \, \cos\theta \, d\Omega \qquad \text{(Spectral flux density} \qquad (8.33) \\ \text{from specific intensity)}$$

$$\left(\frac{\text{W}}{\text{m}^2 \, \text{Hz}}\right) \quad \left(\frac{\text{W}}{\text{m}^2 \, \text{Hz sr}}\right) \qquad \text{(sr)}$$

where the integration is over the angles subtended by the source and $\theta = 0$ specifies the direction normal to the unit area. The cosine factor represents the reduced effective area presented to the incoming flux at angles off the normal; it is a simple version of $\epsilon(\theta, \nu)$. The spectral flux density $S(\nu)$ represents all the power reaching the unit area from the entire range of angles of the source.

If $S(\nu)$ is measured, the expression (33) yields an expression for $I(\nu)_{\text{av}}$. Let the source lie near $\theta = 0$ and let its extent be small so $\cos\theta \approx 1$ over the source. Again use the definition of an average to obtain,

$$I(\nu)_{\text{av}} = \frac{\int_\theta \int_\phi I(\theta, \phi, \nu) \, \cos\theta \, d\Omega}{\int_\theta \int_\phi \cos\theta \, d\Omega} \xrightarrow[\cos\theta \approx 1]{} \frac{S(\nu)}{\Delta\Omega} \qquad \begin{array}{c} (\text{W m}^{-2} \\ \text{Hz}^{-1} \text{ sr}^{-1}) \end{array} \qquad (8.34)$$

The average specific intensity is merely the spectral flux density divided by the solid angle of the source (or of the antenna if the source fills the beam). This also follows directly from the units (or meanings) of the two quantities.

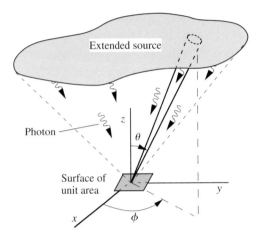

Figure 8.4. A flat surface of unit area receiving flux from an extended (diffuse) source. The spectral flux density $S(\nu)$ (W m^{-2} Hz^{-1}) incident on the surface is obtained from an integration of the specific intensity $I(\phi,\theta,\nu)$ over the angles of the source with the projected area taken into account by the factor cos θ.

Specific intensity of pulsars

If a source is point-like at celestial position θ,ϕ we mean only that its angular size is smaller than the resolution of the instrument imaging it. In this case one measures the flux density S. What can one say about the specific intensity? If the resolution (beam size) is $\Delta\Omega_{\text{beam}}$, one only knows that the solid angle subtended by the source is less than this. This leads to a lower limit to the specific intensity,

$$I(\nu)_{\text{av}} > \frac{S(\nu)}{\Delta\Omega_{\text{beam}}} \qquad\qquad \text{(Point source;} \qquad (8.35)$$
$$\text{W m}^{-2}\text{ Hz}^{-1}\text{ sr}^{-1})$$

It is sometimes possible to limit the size of the solid angle of the source from other evidence. Upon the discovery of isolated neutron stars as radio pulsars, the rapid time variation of the intensity allowed astronomers to deduce upper limits to the size of the radiating regions. For example, the peak width of the repetitive pulse emitted by the pulsar in the Crab nebula is $\Delta t \approx 2$ ms. All parts of a flaring region must physically communicate with each other in this time, so that each part knows that the other parts are flaring. (Otherwise how would one part know when to flare?) Since the communication can occur no faster than the speed of light, this limits the overall size of the flaring region to $D = c\Delta t$ or 600 km in this case. The flaring region must be smaller than this.

At the distance of the Crab, $r = 6000$ LY $\approx 6 \times 10^{19}$ m, this 600-km size yields an angular size for the emission region of $D/r = 10^{-14}$ rad and a solid angle $\Delta\Omega \approx$

$(D/r)^2 = 10^{-28}$ sr. With a measured peak flux density $S = 10^{-23}$ W m^{-2} Hz^{-1} at frequency 430 MHz, one finds that,

$$I(\nu)_{\text{av}} > \frac{S(\nu)}{\Delta\Omega_{\text{timing}}} = \frac{10^{-23}}{10^{-28}} = 10^5 \text{ W m}^{-2} \text{ Hz}^{-1} \text{ sr}^{-1} \qquad (8.36)$$

This specific intensity is so high as to be physically implausible for emission from a body or plasma in thermal equilibrium. It can be shown, see (11.24), that this value of I implies a temperature of $\sim 10^{27}$ K at 430 MHz. At much lower temperatures, $\sim 10^{12}$ K, the particles in a hot gas will have sufficient energy to create mesons, protons and neutrons in profusion. Thus, energy input into a region of limited size will tend to go into particle production rather than into more kinetic energy (i.e., temperature) of the existing particles.

Even for non-equilibrium particle distributions such as those found in synchrotron radiation (Chapter 11), such high intensities are implausible because the high energy electrons collide with, and lose their energies to the very photons they are emitting; this is called *inverse Compton scattering*. This too occurs at an apparent temperature of $\sim 10^{12}$ K. This is called the *Compton limit*. More on these topics will be in *Astrophysics Processes*; see Preface.

The high specific intensity of radiation from radio pulsars was one piece of evidence that other processes were at work. Coherent radiation by groups of relativistic electrons streaming along magnetic field lines in the direction of the observer can result in high observed specific intensities. Similarly, jets of material emerging from a source at highly relativistic speeds, $v \approx c$ relative to, and in the direction of, the observer, can appear to be extremely bright because of relativistic effects.

Surface brightness

We define here the surface brightness and show that it is numerically equal to the specific intensity measured with a distant antenna.

Power emitted from a surface

Heretofore, the specific intensity has been used as a standard way to describe the power received by a telescope. However, the units of specific intensity, W m^{-2} Hz^{-1} sr^{-1}, are also appropriate for the description of the power *emitted from* a surface. Consider a mathematical surface (Fig. 5) of area dA in the upper atmosphere of a star, with its normal in the θ,ϕ direction of the observer. The horizontal "surface" of the star, with normal at $\theta = 0$, is distinct from this surface. The power radiated from unit area of this surface per unit solid angle in the (θ,ϕ) direction in unit bandwidth ($\Delta\nu = 1$ Hz) at frequency ν is traditionally called *surface brightness*,

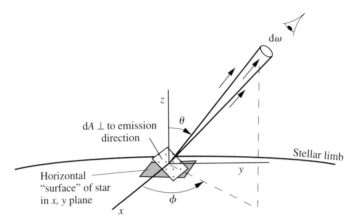

Figure 8.5. Radiation leaving the surface of a celestial object in the direction θ,ϕ toward an observer into solid angle $d\omega$. The surface brightness $B(\theta,\phi,\nu)$ in units of W m^{-2} Hz^{-1} sr^{-1} describes the power emitted per unit solid angle in the direction θ,ϕ and per unit frequency interval at frequency ν from a unit area that is perpendicular to the view direction θ,ϕ.

$$B(\theta,\,\phi,\,\nu) \equiv \text{Surface brightness} \qquad (\text{W m}^{-2}\text{ Hz}^{-1}\text{ sr}^{-1}) \qquad (8.37)$$

(Do not confuse with B magnitudes or with baseline B.)

The surface brightness describes the power coming from the apparent surface of objects such as the sun or the Crab nebula as a function of angle and frequency. The power emitted $\mathscr{P}_{em}$ from a physical surface dA into frequency interval $d\nu$ at frequency ν and into solid angle $d\omega = \sin\theta\ d\theta\ d\phi$ is

$$\blacktriangleright \qquad d\mathscr{P}_{em} = B(\theta,\phi,\nu)\ dA\ d\nu\ d\omega \qquad (\text{W; emitted power}) \qquad (8.38)$$

This serves to define $B(\theta,\phi,\nu)$. Note that the units balance.

Equality of emitted and received intensity ($B = I$)

The units of surface brightness $B(37)$ and of the specific intensity $I(26)$ are identical (W m^{-2} Hz^{-1} sr^{-1}). Here we demonstrate that, in a given observation of a diffuse (resolved) source, their magnitudes are equal, $B = I$. This is an amazing result that tells us that the measurement from earth of a specific intensity gives us directly the actual numerical value of the surface brightness at the surface of the object being observed no matter its distance. This statement is true if one neglects absorption of all kinds and redshifts due to Hubble expansion of the universe.

Consider first the geometry illustrated in Fig. 6a. A very large cloud of atoms, e.g., a distant nebula, emits with a uniform surface brightness B. In Fig. 6b, an antenna at distance r looks out at this sky with a beam that is significantly smaller in solid angle than that of the cloud. The portion of the cloud that is viewed by the

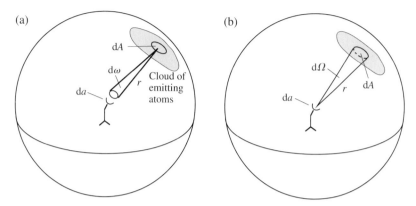

Figure 8.6. Geometry for proof that surface brightness equals the specific intensity, $B = I$. (a) Radiation from an atom in the cloud is directed toward the telescope of area da which subtends a small solid angle $d\omega$. (b) The relatively large beam of the telescope, of solid angle $d\Omega$, views a (large) segment dA of the cloud.

beam of the antenna has area dA. This area subtends a solid angle $d\Omega = dA/r^2$ as viewed from the antenna. The area of the antenna has the smaller value da and faces the source region. Viewed from the cloud (Fig. 6a), the antenna subtends a small solid angle $d\omega = da/r^2$. Each atom of the cloud is presumed to radiate isotropically (equally in all directions).

The power *received* by the antenna is, in terms of the detected specific intensity I and from (29),

$$d\mathscr{P}_{\text{rec}} = I(\theta,\phi,\nu)\, da\, d\nu\, d\Omega = I\, da\, d\nu\, \frac{dA}{r^2} \qquad \text{(W)} \qquad (8.39)$$

where care is required in choosing the correct elements of area and solid angle from Fig. 6. This again is simply the definition of specific intensity.

Each atom in the cloud can radiate toward the antenna, into a cone of solid angle $d\omega$. The power *emitted* toward the antenna by *all* the atoms that are in the antenna beam, within the area dA, is, from (38),

$$d\mathscr{P}_{\text{em}} = B(\theta,\phi,\nu)\, dA\, d\nu\, d\omega = B\, dA\, d\nu\, \frac{da}{r^2} \qquad (8.40)$$

The emitted power $d\mathscr{P}_{\text{em}}$ is defined here to be exactly that which is emitted toward the antenna, all of it and no more. It must therefore be equal to the received power $d\mathscr{P}_{\text{rec}}$ given in (39),

$$d\mathscr{P}_{\text{em}} = d\mathscr{P}_{\text{rec}} \qquad \text{(Equal powers)} \qquad (8.41)$$

Substitute (39) and (40) into (41); the result is $B = I$, as anticipated,

➡ $\qquad I(\theta,\phi,\nu) = B(\theta,\phi,\nu) \qquad \text{(W m}^{-2}\text{ Hz}^{-1}\text{ sr}^{-1}\text{)} \qquad (8.42)$

The import of (42) is that the surface brightness of a diffuse celestial object such as a supernova remnant can be measured directly, without additional calibration factors, knowledge of the distance, etc. The specific intensity at the antenna gives directly the brightness of the celestial surface! Therein lies the importance of the concept of specific intensity; it is the same thing as surface brightness!

Consider an optical CCD image of an optical nebulosity, such as a supernova remnant, the Cygnus Loop. The energy collected by a 1-m^2 telescope and deposited on one pixel corresponding to one square arcsecond is a direct measure of the energy emitted into a $1'' \times 1''$ solid angle by a 1-m^2 segment of the nebulosity which is located in the region that is focused on the pixel. It is not a matter of being proportional; it is an *equality* that yields the numerical value of the surface brightness (in W m^{-2} Hz^{-1} sr^{-1}) of that part of the supernova remnant. The physical conditions of the plasma emitting the radiation can be deduced with the aid of this fundamental quantity.

The ability to measure surface brightness independent of its distance means that a particular nebula will yield the *same* measured specific intensity regardless of its distance from the antenna, as long as its angular size is larger than the antenna beam. The detected power per unit solid angle does *not* depend on the distance of the emitting nebula.

This is not as unreasonable as one might first think. Consider two shells of identical emitting material, one twice the distance from the antenna as the other (Fig. 7), $r_2 = 2r_1$. Each atom of the farther shell will only provide 1/4 the radiation to the antenna compared to an atom in the closer shell, because the flux varies as r^{-2}. However, the diverging antenna beam intersects four times more area of the farther shell than it does of the closer because the intercepted area varies as r^2. Hence, four times as many emitting atoms in the farther shell are viewed by the antenna. Thus the power received by the antenna is the same in the two cases.

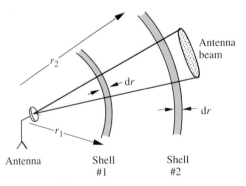

Figure 8.7. Antenna viewing shells of emitting gas at two distances, one twice the other. The specific intensity *I* detected from shell #1 equals that from shell #2.

This gives us another way to justify the relation $I = B$. Consider the specific intensity of a body like the sun, and bring it closer and closer to your detector until the sun's "surface" is right against your detector. We have shown that the measured specific intensity I will not change during this process. When the sun's surface is finally up against the detector, the detector is in fact measuring directly the emitted brightness B. The geometry of Fig. 5 applies; your detector is the surface in the figure with radiation entering its backside. Thus one again finds that $I = B$, as long as absorption and redshifts are not present.

Liouville's theorem

The origin of this equality lies in a fundamental theorem of physics, known as Liouville's theorem. It makes use of the concept of *phase space*, a six-dimensional space with three spatial dimensions, x, y, z and three momentum dimensions, p_x, p_y, p_z. The density in phase space is the number of particles or photons per unit (vol mom)^3, with SI units, m^{-3} (kg m s^{-1})$^{-3}$ = (m^2 kg s^{-1})$^{-3}$.

The theorem states that the density of particles in phase space remains constant even as those particles travel through space under the influence of no forces or of a restricted class of "smooth" forces, of which magnetic fields are an example. Thus cosmic rays (charged particles) traversing the Galaxy, spiraling along the interstellar magnetic field lines, will have the same phase-space density at the earth as they do at their origin, to the extent particles are not removed by collisions en route. (We examine this topic further in *Astrophysics Processes; see Preface*.) The specific intensity (W m^{-2} Hz^{-1} sr^{-1}) is uniquely related to the phase-space density, so it too is conserved as the photons travel through space.

Interesting implications of this equality $B = I$ surround us, for example starting a fire with a magnifying glass by focusing the sunlight. Another ramification is that radiation passing through an optical system has the same brightness before and after its passage. Thus the moon has the same brightness when viewed with binoculars as when it is viewed with naked eyes. The image is bigger but the energy received per pixel of the retina, or per square arcmin, is the same. This is true of the sun also. Thus the filter material safe for naked eye viewing of the sun is also safe for viewing the sun with binoculars as long as it is fully covers both objective lenses.

Energy flow – names, symbols, and units

The names and symbols for the specific intensity and related quantities are hardly standardized in practical use. This is less than helpful to the suffering student (and teacher). However, we try to be consistent within this text. A summary of the names and units of the several energy-related quantities we use are given in Table A4 of the Appendix. Each is directly related to the specific intensity I through an appropriate integration. In fact, the entries immediately following the first line are sequential integrals of I. It is clear from the table that other quantities could be defined, e.g., W Hz^{-1}, by appropriate integration of I.

One often sees the specific intensity (W m^{-2} Hz^{-1}sr^{-1}) designated with a subscript ν as I_ν. The subscript ν is used to designate clearly that the quantity is *per unit frequency interval* (Hz^{-1}), our usual definition of I. Similarly, the subscripted variable I_λ (W m^{-2} m^{-1} sr^{-1} = W m^{-3} sr^{-1}) is used to designate the specific intensity *per unit wavelength interval* (m^{-1}). There is a one-to-one conversion from I_λ to I_ν,

$$I_\nu = \frac{\lambda^2}{c} I_\lambda \tag{8.43}$$

which follows from the condition $I_\nu\, d\nu = -I_\lambda\, d\lambda$; see (11.12).

In this context, the symbol I without subscript is likely to refer to the specific intensity integrated over the entire frequency (or wavelength) band of the measurement. If so, it would have units (W m^{-2} sr^{-1}) and could be called the *integrated intensity*, $I \equiv \int I_\nu\, d\nu$. This quantity is not included in Table A4.

In this text, the subscripts are dropped, and the following correspondences are adopted:

$$I \equiv I_\nu \equiv I(\theta,\phi,\nu) \qquad \text{(Specific intensity)} \tag{8.44}$$

Similarly, the spectral flux density, $S(\nu)$ can have subscripts ν or λ. Again our practice is $S \equiv S_\nu$.

The key to clarity in this sea of definitions is the dimension of the units applied to a given number or symbol. In general, it is dangerous to trust the English language in this matter. On the other hand, *the units tell all*. One must take care to specify the units of symbols used as well as the units of numerical values.

Volume emissivity

A source that appears extended on the sky in the angular directions θ,ϕ is undoubtedly also extended along the line of sight, toward and away from the observer. Such diffuse objects are three-dimensional clouds of material, often gases with some degree of ionization. It is convenient and physical to consider the amount of power emitted by unit volume of an emitting cloud of atoms or ions, namely the *volume emissivity*, j (W m^{-3} Hz^{-1}),

$$j(\boldsymbol{r},\nu) \equiv \text{Volume emissivity} \qquad \text{(W m}^{-3}\text{ Hz}^{-1}) \tag{8.45}$$

where the vector **r** specifies the location of the differential volume element. This is the power emitted from unit volume of the plasma into *all* directions, per unit frequency interval. This is sometimes designated j_ν.

The volume emissivity derives from the actual physical processes that give rise to the emission of photons. If the cloud is transparent (*optically thin*) to its own radiation, one can easily relate j to the surface brightness B of the cloud. We now know that B is equal to the specific intensity I observed by an earth-bound antenna. Thus the observed quantity I can be related to a physical quantity $j(\nu)$, as we now demonstrate.

Relation to specific intensity

Let the emission from a volume element in the cloud be isotropic; the radiation leaves the source in all directions equally. Since all directions constitute 4π sr, the power emitted into one steradian is

$$\text{Power into 1 sr} = \frac{j}{4\pi} \qquad (\text{W m}^{-3}\,\text{Hz}^{-1}\,\text{sr}^{-1}) \qquad (8.46)$$

Further assume that this cloud is optically thin; it does not absorb radiation emitted toward the observer by more distant parts of the cloud. It is also possible to consider *optically thick* clouds that severely absorb their own radiation, or clouds that partially absorb it. (See discussion of radiative transfer in Section 11.5.) The optically thin case is presented here; the surface brightness B of such a cloud may be calculated from the volume emissivity.

Consider a cloud of thickness Λ (m) viewed by an antenna with a beam of solid angle $d\Omega$ (Fig. 8). A volume element of the cloud of area dA (facing the antenna) and thickness dr along the line of sight is shown as an inset to the figure. The emitted power per *unit* volume, and per (Hz sr) is given by (46). Multiply this by the volume of our element $dV = dA\,dr$ to obtain $(j\,dA\,dr)/(4\pi)$ $(\text{W Hz}^{-1}\,\text{sr}^{-1})$, the power emitted from the volume element per (Hz sr). Next divide by the area of the element dA to obtain the emitted power *per unit area* of the cloud as well as per (Hz sr). This, by definition, is the incremental emitted surface brightness dB from one slice dr of the cloud,

$$\blacktriangleright \qquad dB = \frac{j}{4\pi}\frac{dA\,dr}{dA} = \frac{j\,dr}{4\pi} \qquad (\text{W m}^{-2}\,\text{Hz}^{-1}\,\text{sr}^{-1}) \qquad (8.47)$$

This relates dB and the volume emissivity j.

Now, from (42), the detected specific intensity I is equal to the surface brightness B. Thus, for the first shell of thickness dr,

$$dI = dB = \frac{j\,dr}{4\pi} \qquad (8.48)$$

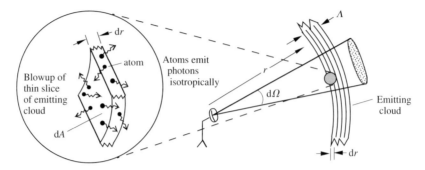

Figure 8.8. Extended transparent cloud containing atoms that are emitting photons in all directions. The cloud is divided into thin spherical shells of thickness dr along the line of sight. Emission from a volume element $dA\,dr$ (inset) leads to a relation between the observed specific intensity I (W m^{-2} Hz^{-1} sr^{-1}) and the volume emissivity j (W m^{-3} Hz^{-1}) of the cloud. The latter quantity is power emitted by unit volume of the gas into unit frequency interval.

Note that in (48), the contribution to the specific intensity dI measured by the antenna is independent of the distance r of the shell. This is in accord with our previous discussion about the distance independence of specific intensity.

The radiation from all the shells may now be summed to obtain the specific intensity (brightness) of the entire cloud, of thickness Λ (Fig. 8). If the volume emissivity j is constant throughout the cloud, each additional thickness dr makes an additional *equal* contribution to the measured specific intensity I because distance to the antenna is not a factor and because the cloud does not absorb its own radiation Thus, one can sum over the slices of cloud,

$$I = \frac{j}{4\pi}\sum_i dr_i = \frac{j}{4\pi}\Lambda \qquad\qquad (j = \text{constant}) \qquad\qquad (8.49)$$

where Λ is the thickness of the cloud.

If j is not constant throughout the cloud but varies with distance r, integration is required,

$$I = \frac{1}{4\pi}\int_0^\Lambda j(r)\,dr \qquad\qquad \begin{array}{c}\text{(W m}^{-2}\text{ Hz}^{-1}\text{ sr}^{-1};\\ j \neq \text{constant})\end{array} \qquad (8.50)$$

In this expression, if j is independent of r,

$$I \xrightarrow[j=\text{const.}]{} \frac{j}{4\pi}\Lambda \qquad\qquad (j = \text{constant}) \qquad\qquad (8.51)$$

which is again (49).

The expression (50) tells us that the detected specific intensity I is a measure of the volume emissivity j summed along the line of sight through the entire thickness of the cloud. It also can be written in terms of the average volume emissivity j_{av} measured through the thickness of the cloud,

$$j_{av} = \frac{\int_0^\Lambda j \, dr}{\int_0^\Lambda dr} = \frac{1}{\Lambda} \int_0^\Lambda j \, dr \qquad (8.52)$$

Use this to eliminate the integral in (50),

➡
$$I = \frac{1}{4\pi} \quad j_{av} \quad \Lambda \qquad \text{(Specific intensity;} \quad (8.53)$$
$$\text{optically thin plasma)}$$

$$\left(\frac{W}{m^2 \, Hz \, sr}\right) \left(\frac{1}{sr}\right) \left(\frac{W}{m^3 \, Hz}\right) (m)$$

This illustrates that the detected specific intensity is simply a measure of the product of the average volume emissivity j_{av} along the line of sight times the thickness Λ of the cloud, if the cloud is transparent (optically thin) to the radiation of interest.

Line-of-sight emissivity (power column density)

According to (53), the specific intensity is proportional to the product $j_{av}\Lambda$ (W m^{-2} Hz^{-1}). This product is, by inspection, the total power generated (per Hz) in a column 1 m^2 in cross section and Λ (m) in length (Fig. 9). In integral form, from (52), it is

$$\int_0^\Lambda j(r) \, dr = j_{av}\Lambda \qquad \text{(Line-of-sight emissivity;} \quad (8.54)$$
$$\text{W m}^{-2} \text{ Hz}^{-1})$$

The integral $\int j \, dr$ is the *line-of-sight emissivity* or the *power column density*. Integrals of this type, namely a density integrated along the line of sight, are quite common in astrophysics and are generally called *column densities*.

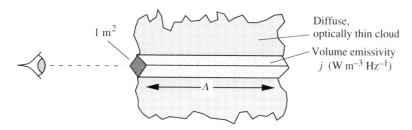

Figure 8.9. A 1-m^2 column through an emitting cloud along the line of sight. The total radiation from a column of cross section 1 m^2 and length Λ is simply the product of the average power emitted per unit volume and the length, $j_{av}\Lambda$ (W m^{-2} Hz^{-1}). The detected specific intensity I turns out to be equal to this product divided by 4π.

The atoms that actually contribute to the radiation detected by an observer do not come from a simple column of the cloud as sketched in Fig. 9, but rather from a much larger area and from a cone of angles as shown in Fig. 8. However, the simple view in Fig. 9 is an easy way to visualize the meaning of the expression (50) or (53) for I. The additional factor $1/(4\pi)$ tells us that only radiation emitted into one steradian from each element of the column contributes, in effect, to the specific intensity. Many measurable quantities in astrophysics are similarly proportional to a source function (like j) integrated along the line of sight.

Problems

8.2 Unresolved point-like sources

Problem 8.21. (a) What is the (partial) luminosity within a narrow 10^4 Hz bandwidth at frequency ν of a radio source radiating isotropically if it is detected at distance 3000 LY with a spectral flux density $S = 1.0$ Jy in this band? Repeat for the broad band 50 to 200 MHz if $S(\nu)$ is constant at 1.0 Jy over this frequency range? (b) Another source at the same distance has a spectral flux density S (W m^{-2} Hz^{-1}) that varies as $S = k\nu^{-2}$ where k is a constant. The (integrated) flux density is measured to be $\mathscr{F} = 3 \times 10^{-20}$ W/m^2 (50–200 MHz). Find: (*i*) the luminosity L in this band (*ii*) the average spectral flux density S_{av} over this band, and (*iii*) the coefficient k. Use the latter to evaluate S at the midpoint (linear scale) of the band and compare to (*ii*). [Ans. $\sim 10^{18}$W, $\sim 10^{22}$ W; $\sim 10^{21}$W, ~ 20 mJy, $\sim 10^{-12}$, ~ 15 mJy]

8.3 Astronomical magnitudes

Problem 8.31. (a) What is the difference of magnitudes for two stars if one is a million times "brighter" than the other? (b) The original Palomar Observatory Sky Survey photographic plates reach to a magnitude of about $m = 21$. Compared to the brightest stars, $m \approx 0$, how "faint" are the 21st magnitude stars, i.e., what is the ratio of fluxes? Repeat for the Hubble Space Telescope (HST) which can reach to $m_V \approx 28$. (c) The bright star Sirius (type A1 V) has $m_V = -1.44$ and is at distance 8.8 LY. Would Sirius be visible to the HST if it were in our sister galaxy Andromeda at distance 2.5 MLY? (d) Would a star identical to the sun (absolute magnitude $M_V = 4.82$) in Andromeda be visible to the HST? [Ans. ~ 15; $\sim 10^{-8}$, $\sim 10^{-11}$;—;—]

Problem 8.32. (a) Use the data in Table 2 to deduce approximately how many photons in the V-band frequency range are incident upon 1 m^2 of the earth each second from a star that is measured to have $m_V = 0$. Remember that the energy of a single photon is $h\nu$. (b) At what magnitude is the photon flux only 1 photon per square meter per second? How does this compare to the faintest stars measured today? [Ans. $\sim 10^{10}$ m^{-2} s^{-1}; ~ 25]

Problem 8.33. In a given observation in the *V* band, Saturn yields a photon flux of 0.86×10^{-11} that of the sun in the *V* band. The *V* magnitude of Saturn is known to be $V_{\text{Sat}} = 1.0$ at the time of the observation. (a) Calculate the apparent *V* magnitude of the sun. (b) How far would the sun have to be removed from the earth to appear as bright as Saturn, i.e., to appear as a first-magnitude star? [Ans. ~ -27; ~ 5 LY]

Problem 8.34. (a) To what extent does the relation between M_{bol} and *L* given in (24) follow from (12)? Hint: consider spectral distribution and isotropy as well as energy vs. photon fluxes. (b) Verify that the relation (25) follows from (24).

8.4 Resolved "diffuse" sources

Problem 8.41. When the moon is full, its magnitude (summed over its entire surface) is $V = -12.7$. The diameter of the moon is about $30'$. (a) The resolution of the human eye is $\sim 1'$. What is the magnitude of the moon per square arcminute? (b) What is the magnitude per square arcsecond; the resolution of a ground-based optical telescope? (c) What is the surface brightness of the moon? [Ans. ~ -6; $\sim +3$, $\sim 10^{-13}$ W m^{-2} Hz^{-1}sr^{-1}]

Problem 8.42. Here we examine the power delivered to the retina of the eye by the sun. (a) The angular resolution of the eye is $\sim 1'$. If one glances momentarily at the sun (not advisable), what is the approximate *V* band power (W) deposited onto one resolution element (area equivalent to one square arcmin) of the retina? The specific intensity from the sun at the *V* band effective frequency is $\sim 2.6 \times 10^{-8}$ W m^{-2} Hz^{-1} sr^{-1} and the *V* passband is 90 THz (Table 2). The radius of the eye's aperture (pupil) is about 1.5 mm. Neglect any absorption by the fluid in the eye. (b) How much power (W) is deposited onto the retina by the entire solar image? The mean angular radius of the sun is $16.0'$. (c) What is the size of the solar image on the retina and what is the energy flux (W/m^2) arriving at the image? The focal length of the eye is ~ 25 mm. (d) Recalculate the flux at the retina directly from the energy flux arriving at the top of the earth's atmosphere from the sun, 1367 W/m^2 which is the "solar constant". Compare to your answer in (c); can you reconcile them? Also compare to the energy per unit area radiated from the 0.01-m^2 surface of a 1-kW electric stove element (burner). [Ans. $\sim 10^{-6}$ W; $\sim 10^{-3}$ W; $\sim 10^4$ W/m^2; $\sim 10^5$ W/m^2]

Problem 8.43. A parabolic (dish) radio antenna of 25-m diameter operating at frequency $\nu = 150$ MHz with a bandwidth $\Delta\nu = 15$ MHz detects radio emission. Consider the efficiency to be unity over the diffraction-limited beam. (a) What are the approximate angular radius and solid angle of the beam? Assume that it is limited by diffraction (see Chapter 5). (b) The power received and recorded at the antenna is 7×10^{-13} W. What is the average detected specific intensity? Under what condition will this equal the actual specific intensity? (Your numerical answer will approximate the true celestial radio background due to synchrotron radiation.)

(c) Suppose you knew from other evidence that *all* the measured radiation came from a small circular nebulosity of angular radius $1'$. What is the specific intensity detected from the source region in this case? (d) Convert the answer, from (c) to wavelength units I_λ. Be sure to state the units. [Ans. $\sim 10^{-2}$ sr; $\sim 10^{-21}$; $\sim 10^{-15}$; $\sim 10^{-8}$, the latter three in SI units]

Problem 8.44.   The Crab nebula is detected as a point source by a broad-beam antenna. The measured spectral flux density at 1 GHz is $S = 7 \times 10^{-24}$ W m^{-2} Hz^{-1} (i.e., 700 Jy). The Crab nebula has an angular size of about $3' \times 5'$. (a) What is the specific intensity I (W m^{-2} Hz^{-1} sr^{-1}) of the detected radiation, averaged over the surface of the nebula? (b) What is the surface brightness of the nebula at 1 GHz, averaged over the surface? (c) What are the linear dimensions (in meters and light-years) that correspond to the angular sizes $3'$ and $5'$ if the Crab is at a distance of 6000 LY? (d) If we assume that the thickness (or depth) of the nebula along the line of sight is midway between the two dimensions obtained in (b), what is the average volume emissivity j_{av} (W m^{-3} Hz^{-1}) of the nebulosity? (Assume that j is a constant along the line of sight.) (e) What is the approximate radio luminosity of the Crab nebula over a band of frequencies of 0.5–5.5 GHz? Adopt the flux density at 1 GHz for the entire band. Compare your answer to the luminosity of the sun while remembering that we included here only a small portion of the entire electromagnetic spectrum radiated by the nebula; see Fig. 11.2. [Ans. $\sim 10^{-17}$; —; ~ 5 and ~ 9 LY; $\sim 10^{-33}$; $\sim 10^{27}$; unspecified units are SI]

9

Properties and distances of celestial objects

What we learn in this chapter

The **information content** in the radiation recorded in observations allows astronomers to derive the **properties** of celestial objects. The **ranges of the values** of these properties are found to be "astronomically" large. **Luminosities** are derived from measured fluxes and distances. The solar luminosity, 3.8×10^{26} W, is a benchmark reference; that of a bright quasar is 10^{13} times larger. The **mass** of the **moon**, **earth**, or of a **galaxy** can be determined by tracking the motion of one or more orbiting objects. The sun's mass, 1.99×10^{30} kg, is also a standard reference; the (Milky Way) Galaxy is $>10^{11}$ times more massive. The **virial theorem** is used to obtain the masses of **clusters of galaxies**. **Temperatures** can be defined for **thermal** sources, wherein the matter and radiation are in, or approximately in, **thermal equilibrium**. The temperatures of a hot gas may be determined in a variety of ways that may yield different values. Thus astronomers refer to **kinetic**, **color**, **effective**, **excitation**, and **ionization temperatures**. The last is obtained from spectral observations with the aid of the **Saha equation**.

The **distance** to a celestial object is not an intrinsic property but it is required to find intrinsic quantities. Ancient astronomers used geometry to learn the **earth size** and **distance to the moon**. The mean earth–sun distance is defined as the **astronomical unit** (AU). 1.00 AU $= 1.496 \times 10^{11}$ m. **Distances to stars** in the solar neighborhood out to about 3000 LY are obtained with **trigonometric parallax**, the apparent motion of stars due to the earth's orbital motion about the sun. **Secular** and **statistical parallax** extend this to longer baselines and greater distances. **Standard candles**, objects of known luminosity, permit distance determinations to greater and greater distances as one moves out the **distance ladder**. Distances to **open clusters** are obtained with the **convergence method**. **Spectroscopic classification** of stars makes them standard candles as does temporal variability of **Cepheid variables** and **RR Lyrae** stars. **Spectral line broadening** in galaxies calibrates their luminosities so they become standard candles also. Other distance indicators are **luminosity functions**, **surface brightness fluctuations**, and **supernovae** of Type 1a. The goal of many distance determinations is to refine the **Hubble law**. One can use this law to find the approximate distance of a distant object from the **redshift** of its spectrum.

9.1 Introduction

The distance to a celestial object is often crucial to determining its size or luminosity. Unlike the angular position of a star or galaxy on the celestial sphere, distances often can be obtained only indirectly. A significant fraction of astronomical history has been dedicated to the determination of distances. Even now, new techniques to determine the distances of the most distant galaxies are an important aspect of cosmological studies.

The values and ranges of the distances, sizes, masses and luminosities of celestial objects tend to be on the large side; this is astrophysics, not atomic physics. It is important to have a feeling for these quantities, but this can be difficult in view of the large values. It is similar to the difficulty in gaining an intuitive feeling for a national debt. It is helpful to concentrate on the nearest power of 10 (the *order of magnitude*) rather than the exact number. For example, you might memorize that there are 10^1 persons in the class, 10^6 in greater Boston, and 10^9 on the planet earth. Then if someone tells you there are $\sim 10^4$ students at MIT, you can put that number roughly in context. Don't visualize people, visualize exponents; think logarithmically!

Nevertheless, the above should not lull the reader into being casual about learning more precise values of some quantities. Astrophysicists use again and again certain values such as the mass, radius, and distance to the sun, and relate other quantities to them. It behooves the reader to learn these to about two-digit accuracy.

In this chapter we discuss the role of luminosities and fluxes, masses, temperatures and finally sizes and distances in astronomy. We present the ranges of values these objects have and outline some of the methods by which these quantities are determined. The largest portion of the chapter deals with the determination of distances.

9.2 Luminosities

The luminosity of an object as discussed in Section 8.2 is the energy it emits per unit time, e.g., joules per second (J s^{-1}) or the equivalent unit, watt (W). It can refer to the energy emitted over the entire electromagnetic spectrum, but it can mean the energy emitted in a more restricted frequency band, e.g., the visible band. Luminosities in optical astronomy usually mean the total energy radiated in the visible and

Table 9.1. *Mass, luminosity, and energy examples*

Source	Mass (kg)	Luminosity (J/s = W)	Total energy radiated^a (J)
Energy used by all humans^b	—	$\sim 10^{12}$	$\sim 10^{22}$
Moon	7.35×10^{22}	—	—
Earth^c	5.98×10^{24}	2×10^{17}	—
Sun^d	1.99×10^{30}	3.845×10^{26}	$\sim 10^{44}$
Rest mass energy of sun ($M_\odot c^2$)	—	—	2×10^{47}
Crab pulsar	$\sim 3 \times 10^{30}$	10^{31}	$\sim 10^{44}$
Galaxy^e	$\sim 7 \times 10^{41}$ ($\sim 3.5 \times 10^{11}\ M_\odot$)	6×10^{36}	$\sim 10^{54}$
Bright quasar	$\sim 10^{38}$	$\sim 10^{39}$	$\sim 10^{54}$

^{a} The total energy takes into account roughly the estimated lifetime of the object and, in the case of the pulsar, the energy loss in all wavebands and the decreasing luminosity during the lifetime.

^{b} For the human race, we estimate very roughly, that the average per-person use of fuels (mostly fossil), including industrial use, is about 1000 W (1 stove burner on continuously), that there are about 10^9 persons, and that this high rate has been going on for about 10^{10} s (300 years). One human, as a *source* of heat, radiates ~ 100 W.

^{c} The luminosity of the earth derives predominantly from absorbed solar energy.

^{d} The mass, luminosity, and radius of the sun are generally represented, respectively, by $M_\odot$, $L_\odot$, and $R_\odot$.

^{e} The (Milky Way) Galaxy mass is an estimate of the luminous and non-luminous mass matter within radius $r = 50$ kLY.

nearby IR and UV regions. The luminosities of some important objects are given in Table 1.

The common astrophysical reference unit is the sun which has a luminosity of

➡ $$1.00\ L_\odot = 3.845 \times 10^{26}\ \text{W} \approx 4 \times 10^{26}\ \text{W} \quad \text{(Solar luminosity)} \quad (9.1)$$

Averaged over its entire mass, the sun emits only 1/5000 W/kg. Since the average density of the sun is 1400 kg/m^3, close to that of water, the average luminosity per unit volume is 0.3 W/m^3, a smaller value than you might have guessed; the light bulb on your desk does a lot better than that!

The luminosity of the Galaxy is about 1.4×10^{10} solar luminosities ($L_\odot$), or about 6×10^{36} W. The most luminous galaxies, *giant ellipticals*, can reach luminosities a factor of 10 greater than the Galaxy. Quasars are extremely luminous nuclei of distant galaxies. Their luminosities can approach 10^{39} W, as does the energy output of the quasar 3C 273. All of these luminosities are many orders of magnitude greater than the total rate of energy used by humans (Table 1).

The estimated total expenditures of energy over the lifetimes of the objects are also given in Table 1. For comparison, the total energy that could arise from the complete *annihilation* of the sun is also given. This is the complete conversion of its mass into energy according to the Einstein relation for the equivalence of mass and energy, $E = M_\odot c^2$. This is called the *rest energy* and has the value:

$$E = M_\odot c^2 = 2 \times 10^{30} \times (3 \times 10^8)^2 \qquad \text{(Rest energy of sun)} \qquad (9.2)$$

$$= 2 \times 10^{47} \text{ J}$$

The total energy expended by the sun and by the pulsar (a $\sim$1 solar-mass neutron star) approach, within a few factors of 10, the total energy that could conceivably be extracted from a 1-$M_\odot$ object. Note also that the large amount of energy used by humans during their short history as high consumers is tiny compared to the output of cosmic energy sources.

The luminosity of a body is typically inferred from the energy flux density $\mathscr{F}$ (W/m^2) detected at the earth. (Colloquially, observed flux densities are referred to as the "brightness" or "intensity" of a star; these descriptive terms should not be confused with "specific intensity" and "surface brightness" defined precisely in Section 8.4 for use in this text.) The relation between luminosity and the flux measured at distance r from the source was given in (8.7) and Fig. 8.1 as,

$$L = \mathscr{F} \, 4\pi r^2 \qquad \text{(W; luminosity)} \qquad (9.3)$$

if the radiation is emitted isotropically. A measurement of $\mathscr{F}$ yields L only if the distance to the star is known.

9.3 Masses

The masses of some important astronomical objects are given in Table 1. The mass of the sun is a unit of mass frequently used in astrophysics; it is called a *solar mass* ($M_\odot$),

$$1.00 \, M_\odot = 1.989 \times 10^{30} \text{ kg} \approx 2 \times 10^{30} \text{ kg} \qquad \text{(Solar mass)} \qquad (9.4)$$

Earth and sun

The masses of the sun and earth may be determined from their mutual motions about their center of mass (*barycenter*). In general for elliptical orbits, the analysis is relatively involved. But if the orbits are circular and if one of the two objects is much more massive than the other, $M \gg m$, it is quite straightforward to find the mass of the heavier object. In this limiting case, the mass M will be essentially at rest while mass m orbits it.

Write the radial component of Newton's second law,

$$F_r = m \, a_r \tag{9.5}$$

Substitute for F_r the radial component of the gravitational force on mass m, and substitute for a_r the inward acceleration, $-\omega^2 r$, of a body (mass m) in circular motion at radius r with angular velocity ω (rad/s),

$$-\frac{GmM}{r^2} = -m\omega^2 r \tag{9.6}$$

Solve for M,

$$M = \frac{\omega^2 r^3}{G} = \frac{4\pi^2 r^3}{P^2 G} \rightarrow \frac{4\pi^2 a^3}{P^2 G} \qquad \text{(kg; mass of central object)} \tag{9.7}$$

The third term is obtained from the relation between the angular velocity ω and the period P of the orbit, $\omega = 2\pi/P$. If the orbit is elliptical, the semimajor axis a takes the place of r (not proven here).

This relation (7) yields, for a circular orbit and $M \gg m$, the mass of the more massive object M in terms of the radius of the orbit and the angular velocity of the less massive object, or in terms of the radius and orbital period. In this limiting case ($M \gg m$), the result does not depend on the mass m! Consider the earth–sun system. These assumptions are quite valid because the sun is about 300 000 times more massive than the earth, and the earth's elliptical orbit has a distance of closest approach to the sun (*perihelion*) only 3% less than the distance at the farthest point (*aphelion*). Thus (7) yields the solar mass if its distance is known (see below).

Similarly, the mass of the earth can be obtained from the moon's motion. In this case, our assumptions are somewhat less valid but still adequate. The ratio of the masses is 81, and the closest and farthest approaches of the moon, *perigee* and *apogee* respectively, differ by 12% (eccentricity 0.0549).

Moon

The mass of the moon is obtained from measurements of the monthly oscillation of the earth as it orbits the barycenter of the earth–moon system every lunar month. The oscillation is revealed by observations of the motions of the planets against the background stars, especially Mars. (This is due to the phenomenon of parallax.) The oscillation indicates that the barycenter is 4700 km from the earth center; compare to the 6400 km radius of the earth. The known distance to the moon leads, by subtraction, to the barycenter–moon distance. This turns out to be 81.3 times greater than the 4700-km earth–barycenter distance. Accordingly, from the definition of the barycenter, the moon is known to be 81.3 times less massive than the earth.

Spiral galaxies and the Galaxy

The mass of the (MW) Galaxy and the masses of other spiral galaxies may be estimated in a manner similar to that used for the earth and sun. Stars in a spiral galaxy rotate in a more or less organized manner about the center at some radius with a corresponding period of rotation. Individual stars can not be resolved except for very bright stars in close galaxies, and the rotational motion of the Galaxy is much too small to be observed directly.

Nevertheless, the orbital speed v of the aggregate of stars in some small region of the galaxy can be obtained from Doppler shifts of the spectral lines from the constituent stars if the inclination (tilt) of the galaxy relative to the line of sight is taken into account. The radius r of the orbit is obtained from the angular displacement of the region from the galaxy center, if the distance from the earth to the galaxy is independently known. The speed and the radius yield the orbital period P. In the crude approximation that most of the mass is at the center of the galaxy, the expression (7) yields the mass.

In fact, a substantial portion of the mass of a galaxy is distributed throughout the galaxy (e.g., the stars themselves). Thus the rotation of the stars observed gives the mass of the galaxy *within* the orbit of the test star observed. The formula (7) is technically valid only if the galaxy is spherical and if the force is strictly an r^{-2} force. These conditions are not met in the typical case because of the disk-like shape of many galaxies and because of the continuous distribution of matter.

Nevertheless, given all the above, one can obtain an estimate of the mass within the orbital radius with (7) if one assumes a perfectly circular orbit that lies in the plane of the Galaxy. Take the speed of our star, the sun, around the center of the Galaxy, $\mathcal{V}_\odot = 220$ km/s, as determined by spectroscopic Doppler-shift observations of distant galaxies. The distance of the sun from the galactic center is $\sim 25\,000$ LY $= 2.4 \times 10^{20}$ m. The result, from (7), for the galactic mass is $\sim 1 \times 10^{11} M_\odot$ for the matter in the disk within the solar radius. Stellar and gas motions at greater radii indicate that all the mass out to about 50 kLY radius is $\sim 3.5 \times 10^{11}\ M_\odot$. Out to 100 kLY, it approaches $10^{12} M_\odot$. Keep in mind that these mass values were obtained solely from the motions of stars in their orbits.

Comparison of these masses to the luminosity of the Galaxy, $\sim 1.4 \times 10^{10} L_\odot$, tells us that much of the matter in the Galaxy has less luminosity per kg than the sun. Indeed, the many stars that are less massive than the sun are less luminous per kilogram than the sun.

The amount of emitted light in our and other galaxies is far too little to explain all the gravitation indicated by stellar and gas motions. There are just not enough stars, even including dead remnants of normal stars (white dwarfs, neutron stars, or black holes), to explain the orbital speeds. The unknown non-emitting material giving rise to this excess gravity is called *dark matter*, as noted previously in Section 4.2.

The masses of galaxies without organized orbital motion can not be obtained in this manner because individual stars can not be resolved and tracked. In these *elliptical galaxies*, the stars orbit the barycenter more or less in random planes and directions. The virial method described below can be used to estimate roughly their masses; the widths of spectral lines (*velocity dispersion*) arising from an aggregate of stars in the galaxy provide information about the galaxy mass.

The masses of the most massive galaxies can range up to $\gtrsim 10^{13}$ solar masses; the smallest "galaxies" have $\sim 10^8 M_\odot$. *Dwarf galaxies* can range down to less than $10^6 M_\odot$. Smaller groupings of stars are called "clusters" (of stars); they are usually associated with galaxies. They are not to be confused with "clusters of galaxies" discussed below.

Clusters of galaxies and the virial theorem

The only objects with substantially greater masses than galaxies, other than the universe itself, are collections of galaxies, known as *clusters of galaxies*. The nearby (~ 55 MLY) Virgo cluster contains about 2500 galaxies. The galaxies move about under the influence of all the other galaxies in the cluster; think of a swarm of bees. The galaxy motions can be used to estimate the mass of the cluster.

For this purpose, astronomers make use of the *virial theorem* which applies to a *gravitationally bound* system of discrete objects ("particles") in *stable equilibrium*. In our case, galaxies are the particles. The theorem states that twice the total kinetic energy of all the particles plus the sum of the potential energies equals zero,

$$2K + V = 0 \tag{9.8}$$

or

$$2 \sum_i \frac{1}{2} m_i v_i^2 - \sum_{\text{pairs}} \frac{G\, m_i m_j}{r_{ij}} = 0 \qquad \text{(Virial theorem)} \tag{9.9}$$

where the summations are over the individual particles and r_{ij} is the separation distance between m_i and m_j. Note that $V = 0$ when the particles are widely dispersed, $r_{ij} \rightarrow \infty$, the usual convention.

The virial theorem $2K + V = 0$ can only be satisfied if the potential energy is negative and twice the value of K, since K is always positive. Particles in free fall from a great distance toward the barycenter would have $K + V = 0$ from energy conservation and would *not* satisfy the virial theorem. In contrast, particles swarming around the barycenter in a stable (unchanging configuration) do obey the virial theorem. In this case, the total energy $E = K + V$ must be less than zero, an indication that such a system is gravitationally bound.

The orbit of a satellite around the earth, or of a planet around the sun, is a simple example of the virial theorem. From Newton's second law, one readily determines that the kinetic energy of the satellite is one-half the negative of the potential energy.

Measurements of the particle separations r_{ij} and speeds v_i, in particular the averages of $v_i{}^2$ and $r_{ij}{}^{-1}$, allow one to find a mass for the entire cluster. The mass determined is that required to provide the gravity necessary to keep particles of speed v_i in a cluster with galaxy separations r_{ij}. At a given typical speed, a more compact cluster (small r_{ij}) indicates more gravity is present. The mass so determined is called the *virial mass*.

Typically, one finds that the virial mass of a cluster is far greater than the visible mass. In other words, it is greater than the total mass of the visible galaxies, even including estimates of their dark matter. The virial mass in clusters is typically 10–50 times greater than the visible mass. Again there is an apparent prevalence of dark matter, but this time on an even larger scale.

9.4 Temperatures

Here we focus briefly on the role of temperature in astronomy. Related material is introduced in Sections 2.3 (temperature), 11.3 (blackbody and thermal bremsstrahlung radiation), and 11.5 (radiative transfer).

Thermal and non-thermal radiation

Temperature is one of the most basic quantities in physics and astrophysics. At its simplest, according to the approximation $h\nu \approx kT$ (2.15), the frequencies of detected photons may be an indicator of the temperature of the originating bodies. For example, photons from the sun at optical frequencies tell us that the surface layers of the sun have temperature of $\sim$6000 K (kelvin).

The infrared emission from a human being indicates a temperature of about 300 K. A pervasive radio radiation from the sky at millimeter wavelengths indicates a temperature of 3 K for the cooled radiation initially emitted in the early years of the universe. X rays of energy 1–10 keV indicate temperatures of $\sim 10^7$ K for the emitting plasmas in the vicinity of very compact gravitational objects, namely neutron stars and black holes. This high temperature represents the high kinetic energies demanded by the virial theorem (8) because the potential energies are large and negative.

The approximate relation between photon frequency and temperature quoted just above is generally valid only for "thermal" bodies where there is an equilibrium between photons and particles. The equilibrium is "perfect" for a *blackbody* in which case the energy spectrum of the radiation will have the characteristic blackbody shape we will encounter in Section 11.3.

It is possible to have *non-thermal processes*, such as *synchrotron emission* (Section 11.3), where the originating particles are not in thermal equilibrium with their surroundings. In this case, the spectrum will have a different shape, a temperature can not be defined, and the emitting particle energy can be very different than $h\nu$.

Temperature measurements

There are several ways to measure temperatures in optical astronomy. The different methods can lead to somewhat different values if the source region is not in perfect thermal equilibrium. They are *color temperature*, *effective temperature*, *excitation temperature*, and *ionization temperature*. Underlying these is the basic laboratory *kinetic temperature*.

Kinetic temperature

In thermal equilibrium, a gas will have a distribution of speeds that is called the *Maxwell–Boltzmann distribution*. The average translational kinetic energy of atoms or molecules in a gas is related to the temperature as

$$\left\langle \frac{1}{2}mv^2 \right\rangle_{av} = \frac{3}{2}kT \qquad \text{(Kinetic temperature)} \qquad (9.10)$$

On earth, we sometimes measure the temperature with a mercury thermometer. Higher molecular speeds impart energy to the mercury causing it to expand and indicate a higher temperature. Unfortunately, it is not possible to directly do this with a distant star.

Color temperature

The color temperature T_c is derived from broadband *UBV* photometry, the measurement of fluxes in the *U, B,* and *V* bands, under the assumption that the observed object has a blackbody spectrum. The relative flux densities in these bands are a measure of the color of the object (Section 8.3). The nature of the blackbody curve is such that, for stellar temperatures, the *B* and *V* filters provide a fair measure of temperature.

The ratio of fluxes in the *B* and *V* bands, $\mathscr{F}_{p,B}/\mathscr{F}_{p,V}$, translates, as noted in Section 8.3, into a difference in the (logarithmic) magnitudes, $B-V$, called the color index,

$$m_B - m_V \equiv B - V \equiv -2.5\log_{10}\left(\frac{\mathscr{F}_{p,B}}{\mathscr{F}_{p,V}}\right) \qquad \text{(Color index)} \qquad (9.11)$$

This follows from the definition of magnitudes (8.12) where $\mathscr{F}_{p,B}$ and $\mathscr{F}_{p,V}$ are the measured photon fluxes (photons m^{-2} s^{-1}) in the two bands. The *difference* in magnitudes is thus a measure of the color. The two colors $B-V$ and $U-B$ are the often quoted results of *UBV* photometry. If a color $B-V$ is measured, it can be described as a color temperature, T_c in kelvin, namely the temperature of an ideal blackbody that would yield the measured value of $B-V$.

Effective temperature

The distribution of light from normal stars roughly approximates the continuum spectrum from a blackbody wherein each square meter emits a flux of σT^4 (W/m^2) and σ is the Stefan–Boltzmann constant, $\sigma = 5.67 \times 10^{-8}$ Wm^{-2} K^{-4}. The luminosity of a spherical star of radius R may therefore be expressed as

$$L = 4\pi R^2 \sigma T^4{}_{\text{eff}}$$ (W; defines effective temperature)   (9.12)

The *effective temperature* T_{eff} is used here because the spectra of stars deviate somewhat from the blackbody shape. The spectra of normal stars show absorption lines at certain frequencies which are not found in true blackbody radiation. This equation defines T_{eff} to be the hypothetical temperature that would yield the true luminosity if the spectra were exactly blackbody in form. The definition is thus

$$T_{\text{eff}} \equiv \left(\frac{L}{4\pi R^2 \sigma} \right)^{1/4}$$ (Effective temperature)   (9.13)

where $4\pi R^2$ is the surface area of the star.

In other words, T_{eff} is the temperature that a blackbody of the same luminosity and radius *would* have. The effective temperature of the sun is 5800 K. For more on blackbody radiation, see Section 11.3.

Excitation temperature

The *excitation temperature* is that deduced from the populations of atomic excited states observed in stellar spectra. The higher the energy of the occupied states, the higher the temperature. For example, the greater the number of hydrogen atoms in the $n = 2$ state relative to the number in the $n = 1$ state, the higher is the excitation temperature. The expression that governs this is the *Boltzmann formula*,

➡ $$\frac{n_j}{n_i} = \frac{g_j}{g_i} \exp\left(-\frac{h\nu_{i,j}}{kT} \right)$$ (Boltzmann formula;   (9.14)
excitation temperature)

which gives the ratio of numbers (or densities) of atoms in the excitation states i (the lower) and j (the higher) as a function of the temperature T, the energy difference between the two states $E_j - E_i \equiv h\nu_{ij}$, and the statistical weights g_i and g_j of the two states. The statistical weight of a state is the degeneracy of the state, i.e., the number of substates with differing quantum numbers that have the same energy. Thus, the $n = 1$ level has $g_1 = 2$ (two s states) and the $n = 2$ level has $g_2 = 8$ (two s states and six p states). The ratio n_j/n_i can sometimes be inferred from the relative strengths of various spectral lines in the spectrum, in which case, the excitation temperature can be obtained from (14).

The exponential factor in (14) tells us that a high-energy state has a lower probability of being occupied than a lower-energy state with the same statistical weight and that the population of the higher state decreases with the increase in the energy difference between the two states. The ratio g_j/g_i tells us that the probability of a state being occupied is proportional to its statistical weight. For example, if the higher state has more substates than the lower, the probability of finding atoms in that state relative to the lower is higher proportionally by the factor g_j/g_i.

The Boltzmann formula follows from considerations of detailed balance when the atoms are immersed in a bath of photons in true thermodynamic equilibrium, i.e., with the blackbody spectrum for the indicated temperature. Basically, one equates the probabilities of upward and downward transitions in photon–atom interactions. Transitions due to particle–particle collisions also yield this relation as long as the system is in true thermodynamic equilibrium.

A famous example is the excitation of cyanogen molecules (CN) in interstellar space by the 2.7-K cosmic microwave background (CMB; see Section 11.3). This was the earliest (1941) detection of this radiation from the early universe. The significance of the measurement was not appreciated until the radiation was detected directly as microwave waves more than two decades later.

The Boltzmann formula is not perfectly valid if, for example, there are particles of a different temperature present. This must be taken into account in the calculation of the temperature of the CMB from the CN measurements. An extreme violation of the Boltzmann formula is the relative population of atoms in a laser or maser. In this case the atoms are externally "pumped" to a high, relatively stable ("metastable") state with applied radiation. This results in an excess population in the higher state which by definition is a violation of the Boltzmann formula. The laser emission is then due to stimulated emission from the higher state.

Ionization temperature

The *ionization temperature* is also determined through spectral studies. It depends upon the extent to which different ionization states (as distinct from the excitation states just discussed) appear in the spectrum, e.g., the ratio of He I to He II, neutral to ionized helium. The relative numbers of ions in the several ionization states at a given temperature and density are obtained from an analysis of equilibrium interactions similar to that for the Boltzmann formula. The equation that follows from this analysis is the *Saha equation*. It can be used to define a temperature if the relative densities of the several ions and of free electrons can be independently known, for example, from the strengths of spectral lines.

The excitation and ionization temperatures represent the physical conditions of the atoms in a region of interest. For the quasi-equilibrium conditions that are typical of local regions of stellar atmospheres, these two temperatures and the kinetic temperature will all be nearly equal.

For a typical star, the temperature varies with height in its interior and also in its atmosphere. For the sun, the temperature is 6400 K at the altitude where the escaping photons can, for the most part, leave the sun without further scatters; this is known as the *photosphere*. The temperature is decreasing in this region. This leads to absorption at certain frequencies by the outer cooler gases so the total radiation observed is less than that expected from a 6400 K blackbody. This is why the effective temperature is only 5800 K. As one moves outward, the temperature arrives at a minimum of ~4200 K. It then begins to increase, eventually reaching 10^6 K in the corona.

Saha equation

The Saha equation allows one to obtain the ionization temperature from measurements of the relative numbers of atoms of a given species that are in adjacent ionization states (Figure 1). The equation follows from consideration of *detailed balance* wherein the probabilities of absorption and emission of an electron by the

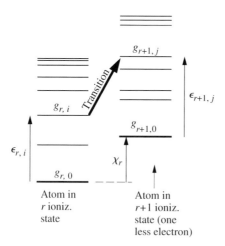

Figure 9.1. Hypothetical energy levels of two states of ionization of a given species of atom, e.g., Ca. The transition between the ionization states can occur between any excitation levels in one state to any excitation level in the other. The statistical weights of the individual states g and the *ionization potential* χ_r are indicated. The latter is the energy between the respective ground states. The Saha equation relates the densities of atoms in the two ionization states to the temperature T and free electron density. [Adapted from *Principles of Stellar Evolution and Nucleosynthesis*, D. Clayton, McGraw Hill, 1968, p. 33]

atom are equated. These transitions are associated with the emission and absorption of a photon respectively. The derivation assumes conditions of perfect thermal equilibrium.

For a given temperature T, the Saha equation yields the quotient $(n_{r+1}n_e)/n_r$, where n_r is the density of atoms (m^{-3}) in the lower (r) of two ionization states, n_{r+1} is the density of atoms in the next higher ($r+1$) ionization state (with one less orbital electron), and n_e is the density (m^{-3}) of free electrons; see Fig. 1. We quote the equation without derivation,

➡ $$\frac{n_{r+1}n_e}{n_r} = \frac{G_{r+1}g_e}{G_r} \frac{(2\pi m_e kT)^{3/2}}{h^3} \exp\left(-\frac{\chi_r}{kT}\right) \quad \text{(Saha equation)} \quad (9.15)$$

where the quantities are:

n_{r+1}	Density of atoms in *ionization* state $r+1$ (m^{-3})
n_r	Density of atoms in *ionization* state r (m^{-3})
n_e	Density of electrons (m^{-3})
G_{r+1}	Partition function of ionization state $r+1$
G_r	Partition function of ionization state r
$g_e = 2$	Statistical weight of the electron
m_e	Mass of the electron $= 0.911 \times 10^{-30}$ kg
χ_r	Ionization potential of state r (to reach state $r+1$)
h	Planck constant $= 6.63 \times 10^{-34}$ J s
k	Boltzmann constant $= 1.38 \times 10^{-23}$ J/K

The partition function $G_r(T)$ is a kind of statistical weight for ionization state r taking into account all the possible bound states it can have. Under many conditions, the upper states do not play a major role. For most astrophysical conditions, the partition functions for neutral and ionized states of hydrogen are closely approximated by $G_r(T) \approx g_{r,0} = 2$ and $G_{r+1}(T) \approx g_{r+1,0} = 1$, respectively. The ionization energy χ_r for the ground state of hydrogen is, of course, 13.6 eV.

The ratio on the left of (15) is large for a high proportion of ionized atoms in the plasma. This is favored for high values of the ratio of statistical weights (first term on right) and the number of phase-space states available to the electron (second term), and if the ionization energy χ_r is not large compared to kT.

9.5 Distances and sizes

Distance ladder

The relative sizes of some astrophysical objects and the distances to them are listed in Table 2 and illustrated on logarithmic scales in Fig. 2. The large ranges mentioned above are very evident here. For example, the ratio of the size of the observable

Table 9.2. *Size (radius) and distance examples*

Object	Radius of object (m)[a]	Distance to object (m)[a]
1-$M_\odot$ black hole	3×10^3 (3 km)	
Neutron star	1×10^4 (10 km)	
Moon	1.7×10^6 (1738 km)	3.84×10^8 (mean)[b]
Earth	6.4×10^6 (6371 km)	
Sun	6.96×10^8	$1.496 \times 10^{11} = 1.00$ AU[c]
Solar system	6×10^{12} (Pluto orbit)	
Nearest star (Prox. Cen)		4.1×10^{16} (4.3 LY)
Crab nebula	5×10^{16} (5 LY)	6×10^{19} (6 kLY)
Center of the Galaxy		2.4×10^{20} (25 kLY)[d]
Galaxy (Milky Way)	~5×10^{20} (50 kLY)	
Andromeda galaxy, M31	~7×10^{20} (70 kLY)	2.4×10^{22} (2.5 MLY)[e]
Virgo cluster of galaxies	5×10^{22} (5 MLY)	4.9×10^{23} (52 ± 5 MLY)[f]
3C273 quasar		2.3×10^{25} (2.4 GLY)[g]
Observable universe		~1.4×10^{26} (15 GLY)[h]

[a] Values are from C. W. Allen, *Astrophysics Quantities*, 3rd Ed., Athlone, 1976 and the 4th Ed., ed. A. N. Cox, Springer Verlag/AIP, 2000 or Zombeck, *Handbook of Space Astronomy & Astrophysics*, 2nd Ed., Cambridge University Press, 1990, unless otherwise referenced.

[b] The moon distance varies by 11% due to its elliptical orbit.

[c] Mean value = semimajor axis of Earth orbit; the earth–sun distance varies by 3% due to the earth's elliptical orbit.

[d] Nominal value used for this text; result from Fig. 5 data is 23 ± 5 kLY.

[e] Freedman and Madore, *ApJ* **365**, 186 (1990).

[f] Jacoby *et al.*, *Publ. Astron. Soc. Pacific* **104**, 599 (1992).

[g] Extragalactic distances are referenced to a Hubble constant of 20 km s^{-1} MLY^{-1} (65 km s^{-1} Mpc^{-1}) and $q_0 = 1$ with relation $D = cz/H_0$. A recent "best fit" value to available data is 71 ± 3 km s^{-1} Mpc^{-1}; Spergel *et al.*, *ApJ* (2003); astroph/0302209.

[h] Hubble distance $R = c/H_0$; approx. distance to the observable limit of the universe.

universe (10^{26} m) to that of a neutron star (10^4 m) is equal to the ratio of the sizes of the earth (10^7 m) and the proton (10^{-15} m), a factor of 10^{22} in each case. Distances can be given in meters, light years, or parsecs.

The sizes of the earth, the sun, the Galaxy, and the observable universe serve as excellent benchmarks as do the distances to the moon, the sun, the nearest star, and the center of the Galaxy. You would do well to memorize them. Try to picture the relation between the sizes and spacings. For example, if the sun were the size of a soccer ball, how big would the similarly scaled earth be, and how distant would it be from the soccer ball? I bet the answers surprise you!

The distance to a celestial object is difficult to obtain, in principle, because the sky appears two-dimensional to us. The angular position of a star on the celestial sphere requires only the two angular coordinates. Indirect means must be used to obtain the

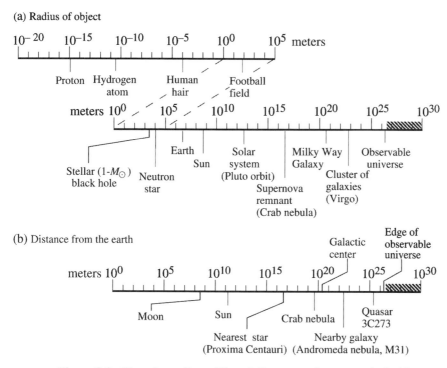

Figure 9.2. Sizes (actually radii) and distances of astronomical objects on loga-rithmic scales. The "parsec" is often used as a distance measure in astronomical literature; in this text we choose to use light years (LY). The logarithmic range of sizes is compared to human and subatomic sizes.

third coordinate, the distance. There is a so-called *distance ladder* whereby different techniques and objects are employed for different regimes of distance.

The ladder begins with the closer objects. Based on these distances, another technique or type of object (a *standard candle*) is used to obtain distances of more distant objects. For example, if the distance to a cluster of stars in the Galaxy is determined, the luminosity of any star in it is easily obtained from the measured flux. If one of the stars is highly luminous and recognizable due to its spectrum or oscillations, that type of star could be used as a standard candle to obtain the distances of galaxies in which the star is found. This is done by comparing the observed flux and luminosity of the standard candle; see (8.9). As one works out to greater and greater distances, the accuracy of the later steps depends on the accuracy of the previous steps. Uncertainties can thus be large.

Moon, earth, and sun

The distance to the moon from the earth center was determined with poor accuracy by the Greek astronomer Aristarchus (~270 BCE) from timing of lunar eclipses.

This method was improved upon by a later Greek astronomer, Hipparchus ($\sim$140 BCE). Ptolemy (second century CE) used a different method; he measured the parallax (Section 4.3) of the moon relative to the distant stars. He noted that the apparent position of the moon in the sky is different for two well separated observers on the earth. A single observer can perform the experiment by letting the earth carry her to a different position as it rotates, but she must subtract the effect of the moon's orbital motion during the interval. The closer the moon is to the earth, the more it will appear to be displaced from its expected track through the stars as the observer rotates with the earth.

Ptolemy obtained a lunar distance of 59 times the earth radius; the actual mean value is 60.3. This is quite good agreement considering that the moon's distance varies by $\sim$12% due to its elliptical orbit. The absolute distance of the moon follows if the radius of the earth is known.

Indeed, the radius of the earth had been found several centuries earlier by Eratosthenes. He assumed that the rays of the sun arrive as a plane wave (as if from a distant point source). Since the earth is indeed spherical, observers at two locations would (in most cases) observe the sun to be at different elevations (angle above the horizon) at the same time. If the observers are directly north–south of one another by a known distance D, the difference between the two measured elevations $\Delta\theta$ of the sun (at the same time) will yield the radius R of the earth according to $R = D/\Delta\theta$.

The distance to the sun is found most directly by the measurement of the speed of the earth in its orbit. This is found by spectroscopic observations of stars in the forward and backward directions. Stars exhibit spectral lines (Section 11.4) that can be measured to obtain the frequency v of the radiation. If, at one point in its orbit, the earth is moving directly away from a star with speed v_r, the spectral line from that star will be Doppler shifted to lower frequency, or toward the red. The Doppler frequency shift Δv is related to v_r for non-relativistic speeds as

$$\frac{v_r}{c} = -\frac{v - v_0}{v_0} = -\frac{\Delta v}{v} \qquad \text{(Doppler shift; } v_r \ll c\text{)} \qquad (9.16)$$

where v_r is the earth speed, c is the speed of light, v is the observed frequency, and v_0 is the rest frequency that would be observed by an observer at rest relative to the emitting or absorbing atom. A measurement of Δv and v thus yields v_r.

The studies of stars in many directions surrounding the earth show red and blue shifts that oscillate on an annual basis due to the earth's orbital motion. (There are also constant shifts due to the sun's motion relative to the other stars; see below.) From the oscillations of stars located in a number of different directions, one can deduce the orbital velocity v_E of the earth as well as the period P_E of the earth's orbit. Of course, the period is also known from observations of the yearly apparent motion of the sun through the constellations and from the annual north–south motion of the sun relative to the earth.

If the earth's orbit is perfectly circular (an approximation) at a distance r_{AU} from the sun, the distance around the orbit is

$$2\pi r_{AU} = v_E P_E \tag{9.17}$$

The two factors v_E and P_E together yield the radius of the earth's orbit about the sun. In fact, though, the elliptical orbit must be taken into account. Measurement of the changing speed v_E and changing angular size of the sun reveal the parameters of the orbit including its semimajor axis a. The semimajor axis of the orbit is known as the *astronomical unit* (AU). As quoted previously (4.7),

$$a = 1.496 \times 10^{11} \text{ m} \tag{9.18}$$

This is the desired distance to the sun.

Trigonometric parallax

The technique of trigonometric parallax yields the distances of the closer stars. As described in Section 4.3, nearby stars appear to move relative to the more distant stars as the earth moves about the sun (see Fig. 4.2). This motion appears as a circle, ellipse, or line depending on the star's location relative to the plane of the earth's orbit about the sun. As noted, the distance r to a star in the direction normal to the ecliptic may be expressed in terms of the semimajor axis of the nearly circular track on the sky, θ_{par}, and the semimajor axis of the earth's orbit. From (4.5),

$$r = \frac{a}{\tan \theta_{par}} \approx \frac{1.5 \times 10^{11} \text{ m}}{\theta_{par} \text{ (rad)}} \qquad \text{(m; parallax distance)} \tag{9.19}$$

where $\tan \theta_{par} \approx \theta_{par}$ for the small angles involved. The distances may be expressed in meters or AU.

The distance r may be expressed in *parsecs*. Rewrite (19) with $a = 1.0$ AU and the angle in arcseconds to obtain the definition

$$r(\text{pc}) \equiv \frac{1}{\theta_{par} \text{ (arcsec)}} \tag{9.20}$$

which is in accord with our discussion prior to (4.8) wherein the parsec was defined as the distance at which the parallax is $1.00''$. Traditionally, astronomers define "parallax" $\pi = \theta_{par}$ (arcsec) so that $\pi(\text{arcsec}) = [r(\text{pc})]^{-1}$. Thus if one has $\pi = 0.10''$, the distance is $r = 10$ pc.

Traditional trigonometric parallax with ground-based telescopes has good precision out to a distance of $\sim$300 LY. Here the angular displacements become so small

(0.01″ at 100 pc) that the errors become significant. The centroids of stellar images of size ~1″ must be measured to these precisions. Typically ~40 images taken over a period of ~5 years will be measured. Satellite observatories do much better. The Hipparcos satellite (launched 1989) was dedicated solely to astrometry and yielded ~10^5 stellar positions more precise than 1 milliarcsec (mas). The Hubble Space Telescope carries out limited astrometry to accuracies of ~0.5 mas. At 1 mas, one reaches to about 3000 LY.

Note that the stellar distances are obtained in units of the sun–earth distance (1 AU). Thus the determination of the magnitude of the astronomical unit (see above) could be called the first step on the distance ladder, and the distances to other stars with trigonometric parallax would be the second step.

Distances to open clusters

The distribution of stars on the sky shows a number of open clusters; these are groups of stars that presumably formed together from the same collapsing cloud at a given location within the Galaxy at some time in the past. A typical open cluster might contain ~100 cataloged stars and be ~20 LY in diameter. Such groups eventually disperse due to the random motions of the individual stars. The open clusters now in the sky are significantly younger than the Galaxy; they have not yet had time to disperse. It turns out that a number of these clusters are moving at a significant speed through the Galaxy, relative to the sun. It is this feature that makes it possible to obtain a distance to a cluster by purely geometric means called the *moving cluster method*, which we now describe.

Convergence

The motion of the cluster as a whole typically turns out to be much greater than the motions of individual stars within the cluster. Thus one can visualize a cluster of individual stars moving through space as a group, each with the velocity of the cluster. If the radial portion of the motion is such that the cluster is receding from us, the cluster will become smaller and smaller in angular size as time progresses. The azimuthal and radial motions of the cluster cause the tracks of the individual stars on the sky (as seen by the observer, Fig. 3a) to eventually converge to a single point, just as two rails of a railroad track converge toward a single point at great distances. This is strictly true only for our ideal picture of stars motionless with respect to the cluster.

The celestial locations of selected stars of the Hyades cluster at time t_1 are at the tails of arrows in Fig. 3a; star A is an example. The arrows indicate the *apparent* directions and angular speeds of motion $d\theta/dt$. This motion, the *proper motion* (Section 4.3), is observed from the comparison of sky images taken at

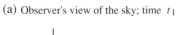
(a) Observer's view of the sky; time t_1

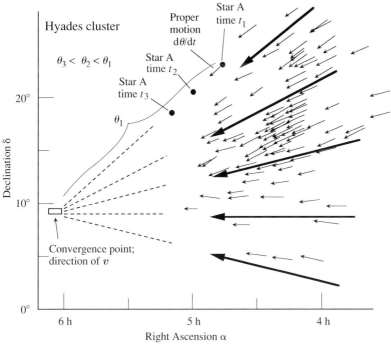

(b) Views normal to line of sight and to star vector velocity v

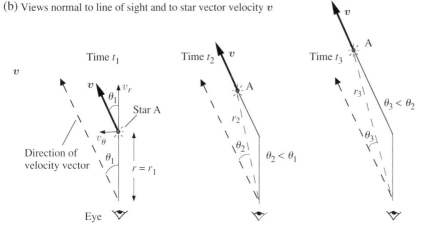

Figure 9.3. The moving cluster method of obtaining the distance to an open cluster. The motion of the stars relative to the center of mass of the cluster is assumed to be small. (a) The individual stars of the Hyades cluster plotted on the sky with the directions and relative magnitudes of their proper motion shown (arrows) as derived from data from 1908 to 1954. All of the stars appear to be headed toward the *convergence point* at $\alpha \approx 6$ h and $\delta \approx +9°$. The angle θ_1 for star A at time t_1 is shown. Anticipated positions of star A at times t_2 and t_3 are also shown. (b) Sketches in plane of view direction and velocity vector of star showing velocity of star A at three times. The angular size of the cluster and the angle θ both decrease as time progresses. [(a) After J. A. Pearce, *PASP* **67**, 23 (1955) and in O. Struve, B. Lynds, and H. Pillans, *Elementary Astronomy*, © Oxford University Press, 1959]

intervals of several years. The *convergence point* is immediately apparent; all the stars seem to be heading toward one position on the celestial sphere at $\alpha \approx 6\,\text{h}$ and $\delta \approx +9°$. The motion, $d\theta/dt$ (radians per second), is a very small quantity, of magnitude $0.075''/\text{yr}$ for a typical star in the Hyades cluster.

Recall that a vector direction in three-dimensional space may be represented as a point on the infinitely distant celestial sphere. Thus the convergence point represents a vector direction in space. This direction is the direction of the velocity vector v of our sample star or any other star in the cluster. (Remember that in our model all the cluster stars have the same velocity v.) The situation is illustrated with the three sketches of Fig. 3b wherein a star at a distance $r = r_1$ from us has velocity vector v parallel to the dashed arrow. As the star recedes, the perceived angle θ between the star and the dashed line gradually decreases: $\theta_1 > \theta_2 > \theta_3$. In the limit of very long times, the angle approaches zero. Thus the star's position converges to the point on the celestial sphere which represents the direction of the velocity vector, as stated.

The utility of the convergence-point determination is that the angle θ between the current star position and its velocity vector (convergence point) is a measurable quantity. This angle θ is shown in both figures (Figs. 3a,b) for star A at time t_1. It is a large angle, of order $20°$. In our limited lifetime, star A in the Hyades cluster moves hardly at all, only about $4''$ in 50 years, or $\sim 10^{-4}$ the distance to the convergence point.

Distance to a cluster star

The velocity vector v has two components, a radial component v_r and a tangential component in the plane of the sky, v_θ (Fig. 3b, time t_1). The radial velocity v_r is readily obtained from spectroscopic measures of line frequencies. The frequency shift due to the Doppler effect yields directly v_r according to (16).

The tangential component v_θ is related to the measurable proper motion $d\theta/dt$,

$$v_\theta = r\frac{d\theta}{dt} \qquad \text{(m/s; transverse velocity)} \qquad (9.21)$$

and to the observable θ (Fig. 3b),

$$\tan\theta = \frac{v_\theta}{v_r} \qquad (9.22)$$

These two equations include two unknowns, v_θ and r. Thus one can solve for the distance r,

$$\Rightarrow \quad r = \frac{v_r \tan\theta}{d\theta/dt} \qquad \text{(Distance of star;} \qquad (9.23)$$
$$\text{moving cluster method)}$$

Thus, with knowledge of the measured quantities, θ (rad), v_r (m s^{-1}) and $d\theta/dt$ (rad s^{-1}) the distance r (m) to the star is obtained.

Distance to the cluster

The distances of about 40 stars in the Hyades cluster may be obtained in this manner. These distances, together with the celestial (angular) positions, locate all the measured stars in three-dimensional space. Thus the distance to the cluster is obtained. The distance to the Hyades cluster is found from this method quite precisely, $r = 144 \pm 3$ LY. Such distances are the underpinning of other distance scales because both cepheid variables and main-sequence stars found in clusters serve as standard candles (see below). The cluster distance serves to calibrate their luminosities. Other clusters sufficiently nearby (so the required measurements can be made) include the Ursa Major group, the Scorpio–Centaurus group, the Praesepe cluster, and the double cluster h and χ Persei. The latter is at quite a large distance, 7600 LY.

Secular and statistical parallaxes

The use of parallax motions can be extended to larger distances by making use of the sun's motion relative to the barycenter of the stars in the nearby solar region of the Galaxy. These stars define the *local standard of rest* (LSR), the frame of reference in which their peculiar motions (vector velocities in three-dimensional space) average to zero. The solar motion in this frame of reference amounts to 4.1 AU/yr, as presented previously in Section 4.3.

Secular parallax

In traditional parallax determinations of distances, the baseline motion of the earth is limited to ± 1 AU. The annual oscillatory motion of the earth leads to an oscillatory motion of the apparent track of the star under study. It is thus easy to identify unambiguously the apparent stellar motion (parallax) that is due to the earth's orbital motion.

The sun carries the earth over much greater distances than 1 AU over the years, for example $20 \times 4.1 = 82$ AU in 20 years relative to the LSR. One would hope to use this larger motion to make parallax measurements of more distant stars than is possible with traditional parallax. Unfortunately, though, in this case, there is no telltale oscillatory motion of the star track that identifies the part of the apparent motion due to the sun's motion.

The sun moves steadily along a more or less straight line at constant speed since the time between stellar encounters is very long. Similarly, another star whose distance we hope to learn, will have its own peculiar motion relative to the LSR, and this is also approximately straight line motion at constant speed. The apparent motion of the star against the more distant "background" stars according to a hypothetical observer at the solar system barycenter is thus a straight line due to the motions of *both* the sun and the star. Since the motion of the star is generally not independently known, one can not trivially extract the portion of the apparent motion due solely to the sun's (actual) motion, which is needed to arrive at a parallax distance. However, it is possible to extract the apparent motion due to the sun's motion alone with the appropriate statistical average over a large number of stars.

The *average* peculiar motion of a large group of stars is zero relative to the LSR, by definition. Thus, if the apparent motions of many stars in our local region are measured relative to the distant background stars, their *average* apparent motion must be due solely to the sun's motion. The average apparent motion would be a drift in the backward direction, opposite to the sun's motion in the LSR. This is the parallax motion that can be used to obtain distances. This method is called *secular parallax*. In the following discussion, we illustrate how this is done while considering the observer to be at the barycenter of the solar system. In practice, one can subtract out the wobble in the observed tracks due to the earth's orbital motion.

Consider a large sample of stars that are known to be all of approximately the same luminosity because they all are of the same "spectral type". If, in addition, we select from these only those that exhibit the same apparent magnitude or flux, the sample will then consist of stars that are all at about the same distance from the sun. Further, among these "constant-distance" stars, consider, for simplicity, only those that lie in a direction roughly normal to the solar velocity vector; that is, $\theta_v \approx 90°$ (Fig. 4a). In the secular-parallax method, one measures the component of the proper motion that lies parallel to the sun's velocity vector, i.e., along the great circle passing through the star and the apex of the solar motion (Fig. 4a). This is called the *upsilon component* $(d\theta/dt)_v$ of the proper motion.

To obtain a long baseline of solar motion, one measures the changing positions of all the stars in the selected sample relative to the background stars on plates taken over a period of, say, $\sim$20 years. The *average* of all the upsilon motions is presumed to be due to the sun's motion relative to the LSR because the peculiar motions of the sample should average to zero relative to the LSR.

For normal parallax, we used the relation (19), $r \approx a/\theta_{par}$, where a is the semi-major axis of the earth's orbit. In our case, for plates taken 20 years apart, the sun would have moved the above quoted $a' = 82$ AU, and the stars in our sample at $\theta_v \approx 90°$ would have moved, on average, some angular distance θ_{par}' toward the antapex; see stars in Fig. 4b. These two values, a' and θ_{par}' can be used in place of a and θ_{par} to calculate the distance: $r = a'/\theta_{par}'$ of the stars in our constant-distance sample.

Equivalently one can use the motions that occur in a 1-s time interval, averaging over the 20 years. Replace a' with the distance of travel of the sun per second, $v_\odot$, and replace θ_{par}' with the *average* upsilon motion in unit time (radians per

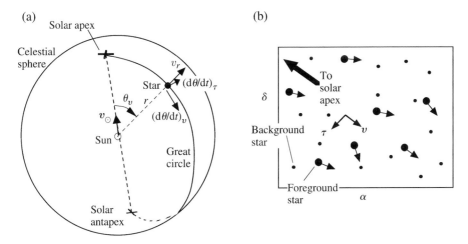

Figure 9.4. (a) The three measurable components of the motion of a star relative to the sun, shown for a star on the celestial sphere: the radial component of *linear* velocity v_r measurable through spectroscopy, and the two transverse components of *angular* velocity in the plane of the sky measurable through imaging. The upsilon component, $(d\theta/dt)_\upsilon$, is the motion of the star parallel to the great circle connecting the star and the apex of the solar motion relative to the local standard of rest. The tau component $(d\theta/dt)_\tau$ lies perpendicular to the great circle. (b) Observer's view of the sky. The closer stars (large dots) exhibit a general backward drift due to the sun's motion toward the apex as well as their own random motions. Caution: the upsilon and velocity symbols are similar.

second), $\langle(d\theta/dt)_\upsilon\rangle_{\text{av}}$,

$$\Rightarrow \qquad r = \frac{v_\odot}{\langle(d\theta/dt)_\upsilon\rangle_{\text{av}}} \qquad \text{(m; secular parallax; distance to the} \qquad (9.24)$$
$$\text{sample of stars normal to solar apex)}$$

This is the desired distance to our sample of constant-distance stars. The numerator is the speed of sun in the LSR, 19.5×10^3 m/s $= 4.1$ AU/year. This illustrates the principle of the method. In practice, one would make use of stars at other angles θ_υ and would take into account a modest spread of types or luminosities, i.e., distances, in the sample.

Statistical parallax

The method of *statistical parallaxes* makes use of the other component of the proper motion in the plane of the sky, the *tau component* $(d\theta/dt)_\tau$ (Fig. 4). This motion is perpendicular to the solar motion and is unaffected by it. Only the peculiar motion of the star affects the tau component, and this, by definition, should average to zero in the LSR and also in the frame of the sun, for a large sample of nearby stars. The tau component, like the upsilon component, is an *angular* velocity.

Again consider a sample of stars at a common distance and in a direction normal to the solar velocity. For stars in these particular (normal) directions, the radial velocity v_r (component along the line of sight) is also unaffected by the solar motion; hence it too is due only to peculiar motions. Furthermore, one expects the absolute value of the speeds in the radial and tau directions, averaged over the stars in the sample, to be equal because all directions of the motions are assumed to be equally probable.

The radial velocity v_r may be obtained from the spectroscopic Doppler shifts (16) of the spectral lines, and the average of the absolute values $|v_r|_{av}$ can be calculated. The tau components of the stars' angular motions can also be measured and the absolute values averaged, $|(d\theta/dt)_\tau|_{av}$. It is necessary to take the absolute values before averaging; a straight average would yield zero velocity. Alternatively, one could calculate the root-mean-square values of the τ components.

The relation between linear azimuthal velocity in the τ direction v_τ and the angular velocity in the same direction $(d\theta/dt)_\tau$ for a single object at distance r is simply $v_\tau = r(d\theta/dt)_\tau$, from the definition of the radian. The same relation will hold for the averages of the linear and angular τ motions for a large number of stars. If the averages were based on a single object, the relation would surely hold. Or you can prove this by carrying the relation through the averaging process. The relation with the average values could be solved for the distance r.

In our case, we do know the values of $(d\theta/dt)_\tau$ and hence also the average $|(d\theta/dt)_\tau|_{av}$ but we do not know the linear velocities v_τ. Nevertheless, as noted, we presume these to be equal, on average, to the radial velocities we did measure. Thus our two measured (average) values can yield the desired distance r to our sample objects,

$$\Rightarrow \quad r = \frac{|v_r|_{av}}{|(d\theta/dt)_\tau|_{av}} \quad \text{(m; statistical parallax; distance to a group of stars)} \quad (9.25)$$

In practice, one must again take into account the effect of stars in the sample having a spread of distances as well as projection effects for stars not normal to the solar apex. This technique has yielded distances of stars out to ~ 3000 LY.

Standard candles

Subsequent steps in the distance ladder involve the identification of *standard candles*. The calibration of the standard candles is a step-by-step process. First the distance to a nearby object is obtained, say by trigonometric parallax. This distance, together with the measured flux, yields the luminosity L according to (3). If the object is a recognizable member of a class for which all members have the same luminosity, the members of the class can serve as standard candles out to greater distances. The luminosity must be sufficiently high so that it can be detected at these larger distances, and it must have spectral or temporal characteristics that make it recognizable. Whenever another such object is found, its distance follows directly from its measured flux and the known luminosity of the class.

If one of these objects is located in a cluster of objects on the sky, the distance of the cluster thus also becomes known. One of the *other* cluster objects may be a more luminous object of another type which may well be suitable as a new standard candle. With knowledge of its distance and measured flux, its luminosity is known. With its greater luminosity, it can be used out to even greater distances. Of course, it must also exhibit easily recognizable spectral or temporal characteristics so other members of the class can be identified. This second standard candle would be yet another step in the distance ladder.

Luminous normal stars and cepheid variables (see below) are important standard candles. Also, planetary nebulae, supernovae, and galaxies themselves are used as standard candles. The determination of the distances to distant galaxies plays a major part in the effort to determine the past and future history of the universe.

Spectroscopic classification

All normal stars can be classified according to their spectral types. Each type is characterized by a particular luminosity and a unique combination of spectral features that follow from the mass of the star and its stage of evolution. Among the stars burning hydrogen in their core (like our sun), the low-mass stars are generally cool (color red) with luminosities as low as $10^{-3} L_\odot$ whereas high-mass stars are white hot in color and have luminosities up to about $5 \times 10^5 L_\odot$. Stars of the various types can therefore be used as a whole series of standard candles. Since these standard candles are based on their spectral type and since they yield distances, the process is sometimes called *spectroscopic parallax.*

The luminosities of individual stars obtained with the parallax method may be used to calibrate this system. However, the stars sufficiently close to the earth to be measured directly via parallax are mostly low-mass stars that are not particularly luminous. Thus the distance to which they can be used as standard candles is limited. The rarer more luminous stars must be calibrated indirectly.

The luminosities of the more luminous stars were first calibrated from observations of the stars in the Hyades cluster, the distance of which is known (see above). The Hyades cluster does not contain the most luminous stars because they have used up their nuclear fuel (burned out). However, younger clusters do contain them. The already-calibrated less-luminous stars or cepheid variables (see below) in a younger cluster give its distance. This allows the luminosities of the rare

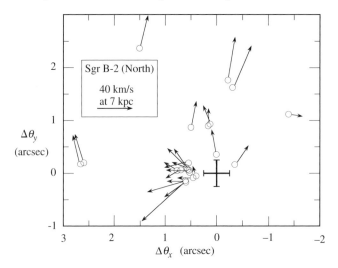

Figure 9.5. Cluster of H_2O masers in radio source Sgr B-2 near the center of the Galaxy. The arrows represent directions and speeds of masers on the sky. Note that the masers are clustered within an area of only 4″ extent. The masers are expanding from a point within the region indicated by the dashed cross, presumably the location of a newly formed star. Modeling of these data and also radial velocity data yields a distance to the galactic center of ~23 000 LY. [From M. Reid *et al.*, *ApJ* **330**, 809 (1988)]

luminous stars it contains to be determined. In this way they too become standard candles.

Galactic center and Crab nebula distances

The distance to the center of our galaxy has been obtained by radio astronomers with a method that made use of a cluster of H_2O masers in a star-forming region in the close vicinity of the galactic center (Fig. 5). The masers are probably condensations of interstellar gas and dust in the vicinity of a luminous newly formed star. They operate similarly to a laser but radiate in the microwave frequency range; they emit at precise frequencies characteristic of a constituent molecule, H_2O in this case.

The angular positions of the individual masers are measured repeatedly with very long baseline interferometry to obtain their angular velocities. Radial linear velocities are obtained from Doppler shifts of the spectral lines. The angular motion of the masers on the sky (arrows in Fig. 5) shows them to be moving away, more or less, from a common point on the sky, probably an (unseen) star. If a spherical expansion is assumed (e.g., due to a stellar wind), these data can be modeled to give a distance to the source.

Consider a maser to be part of the close edge of a spherical shell of gas moving outward from the central star. It will be directly approaching the observer, so the measured Doppler shift will yield the velocity magnitude $v = v_r$ (16). Consider next another maser on the right edge of the same shell as viewed by the same observer. This maser moves in the plane of the sky normal to the line of sight and will have only a tangential component of velocity $v_\theta = r(d\theta/dt)$, where $d\theta/dt$ is the directly measured angular displacement per unit time. In our model of spherical expansion, the two masers would have the same speed. The distance to the star follows from equating the two speeds,

$$v_r = v_\theta = r\frac{d\theta}{dt} \qquad (9.26)$$

➡ $$r = \frac{v_r}{(d\theta/dt)} \qquad (9.27)$$

The complete analysis will include all the masers associated with this object, the results being appropriately weighted and averaged. The published result from this experiment is 23 ± 5 kLY which is substantially less than the value used for the distance to the galactic center for many years previously, $32\,600$ LY ($= 10\,000$ pc). In this text, we adopt $25\,000$ LY as the nominal value to the galactic center. Note that the underlying principle of the analysis is similar to that discussed above for the statistical parallax, namely the simultaneous measure of angular and linear velocities.

The same method has been used by optical astronomers to obtain the distance to the Crab nebula which is the remnant of a supernova explosion several arcmin in size with a bluish synchrotron nebula and an expanding shell of red hydrogen-emitting filaments (Fig. 1.3). The radial and angular velocities of the filaments are measured to yield a distance of ~ 6000 LY. The expansion is not perfectly spherical which leads to some uncertainty in this distance.

Cepheid and RR Lyrae variables

Cepheid variables are luminous stars that are 300 to 30 000 times the luminosity of the sun (300–30 000 $L_\odot$). Their luminosities vary periodically with periods of days (1–100 d); see Fig. 6a. The luminosity variation is due to oscillation in the size (height) and temperature of the stellar atmosphere associated with variation in the degree of ionization of the atmospheric gases. The period of the oscillations is highly correlated with the average luminosity (Fig. 6b). A cepheid variable with a period of 100 days is roughly 40 times more luminous than a cepheid variable with a 1 day period. These stars are sufficiently luminous to be found in galaxies out to ~ 60 MLY with the Hubble Space telescope, e.g., in the Virgo cluster of galaxies.

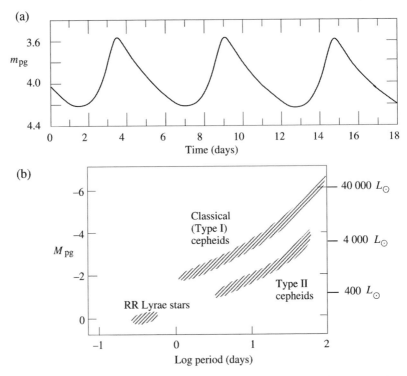

Figure 9.6. (a) Typical variation of flux from a cepheid variable star as a function of time. The ordinate is "apparent photographic magnitude" ($m_{pg} \approx m_B$) which is proportional to the logarithm of the detected flux $\mathscr{F}$ (see Section 8.3). In this case, the flux changes by a factor of about two, and the period is about 5.5 d. (b) Average luminosity (right hand labels) of cepheid and RR Lyrae stars as a function of the period of oscillation. The left hand axis gives the absolute photographic magnitude, a measure of luminosity. [Adapted from Abell, *Exploration of the Universe*, 3rd Ed., Holt Rinehart Winston, 1975, p. 407, p. 411]

There are two types of cepheid variables recognizable by the shape of their light curves, the *classical cepheids* (Type I) and the less luminous *Type II cepheids* (Fig. 9.6b). The former are young stars with relatively short lifetimes, $\sim 10^8$ yr, and the latter are older stars of ages $\sim 1.5 \times 10^{10}$ yr. The former are found in galaxies with star-forming regions such as the Magellanic clouds and spiral galaxies, while the latter are found in places with older star populations such as the halo of the (MW) Galaxy and globular clusters.

There is also a class of *RR Lyrae variables* that are less luminous and have periods ranging from 0.2 to 0.9 d. They are found in, and yield distances to, globular clusters as distant as our neighboring sister galaxy, the Andromeda nebula (M31) at a distance of 2.5 MLY.

The period of the oscillations of such a variable star is directly measured, and this yields the luminosity from Fig. 6b, or its equivalent. Comparison to the measured mean flux then yields the distance.

Knowledge of the absolute mean luminosities of the cepheid variables is required if they are to be used as distance indicators. In other words, the luminosity must be calibrated. This was a difficult observational problem. Historically, the period–luminosity relation was first noted in 1912 by Henrietta Leavitt. She identified 25 cepheid variables in the Small Magellanic Cloud (SMC), a small satellite galaxy of the (MW) Galaxy, and noted that the brighter stars had longer periods. All stars in the SMC are presumed to be, more or less, at the distance of the SMC which was unknown at that time. (It is ~200 000 LY.) This allowed a curve such as Fig. 6b to be constructed but without an absolute calibration of the luminosity.

The luminosity calibration was accomplished through measurements of the nearer brighter cepheid variables in the Galaxy. This could not be done with trigonometric parallax studies because the cepheids were too distant. However, some cepheids lie in open clusters of known distance. Also the statistical methods described above were used. The RR Lyrae stars are also calibrated with such methods.

The relationships in Fig. 6b were well established by ~1939 for the classical cepheid variables. Prior to this, in 1923, Edwin Hubble identified several cepheid variables in the Andromeda nebula (M31). Their extreme faintness established for the first time the great distance of this nebula, now known to be ~2.5 MLY. This demonstrated its great size and luminosity and hence that it is another galaxy like our own. Previously, it was not clear whether such objects were smaller nebulae within the Galaxy or large nebulae outside it.

The cepheids are so luminous that, as noted, they can be identified in very distant open clusters that contain the rare high-mass stars in the Galaxy and in nearby galaxies. In turn this gives us the luminosities of these high-mass objects.

The physics that governs the oscillations of size and luminosity of cepheids is interesting; it is sketched briefly here. The degree of ionization of hydrogen and helium near the surface of the star controls the opacity to radiation and hence the rate of energy radiated into space from the star. The periodic variation of stellar radius (atmospheric height), density and temperature of the cepheid variables affects these ionization levels leading to periodic opacity changes. In cepheid variables, the periodic opacity releases or holds in energy at phases that cause the oscillations to be reinforced rather than damped.

It is not unlike a child on a swing who pumps energy into the swinging motion or takes energy out depending on the phase with which he leans back and forth. In a cepheid star, the gas elements of the stellar atmosphere serve as *heat engines* that do work on their surroundings. The work varies with the phase of the oscillation so as to sustain it. Many stars enter this state of instability during a late stage of their evolution.

Hubble law

A new powerful method of determining distance was discovered in 1929 by Edwin Hubble. It had been known since 1912 that the spectral lines in the spectra of galaxies generally are found at lower (redder) frequencies than the frequency measured in the laboratory (Fig. 1.11). This shift is known as a *redshift*. Hubble found that the faintest galaxies have the greatest redshifts and was able to conclude that the redshift was proportional to distance.

Receding galaxies

The redshift may be interpreted as due to the Doppler effect (16) brought about by high recessional velocities; the emitting galaxies are moving rapidly away from the (MW) Galaxy. In this interpretation, Hubble's result tells us that the galaxy recessional speed v is proportional to its distance r from the Galaxy. This revealed to us that we live in an expanding universe. The speed-distance proportionality is known as the Hubble law,

$$v = Hr \hspace{3cm} \text{(Hubble law)} \hspace{1cm} (9.28)$$

where it is evident that H, the *Hubble constant*, has units of s^{-1} if the recessional speed and distance are given in SI units, (m s^{-1}) and (m) respectively.

Astronomers use the mixed unit (km s^{-1} Mpc^{-1}) for H, which has a physical connection to the recessional velocity as a function of distance if one thinks of distance in terms of parsecs. In this text we use (km s^{-1} MLY^{-1}), which gives the recessional speed in km/s of a hypothetical galaxy at distance 1 MLY.

In general relativity, the Doppler interpretation of the redshift is not the correct view. One should think of the electromagnetic waves being stretched to longer wavelengths as the universe expands while they are en route from the source to the observer. In either view, the redshift is a consequence of an expanding universe.

Expanding universe

In the simplest picture of an expanding universe, the speed of the individual galaxies do not change as they fly away from each other. Think of the galaxies as raisins in a bread being baked in an oven. As the bread expands (linearly with time), the raisins move farther and farther from one another. The recession speed of one raisin relative to another is proportional to the spacing between them.

This simple picture has no chosen center of expansion; no one galaxy is special. This is philosophically appealing in that it obeys the *cosmological principle* which states that the universe is similar at all locations at a given epoch. The necessary consequence of such an expansion is that it gives a distance–speed relation that is linear, just as Hubble discovered! One would hardly have expected anything different.

The Hubble proportionality "constant" is appropriate at a single stage (*epoch*) of the expansion. As a function of time in our constant velocity picture, it is *not* constant. Consider a given galaxy moving away from the Galaxy at a constant velocity, so that much later, it will be at a larger distance r with the same v. At the later time, H will thus have a smaller value, $H(t) = v/r$. The value of the constant *today*, "at the current epoch", is designated H_0. This is the constant that relates the galaxy speeds to their distances in our time,

$$v = H_0 r \qquad \text{(Hubble law; current epoch)} \qquad (9.29)$$

A specific time or "epoch" is that agreed to by a number of *fundamental observers* (FOs) distributed throughout the universe, each at rest relative to the local matter. Each FO can set his watch according to a pre-agreed-upon local matter density which is decreasing due to the continuing expansion of the universe. Thus, at a given epoch, the universe appears the same locally to all FOs.

Value of Hubble constant

The value of H_0 is obtained from measurements of v and r for a large number of galaxies; it depends upon careful distance calibrations. This value has been and is still a much discussed number. Proponents have argued for values ranging from 15 to 30 km s^{-1} MLY^{-1} (or, multiplying by 3.26 LY/pc, $\sim$50–100 km s^{-1} Mpc^{-1}). Recent measurements are beginning to bring about a consensus that the value is approximately 65 km s^{-1} Mpc^{-1} (within about 15%), or $\sim$20 km s^{-1} MLY^{-1}. Very recent results (2003) from the satellite observatory, the Wilkinson Microwave Anisotropy Probe (WMAP), suggest a value of 21.8 ± 1 km s^{-1} MLY^{-1} (71 km s^{-1} MLY^{-1}) for the currently most favored expansion model. In this text, for convenience, we use the round-number value,

$$\begin{aligned} H_0 &= 20 \text{ km s}^{-1} \text{ MLY}^{-1} \ (= 65.2 \text{ km s}^{-1} \text{ Mpc}^{-1}) \qquad \text{(Hubble} \\ &= 2.11 \times 10^{-18} \text{ s}^{-1} \qquad\qquad\qquad\qquad\qquad \text{constant)} \end{aligned} \qquad (9.30)$$

When astronomers publish results that depend on the value of the Hubble constant, they often include a dimensionless adjustment factor in their mathematical expressions, e.g., $h_{65} \equiv H_0/65$. This allows for a value of H_0 other than the one they adopted for their result. In this text, we simply use the value given in (30).

Once the value of H_0 is known, even approximately, it can be used to find the approximate distance of any anonymous galaxy or quasar, as long as the redshift of its spectral lines can be measured. The measured redshift provides the distance through (29). For example, if a Doppler velocity of 600 km/s is measured, one could conclude that the distance was about $600/20 = 30$ MLY.

Redshift

The *redshift z* is defined as

$$z \equiv \frac{\lambda - \lambda_0}{\lambda_0} \qquad \text{(Redshift)} \qquad (9.31)$$

where λ is the measured wavelength of a given atomic spectral line observed in a galaxy and λ_0 is its laboratory ("rest") wavelength. From the classical Doppler relation (16), we find that z is proportional to the recession speed v,

$$\frac{v}{c} = \frac{\Delta\lambda}{\lambda} = z \qquad \text{(Doppler shift)} \qquad (9.32)$$

where $\Delta\lambda = \lambda - \lambda_0$. Eliminate v from (29) and (32),

$$\blacktriangleright \qquad r = \frac{c}{H_0}z \qquad \text{(Distance for } z \ll 1\text{)} \qquad (9.33)$$

This linear relation is valid only for non-relativistic recessional speeds ($v = cz \ll c$), i.e., for $z \ll 1$. At large values there are additional terms, and furthermore one must specify which of the several distance parameters in general relativity is being used. At the small recessional velocities (compared to c) encountered by Hubble, such distinctions were not important. Modern work must take them into account.

Use (33) to calculate the distance of a galaxy that yields a measured redshift of $z = 0.002$,

$$r = \frac{3 \times 10^8 \text{ m s}^{-1}}{2.11 \times 10^{-18} \text{ s}^{-1}} 0.002 = 2.84 \times 10^{23} \text{ m} = 30 \text{ MLY} \qquad (9.34)$$

which is the same example we just gave above. Equivalently, in astronomical units,

$$r = \frac{3 \times 10^5 \text{ km s}^{-1}}{20 \text{ km s}^{-1} \text{ MLY}^{-1}} 0.002 = 30 \text{ MLY} \qquad (9.35)$$

At great distances where the recession speeds approach the speed of light, the Hubble law will deviate from the linear expression (33) in a way that depends upon the exact nature of the expansion of the universe. This will change the form of the distance–redshift relation at these large distances. For example, the expansion could be slowing due to the mutual gravitational attraction of the galaxies that make up the universe, or it might be accelerating due to "dark energy" as recent data suggest. In these cases, the starlight left distant galaxies at much earlier times, so the changing speed should become apparent in comparisons of the redshifts of close and distant galaxies. Currently, much effort is being expended to obtain distances independent of the Hubble law, e.g., with galaxies and supernovae as luminous standard candles (see below), in order to determine these deviations.

Size and age of the universe

Returning to the simplest model wherein the expansion speed of any given galaxy is constant in time, one can find an approximate "size" of the "observable" universe. We presume this distance to be that where the expansion speed is approaching the speed of light. At this distance, one could not observe a celestial object because, at this recession speed, signals from it would be redshifted to nearly zero frequency. In the limit of recessional speed c, the photons would have no energy.

To find this distance, substitute c for v in (29) to find

$$\Rightarrow \qquad r_{univ} \approx \frac{c}{H_0} = \frac{3 \times 10^5 \text{ km s}^{-1}}{20 \text{ km s}^{-1} \text{ MLY}^{-1}} = 15 \text{ GLY} \qquad \text{(Radius of} \qquad (9.36)$$
$$\text{observable universe)}$$

the distance we quote in Table 2.

In this simple model, the material at the 15 GLY distance would be receding at speed c, and hence would have left our location 15 Gyr previously. This is an approximate age of the universe T_{univ}. Formally one would write,

$$T_{univ} \approx \frac{r_{univ}}{c} = \frac{1}{H_0} \equiv t_H \qquad \text{(Hubble time)} \qquad (9.37)$$

where the time $t_H = H_0{}^{-1}$ is called the *Hubble time*. Substitute the numerical value for H_0 (30) into (37) to find the anticipated age,

$$\Rightarrow \qquad T_{univ} \approx t_H = 15 \text{ Gyr} \qquad \text{(Approximate age of} \qquad (9.38)$$
$$\text{the universe)}$$

The WMAP observers report an age of 13.7 ± 0.2 Gyr.

Extragalactic "standard candles"

The quest for standard candles as independent indicators of distance continues to this day. Modern sensitive instrumentation and the ingenuity of contemporary astronomers are bringing about important advances. The objects now used as standard candles include supernovae at their peak brightness, galaxies of the spiral and elliptical types, globular clusters, novae, and planetary nebulae. Each of these can be very luminous and hence can be seen to very large distances.

These classes would not seem to be useful standard candles because they come with differing luminosities. Nevertheless, sometimes there are sufficiently large numbers of a given type of object in a given locale to use global properties of the entire sample as a distance indicator. The *luminosity function* discussed below is an example of such a global property.

Astronomers have also identified observable characteristics (e.g., spectral line broadening) of extragalactic objects that are related to the luminosity. Such objects can then be used as standard candles. This is completely analogous to cepheid or RR Lyrae variables which exhibit different luminosities, but the period of oscillation specifies the luminosity. It is also analogous to the spectroscopic classification of stars; the spectrum classifies the star and thus its luminosity is known. Such an approach is used with supernovae and galaxies as we now discuss.

Luminosity functions

A hypothetical example of the use of global properties as a distance indicator would be to determine the average luminosity of all the globular clusters in a distant galaxy. Globular clusters are compact collections of $10^5 - 10^6$ stars of which several hundred surround the center of the Galaxy (Figs. 1.9 and 3.3). One can argue that the average luminosity of globular clusters should be the same from galaxy to galaxy, and that this average value could then be used as a standard candle.

In practice, one uses the entire shape of the *luminosity function* rather than simply the average luminosity. This function is the distribution of luminosities of a given type of object such as globular clusters in a given galaxy. Specifically, it is the number of objects per unit luminosity interval as a function of luminosity. As shown in Fig. 7, it is usually plotted on a log–log plot, in this case, log relative number per unit absolute magnitude vs. absolute magnitude; recall that magnitude is a logarithmic quantity.

It turns out that globular clusters in a typical galaxy have a distribution that peaks at some luminosity and has a characteristic width. The luminosity function can be characterized with these two parameters (peak luminosity and width). The two parameters serve as a standard candle, if they are calibrated to some known distance.

The luminosity function of planetary nebulae is also an important distance indicator. Planetary nebula are gas clouds ejected from stars of high temperature and luminosity late in their evolution. The ejected clouds of gas are illuminated by the ultraviolet radiation of the central star (Fig. 1.8). The cloud effectively absorbs and reradiates much of the luminosity, mostly as spectral lines in the optical part of the spectrum. The luminosity in these spectral lines can be seen from great distances, and a given galaxy might contain several hundred detectable objects.

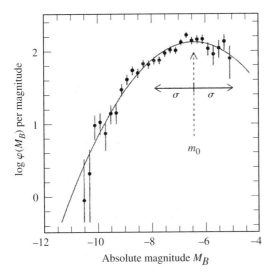

Figure 9.7. Luminosity function of globular clusters from ~2000 clusters in four elliptical galaxies in the Virgo cluster of galaxies. The ordinate is the log of the relative number of clusters in unit interval of absolute B-band magnitude, and the abscissa is the absolute magnitude in the blue band. The distribution is characterized by the magnitude, m_{peak} and a width 2σ which are used together as a standard candle for other galaxies. [From G. Jacoby, *et al.*, *PASP* **104**, 599 (1992); courtesy W. Harris]

The luminosity function or other characteristic of a distance-indicating object may vary with the type of galaxy with which it is associated. Galaxies come in a wide range of types from *spiral* to *elliptical* (Hubble types) with differing ages of stars and colors (reddish to bluish) and metal content. Distance indicators can, in principle, vary slowly with galaxy type. This effect can be calibrated with observations of the closer galaxies.

Supernovae

Supernovae are violent outbursts from stars that gravitationally collapse. The luminosity can approximate or exceed that of an entire galaxy in which it resides for a short time (days). One type of supernova (Type Ia) is believed to arise from a dense white-dwarf star that gradually accretes matter from a companion until it reaches its maximum allowed mass. At this point it undergoes runaway nuclear burning and incinerates itself. The nature of the triggering mechanism (slow accretion) suggests that every such event would release the same amount of energy; that is, the peak luminosity of the outburst should be a standard candle. This method is proving to be a valuable distance indicator for very large distances. In fact, as described below, it is now yielding rather surprising results.

Line-broadening in galaxies

Galaxies themselves have been used as standard candles. It turns out that large *spiral galaxies* are more luminous if they are more massive. This is not surprising because a greater mass indicates more stars. Also, the more massive galaxies cause the gas in their outer regions to rotate around their centers faster than for lower-mass galaxies. This follows directly from $F = ma$ and Newton's gravitational law.

This speed of rotation shows up as a Doppler broadening of the spectral lines emitted by the rotating hydrogen gas at the radio wavelength of 21 cm (1420 MHz) if the spectrum includes radiation from the entire galaxy. In this case the receding and approaching parts of the galaxy will yield, respectively, red and blue shifted lines. Taken together in one spectrum the 21-cm line thus appears broadened. Because both the luminosity and the broadening are correlated with the mass, the line broadening specifies the luminosity of the galaxy in question. In short, a spiral galaxy with a known Doppler broadening is a standard candle. This method for the determination of distances is known as the *Tully–Fisher method*.

A similar relation is used for galaxies which are lacking in interstellar gas and for which there is no organized rotation; the angular momentum is low. In these *elliptical galaxies*, the stars move around in the gravitational well with more or less random velocities. Again, if the total mass is greater, the star velocities and the overall galaxy luminosity are greater. The absorption lines in the optical spectra of the individual stars contribute to the overall galaxy spectrum. This overall spectrum exhibits broadened spectral lines because of the random velocities of the stars contributing to it.

Thus, the line widths are again a measure of the luminosity. Elliptical galaxies with a specific line width are thus standard candles of a known luminosity. This is known as the *Faber–Jackson method*. It has been extended to include an angular-diameter parameter, and is now called the D_n–σ *method*, where the two symbols refer to the angular diameter (containing a specified average surface brightness) and the line width.

Surface brightness fluctuations

A very different method invokes pixel to pixel fluctuations of intensity in the images of elliptical galaxies. A distant galaxy will generally appear as a diffuse blob due to the thousands of unresolved stars in each image element, that is, in each pixel of a CCD. (Charge-coupled devices are described in Section 6.3.) However, a relatively small number of extremely luminous stars (*giants*) in the galaxy will give the galaxy a roughened or mottled appearance because each pixel contains only a few such objects.

Since there are typically several such stars in a pixel, one can not use a single such object as a standard candle. Happily though, it turns out that the flux from a single "average" star is directly obtainable from the pixel-to-pixel fluctuations of light in the CCD image. Thus these stars can be used as standard candles if their average luminosity is known from studies of closer examples.

Consider the situation of Fig. 8a wherein a galaxy is imaged onto a CCD at the focus of a telescope. Each cell in the figure represents a single pixel, and the dots are the giant stars in the galaxy. The galaxy is presumed to be larger than the pixel arrays shown. If the galaxy is removed to twice the distance, the galaxy will be reduced in angular size and likewise the angular spacing between giants will be reduced. In the image plane, the giants will thus be closer together, and more of them will be imaged onto each pixel; see Fig. 8b.

The data from such observations consist of a single number from each pixel representing the total number of photons recorded in that pixel during the exposure. In Figs. 8c,d, we represent this number with a shading. The shading takes into account the numbers of giant stars in each pixel as well as the fact that the stars at the farther distance each yield only 1/4 the flux of those at the closer distance. It is important to note that the calibration of the shading with photon number is the same for Figs. 8c and 8d.

The number of photons recorded per pixel in Fig. 8, averaged over many pixels, is independent of distance to the galaxy because the larger (average) number of giant stars in each pixel at the farther distance exactly cancels the decrease in flux from each such star. This is just the distance independence of specific intensity discussed in Section 8.4. We thus argue that the "average" shadings in the two frames (Figs. 8c,d) are the same. The *average* of the pixel values provides no distance information.

However, there is an obvious and important difference in the two cases of Figs 8c,d. In the closer case, the lesser number of giant stars in each pixel yield greater fluctuations about the mean flux. The fluctuations are thus a strong distance indicator. We now show quantitatively how Poisson fluctuations provide the desired distance information.

If the giant stars all have similar spectra, the photon number recorded in some bandwidth in each pixel is a measure of the energy recorded in each pixel. For a single point source, this is a measure of a flux density (W m^{-2} Hz^{-1} in SI units) because we know the effective area and bandwidth of the telescope system. We therefore designate the energy per pixel from a single star to be the "flux" f.

Let us further designate the expected (or mean) number of bright giants per pixel to be m. The expected total energy recorded in each pixel is thus the product fm. Here, for our purposes, we assume all the stars have the same flux. Since each pixel corresponds to a known solid-angle element on the sky, the product fm is thus a measure of the specific intensity (W m^{-2} Hz^{-1} sr^{-1} in SI units). Recalling that

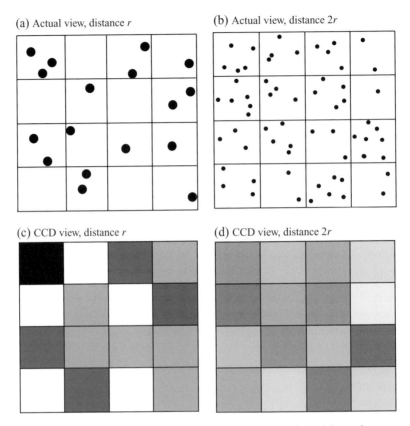

Figure 9.8. Giant stars (black circles) in a galaxy as viewed by a charge-coupled device (CCD). Each cell is a pixel. (a,b) Stars for two distances of the galaxy, one twice the other. Note that the star patterns of (a) are in the central 4 boxes of (b). (c,d) Shadings indicating the number of detected photons recorded in each pixel for the two cases (a,b). The calibrations of shading to photon number are the same for (c) and (d). The recorded photon number averaged over all pixels is almost the same (within statistics) for the two cases, but the fluctuations about the mean are much greater for the closer case (c). Table 3 applies directly to this example. [Adapted from J. Tonry, personal communication; also see G. Jacoby *et al.*, *PASP* **104**, 599 (1992)]

$I = B$ (8.42), we call the product a surface brightness b,

$$b = fm \qquad \text{(J/pixel; mean surface brightness)} \qquad (9.39)$$

This is the *expected* (or mean) surface brightness of the galaxy measured in units of recorded joules per pixel.

The fluctuations of this brightness from pixel to pixel reflect the fluctuations in the number of giants per pixel. The fluctuation in this number will obey the Poisson distribution which has an rms deviation of $m^{1/2}$ (6.8). This latter number must be multiplied by the flux from a single star to obtain the rms deviation in the surface brightness b,

$$\sigma_b = f m^{1/2} \qquad \text{(W; rms dev. in brightness)} \qquad (9.40)$$

Eliminate m from (39) and (40) and solve for f, the flux from a single giant star in the galaxy,

➥ $$f = \frac{\sigma_b^2}{b} \qquad \text{(J/pixel; flux from single giant star)} \qquad (9.41)$$

Thus the two measured quantities σ_b and b yield the flux density f (J/pixel) from a single star. As noted this can be converted to SI units. As we know, the flux density from a single star yields its distance, given independent knowledge of its luminosity. Thus (41) can be used to find the distance to the galaxy containing the giant stars. In practice the giants will have a spread of luminosities which must be taken into account in the analysis.

The quantity b in (41) is the *expected* (or mean) surface brightness. The mean value is obtained by averaging the values recorded in the pixels of the image. Similarly σ_b is obtained by taking the rms deviation of these same recorded values. As noted, the average surface brightness b is independent of distance to the galaxy (more stars per pixel but less flux from each star). In contrast, the rms brightness deviation σ_b decreases with distance since the flux f in (40) decreases faster with distance than the rms star number deviation $m^{1/2}$ increases. The flux f from a single star (41) thus decreases with distance, as it must.

The detected giant stars in elliptical galaxies turn out, from independent determinations, to have an average luminosity L_{av} (per giant star) that is quite constant from galaxy to galaxy. Thus indeed they can serve as standard candles. The flux recorded by a pixel from a point source of luminosity L_{av} at distance r from earth is $f = (L_{av}/4\pi r^2)A$, where A is the effective area of the telescope mirror and where we adopt a non-relativistic Euclidean universe. We use the area A of the mirror because all the light that impinges on this area from a point source is deposited on the pixel where the point source is imaged. This expression can be equated to (41) to yield the distance r to the average giant star in terms of known quantities,

➥ $$r = \left(\frac{b L_{av} A}{4\pi}\right)^{1/2} \frac{1}{\sigma_b} \qquad \text{(Distance to source)} \qquad (9.42)$$

This discussion illustrates the principle of this fluctuation method. It has in fact proved to be a practical method of finding distances independent of the redshift. In turn these yield values of the Hubble constant, from $H_0 = v/r$ where v is obtained from the redshift. The method requires that the galaxy under study be sufficiently large in angular extent to extend across a substantial number of telescope resolution elements. The average luminosity L_{av} of a single giant star may be obtained from cepheid distance indicators in nearby galaxies, e.g., Andromeda (M31).

Table 9.3. *Surface brightness fluctuations, example*[a]

Parameter	Distance r	Distance $2r$
Mean no. stars per pixel	m	$4m$
Flux from one star per pixel	f	$f/4$
Surface brightness b	fm	$(f/4)\,4m = fm$ (same)
Surf.-brightness fluct. σ_b	$f\sqrt{m}$	$\frac{1}{4}f\sqrt{4m} = \frac{1}{2}f\sqrt{m}$
Flux $\sigma_b{}^2/b$ (41)[b]	$\frac{(f\sqrt{m})^2}{fm} = f$	$\left(\frac{1}{2}f\sqrt{m}\right)^2 (fm)^{-1} = f/4$

[a] Refer to Fig. 8; courtesy of J. Tonry (personal communication).
[b] The flux in line 5 agrees with that in line 2.

The illustration of Fig. 8 can be put in the context of the above derivation with Table 3, which shows how the several quantities change when the distance is increased a factor of two. The last row shows that indeed the expression (41) yields the flux from a single star, in agreement with line 2.

Ultimate goals

Extragalactic distances depend directly on distance determinations lower on the ladder. One might simplistically describe the ladder sequence as follows: (*i*) the astronomical unit, (*ii*) parallax and open clusters, (*iii*) cepheid distances, and finally (*iv*) extragalactic standard candles. In fact, the numerous methods of the actual ladder link and overlap in a variety of ways. A version of the ladder is shown in Fig. 9. We do not attempt to explain each of the methods in the figure although many of them have been discussed or alluded to in this chapter.

There are two other approaches to determining the Hubble constant that do not depend directly on the ladder, though they are less well developed. One is based on *gravitational lensing* of distant quasars and another is based on fluctuations in the cosmic microwave background due to photon scattering in the hot plasmas of clusters of galaxies, the *Sunyaev–Zeldovich* effect.

A prime objective of these several methods is to obtain absolute distances to the most distant galaxies. This directly provides the luminosity of any such galaxy and thus provides the absolutely calibrated luminosity functions for the several types of galaxies. In other words, it yields the power output of these galaxies, a fundamental quantity indeed. The distances to the galaxies together with their measured recessional speeds (from spectral redshifts) yield H_0, the Hubble constant at the current epoch.

Deviations of the distances from those predicted by the linear Hubble law, $r = v/H_0$ (29), indicate there are motions superimposed on the "Hubble flow". These are

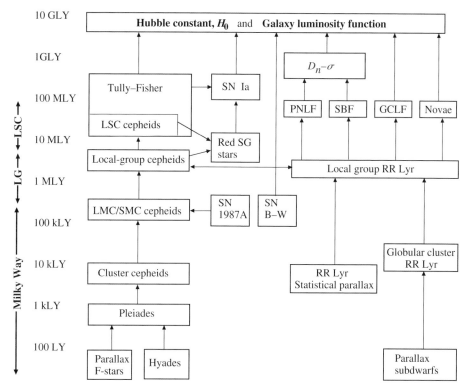

Figure 9.9. Distance "ladder" showing how distance determinations of close objects (e.g., through parallax) are the basis of distance determinations out to greater and greater distances, e.g., through cepheid and RR Lyrae variables to extragalactic indicators such as line broadening in the spectra of galaxies. Abbreviations: B–W – Baade–Wesselink; GCLF – globular-cluster luminosity function; D_n–σ – Diameter–line width extension of Faber–Jackson method; LG – local group (of galaxies); LMC/SMC – Large (Small) Magellanic Cloud; LSC – local supercluster (of galaxies); PNLF – planetary-nebula luminosity function; SBF – surface-brightness fluctuations; SG – supergiant; SN – supernova. [From G. Jacoby *et al.*, *PASP* **104**, 599 (1992)]

due to gravitational attractions between galaxies and between clusters of galaxies. Studies of this tell us how much gravitational mass there is in the universe.

Systematic deviations from the linear Hubble law at large distances and in all directions in principle can reveal whether the expansion rate is decreasing, constant or increasing. Light from distant galaxies left those galaxies long ago so observations probe the expansion at early times. Slowing of the expansion is expected from mutual gravitational attraction; it is a natural outcome of Einstein's general theory of relativity. It is the same as a ball thrown upward; it will decelerate as it moves upward due to the gravitational attraction between it and the earth.

A major complicating factor in using distant galaxies (at large look-back times) as standard candles is that they evolve with time. Their colors and luminosities change as their constituent stars form, age, and die. Thus measurements have not been able to resolve whether or not the universe is actually slowing as gravity would suggest.

Recent studies that use distant Type Ia supernovae as standard candles are now indicating that the expansion might, in fact, be accelerating, exactly counter to our expectation. If so, this could be due to a weak repulsive force that acts preferentially over very large distances. Such a force is implied by a non-zero value of the *cosmological constant* in Einstein's theory of general relativity. The mysterious energy source for this acceleration has been called *dark energy*. Generally, the nature of the expansion is of great interest to astrophysicists because it is fundamental to the theory behind the evolution of the universe.

Problems

9.1 Introduction

Problem 9.11. Find the following ratios to one place accuracy. Give the values that go into the ratios in convenient units (not necessarily SI) and make some comment on each that will help you visualize them. Think about them; get them into your bones.

A. Size and distance
  (a) radius of earth *to* radius of a marble.
  (b) distance to sun *to* radius of sun.
  (c) distance to center of the (Milky Way) Galaxy *to* distance to nearest star.
  (d) distance to nearest star *to* diameter of a typical star like the sun.
  (e) distance to neighbor galaxy, M31, *to* diameter of the (MW) Galaxy.
  (f) distance between NYC and Boston (400 km) *to* diameter of Boston and its inner suburbs out to about 7.5 km; compare to (d) and (e).
  (g) distance to "edge" of universe *to* distance to M31.
B. Energy
  (h) 10-MeV gamma ray *to* a 3-keV x ray.
  (i) 3-keV x ray *to* a 500-nm visible photon.
  (j) 500-nm (optical) photon *to* a 10-m radio wave.
  (k) gravitational potential energy of sun ($\sim GM^2 R^{-1}$) *to* rest-mass energy of sun ($M_\odot c^2$).
  (l) same as (k), but for a neutron star of 1.4 $M_\odot$ and radius 10 km.
C. Power (energy per second)
  (m) solar power intercepted by earth *to* luminosity of sun.
  (n) luminosity of the Galaxy *to* the luminosity of the sun.

9.2 Luminosities

Problem 9.21. (a) Assume that the earth (with its atmosphere) totally absorbs the solar radiation that impinges upon it. If this radiation is reradiated by the earth at the same rate it is absorbed, what is the luminosity of the earth due to this reradiated energy? (b) If the reradiation occurs uniformly over the entire spherical surface of the earth, what is the temperature T of that surface? Assume the radiation is blackbody radiation which has the characteristic that $1\ m^2$ of the surface radiates a total of σT^4 watts over a broad band of wavelengths, where $\sigma = 5.7 \times 10^{-8}$ W $m^{-2}\ K^{-4}$. (c) What is the approximate wavelength at peak power of this reradiated power? Hint: refer to (2.13). In what wavelength band (radio, IR, optical, etc.) is this wavelength? [Ans. $\sim 10^{17}$ W; ~ 300 K; $\sim 20\ \mu m$]

Problem 9.22. (a) If the average luminosity (10^{39} W) of a quasar (Table 1) arises because the presumed black hole at its center accretes a star of $\sim 1\ M_\odot$ every now and then, about how often should it have one of these "star" meals? Assume that about 10% of the rest-mass energy mc^2 of the eaten star is converted to the radiation we observe. (b) Find (classically) an expression for the distance from a central mass M to which a mass m must fall from infinity, such that the loss of potential energy of m equals its entire rest mass energy mc^2. (It is impossible to extract more energy than this.) Twice your answer will be the famed Schwarzschild radius, the event horizon for a non-rotating black hole. What is the Schwarzschild radius for a quasar mass of $10^8\ M_\odot$? What is it for the sun ($1\ M_\odot$)? (c) Write in terms of $\dot{m}$ an expression for the maximum luminosity one could conceivably obtain from accretion at the rate of $\dot{m} = dm/dt$ (kg/s). Discussion topics. Why is a large quasar mass required to support a large quasar luminosity? (Hint: look up *Eddington luminosity*.) What is the meaning of the Schwarzschild radius? How might it affect the luminosities from accretion onto a black hole. [Ans. ~ 7 months; ~ 1 AU, ~ 1 km; —]

9.3 Masses

Problem 9.31. Substitute into (7) to obtain values for the masses of the sun and earth. The sidereal (relative to the stars) orbital periods of the earth and moon are 365.256 d and 27.32 d respectively, and the semimajor axes of 1 AU $= 1.50 \times 10^{11}$ m and $\sim 3.84 \times 10^8$ m, respectively. One day is 86 400 s. Comment on the relative degrees of agreement of your values with those in Table 1.

Problem 9.32. Demonstrate that a star in a circular orbit about the center of a galaxy has a greater velocity than it would orbiting a less massive galaxy at the same radius. Assume that all the mass of the galaxy is near its center.

Problem 9.33. Find an expression for the mass M of a cluster of galaxies consisting of N galaxies ($N \gg 1$) of mass m in terms of the mean square speed $\langle v_i^2 \rangle_{av}$

of the galaxies and the average of the reciprocals of the pair separations, $\langle r_{ij}^{-1} \rangle_{\mathrm{av}}$. Start with the virial theorem (9).

$$\left[\text{Ans. } M \approx \frac{2}{G} \langle v_i^2 \rangle_{\mathrm{av}} \langle r_{ij}^{-1} \rangle_{\mathrm{av}}^{-1} \right] \tag{9.43}$$

9.4 Temperatures

Problem 9.41. Cyanogen molecules (CN) in interstellar space are immersed in a sea of photons with a blackbody spectrum. The molecules have low-level rotational states, the lower two of which are separated by 4.72×10^{-4} eV. The upper state has spin 1 (statistical weight 3) and the lower spin 0 (statistical weight 1). The relative numbers of molecules ("populations") in these two states can be determined from spectral absorption lines in the light of a bright background star as follows. The optical photons (~ 3.2 eV) are absorbed from these two states as they excite the CN to much higher states, thus creating several absorption lines. The frequency of a given absorption line indicates from which low-lying state and to which high state the transition occurred. The relative strengths (equivalent widths) of the absorption lines thus indicate the relative numbers of atoms in the two low-level rotational states, in their equilibrium state. Observation of ζ Oph indicates that the ratio of CN atoms in the higher low-level state to those in the lower low-level state is about 0.40. What temperature would you deduce for the blackbody radiation in which the atoms are immersed? Assume perfect thermodynamic equilibrium, so the Boltzmann formula (14) is valid. [Ans. ~ 3 K]

Problem 9.42. Consider the Saha equation (15). (a) For a pure hydrogen plasma (so that $n_{r+1} = n_{\mathrm{e}}$), what, qualitatively, is the effect on the ionization ratio n_{r+1}/n_r (*i*) of increasing, from an external source, the electron density n_{e} if T is held fixed, (*ii*) of increasing the temperature T, and (*iii*) of increasing the total number density n of hydrogen atoms at fixed T? Assume in each case that the statistical factors and the ionization potential do not change. (b) What is the effect of additional species (types of atoms) being present for each of these three cases? (c) Qualitatively argue that each of your answers is physically reasonable.

Problem 9.43. The photosphere of the sun is in quasi thermal equilibrium at $T = 6400$ K. Consider that the gas is pure hydrogen, so $n_{\mathrm{e}} = n_{r+1}$. Spectral lines indicate the plasma is quite weakly ionized, $n_{r+1}/n_r = 4 \times 10^{-4}$. Use the Saha equation to find the total density of hydrogen atoms at the photosphere. Neglect the effect of other atomic elements. See text for needed parameters. [Ans. $\sim 10^{23}$ m^{-3}]

9.5 Distances and sizes

Problem 9.51. You are a resident of Syene (now Aswan) in southern Egypt in ~ 225 BCE. You notice that sunlight strikes the bottom of a water well at local noon on the summer solstice (June 21); the sun is at the zenith. On the same date, your friend Eratosthenes who lives in Alexandria in the north of Egypt observes that the sun is somewhat south of the zenith at local noon (when the sun is at its highest point and hence due south of the zenith). He measures this angle to be 1/50th of a circle (7.2 degrees). You know that a camel caravan traveling at about 100 stadia a day takes 50 days to reach Syene from Alexandria. Assume that the earth is a sphere and that Syene is due south of Alexandria. (a) What is the radius of the earth in stadia and in meters adopting 185 meters per stadium? How does your result compare to the true value 6370 km (mean radius). (b) Comment on the sources of error. (c) The latitude and longitude of Alexandria are 31° 13′ N, 29° 55′ E, of Syene 24° 05′ N, 32° 56′ E. The sun is at declination +23° 27′ N at the summer solstice; its angular diameter is 32′. What can you say about the depth of the well in Syene if its diameter (or N–S width) is $w = 1.0$ m?

Problem 9.52. Find the distance to the moon from the earth center using the Ptolemy principle, as follows. One night on the equator, you notice the moon directly overhead and note carefully where it is relative to the distant stars. When the earth has rotated *exactly* 90° relative to the stars, approximately 6 hours later, you note that the moon has moved relative to the stars to a position 2.3343 degrees eastward from its earlier position. For simplicity, assume the moon's orbit is circular and in the equatorial plane of the earth, and that the earth–moon barycenter is at the earth center. The sidereal day is 86 164.1 s (earth rotation period relative to stars), and the sidereal orbital period of the moon (time to return to its original position relative to stars according to a hypothetical observer at the earth center is 27.322 d (where 1 d = 86 400 s). The moon's orbital motion causes it to move eastward through the sky. (Hint: correct for this.) The equatorial radius of the earth is 6.38×10^6 m. Give your answer in earth radii and meters. [Ans. $\sim 4 \times 10^8$ m]

Problem 9.53. The speed of the earth in its nearly circular orbit about the sun is found to be 29.8 km/s using spectroscopic methods. The sidereal period of the earth's orbit is 365.256 d. (1 d = 86 400 s) What is the distance to the sun? [Ans. $\sim 10^{11}$ m]

Problem 9.54. Consider the receding Hyades cluster. (a) Find the radial (line of sight) velocity v_r (m/s) of Star A at time t_1 (Fig. 3a). Use the numerical values in the text for r and $d\theta/dt$ together with a measurement from Fig. 3a. (b) What is the tangential velocity component v_θ for Star A? (c) How long does it take for the cluster to reach the convergence point? [Ans. ~ 40 km/s; ~ 15 km/s; —]

10

Absorption and scattering of photons

What we learn in this chapter

Our knowledge of celestial objects must take into account **absorption** and **scattering** of photons as they travel to earth observers. These processes are highly **frequency dependent** and thus affect some bands more than others. Photon–electron interactions include **Rayleigh**, **Thomson** and **Compton scattering** which explain, respectively, the **blue sky**, light from the **solar corona**, and a distorted spectrum of 3-K background radiation in the direction of x-ray emitting clusters of galaxies (**Sunyaev–Zeldovich effect**). Photons of very high energy, $\gtrsim 10^{15}$ eV, are absorbed through **pair production** interactions with photons of the **cosmic microwave background**. Photons with energies from 13.6 eV (ultraviolet) through ~ 2 keV ("soft" x ray) are absorbed by atoms in interstellar space through the **photoelectric effect**. Optical light from stars in the plane of the Galaxy is absorbed (**extinction**), reddened (**color excess**), and polarized by **interstellar grains** (**dust**). The polarized starlight maps out interstellar **magnetic fields**. A useful correlation exists between the locations of **dust** and **hydrogen** in the Galaxy.

The beam intensity that survives passage through a uniform absorbing medium **decreases exponentially** with distance traveled. The rate of decrease depends upon the **cross section** (m^2 per absorbing atom) or **opacity** (m^2 per kg) of the absorbing medium. Photoelectric absorption in the **interstellar medium (ISM)** depends strongly on the composition of the interstellar gases (**cosmic abundances**) and is a strong function of photon energy. X-ray (>2 keV) and infrared astronomy are well suited for studies of **stellar systems** in the **plane of the Galaxy** because neither is particularly susceptible to absorption by atoms or by interstellar grains over galactic distances.

10.1 Introduction

Many photons en route to the earth from distant objects do not arrive safely. They may encounter an electron, an atom, a molecule, a dust grain, a star or planet, or even gravitational fields and be absorbed or scattered into different directions of propagation. Scattered photons may appear to an observer as a faint background glow around an object, as a distorted or displaced image, or as a general glow from interstellar space; they no longer arrive directly from the object from which they emanated.

This removal of photons from the image of an object can dramatically affect its appearance, spectrum or temporal variability. An astrophysicist routinely takes these processes into account when studying celestial objects. In fact, the absorption of photons can be quite helpful; it allows detection of diffuse gases between the stars and galaxies. The bending of light beams by gravitational fields allows us to detect dark matter in galaxies and clusters of galaxies.

Interstellar or intergalactic space is not the only place where the loss of photons takes place. It turns out that photons from the interior of a star can not reach the surface; we observe only photons from the star's atmosphere. We infer the physical processes taking place within a star from the indirect evidence offered by its surface. In an attempt to overcome this problem in stars, physicists and astronomers in recent years have been studying *neutrinos* created in the center of the sun. Neutrinos are created in energy-generating nuclear reactions deep within the sun. They have such a low probability for absorption that they can travel freely through the sun and escape from its surface.

Absorption or scattering, in general, will make a distant object appear fainter, just as a distant street light appears dimmed on a foggy night. If the brightness of a star is the primary clue to its distance, a star so dimmed would appear to be more distant than it really is. Stars in the galactic plane are strongly affected in this way. These processes often depend strongly upon the frequency of the radiation being absorbed. For example infrared radiation can emerge from deep within star forming regions (e.g., the Orion nebula) whereas optical light can not.

The amount of absorption can also vary with angular position over an extended object (if the intervening matter is unevenly distributed) or with time (matter occasionally intervenes). An example of the latter is the periodic partial obscuration of x rays from a neutron star in an accreting binary system by gas in the *accretion disk* that comes into the line of sight once each orbit.

Fortunately, the physics of the absorbing and scattering processes is usually quite well known from laboratory experiments and theory. Thus it is often possible to reliably infer information about the distant emitting objects. Furthermore, since the absorbing medium is sometimes a part of the system being observed, absorption can tell us about the nature of the system itself.

This chapter will introduce some of the processes involved in absorption and scattering, specifically photon–electron and photon–photon interactions, the extinction of starlight, and the photoelectric effect. The concepts of cross section and opacity are also presented. The fate of photons of various wavelengths (radio, optical, x-ray, etc.) as they travel is a powerful probe of interstellar and intergalactic spaces. These processes apply equally well to the fate of photons as they traverse the earth's atmosphere (Section 2.4).

10.2 Photon interactions

In this section, we present an overview of processes that lead to absorption and scattering of photons by electrons, photons, atoms, and, in the following section, larger aggregates of matter. Physical derivations of these processes, for the most part, will be found in *Astrophysics Processes* (see Preface).

Photon–electron interactions

The interactions wherein an electron disrupts the serene straight-line travel of a photon are usually distinguished by the different energies of the incident photon. Usually, one may think of the electrons as being nearly at rest, but in certain cases, the electron may be very energetic and even relativistic such as in a hot plasma. The electrons may be free or part of an atom. But, in each of the three cases discussed here, Rayleigh, Thomson, and Compton scattering, the interaction is primarily between the photon and electron.

Rayleigh scattering

A relatively low-energy photon may be absorbed and immediately re-emitted by an atom without causing the atom or molecule to change its energy state. This is known as *Rayleigh scattering* which may be understood classically as the interaction of the photon and one of the electrons. Consider the electron to be bound by a mechanical spring to the atom or molecule and approach the problem as a classical oscillator. The electric field of the incoming electromagnetic wave (photon) causes the bound electron to oscillate. The electron reradiates energy in the form of electromagnetic waves, as any accelerating charged particle must.

The radiated power of an accelerated charge is, you may recall, $P = e^2 a^2/(6\pi\varepsilon_0 c^3) \propto a^2$. For frequencies well below resonance, the acceleration of the oscillator is $a \propto \omega^2$, from the second derivative of $x = x_0 \cos \omega t$. Thus the radiated power varies as $P \propto \omega^4$. Since this radiated power at frequency ω must equal that absorbed from the incoming beam, the scattering is highly frequency dependent, with high frequencies being scattered the most. The scattering of sunlight from molecules in our atmosphere is an example of this. The blue light is scattered more than the red, which gives rise to a blue sky.

Thomson scattering

If the photon energy is sufficiently high, substantially higher than the ionization energy, then the target electrons will appear to be "free", and *Thomson scattering* applies. Like Rayleigh scattering, this too can be described with classical physics. In fact, Thomson scattering is the high-frequency limit of Rayleigh scattering. It describes the scattering of photons by *free* electrons, namely those that are not bound to an atom. However, as noted, it also applies to those that appear to be free because their binding energy is much smaller than the photon energy.

In this (classical) process, the photon is absorbed and immediately reradiated by the electron into a different direction, but it retains essentially all of its initial energy. This is because the photon is assumed to have a much smaller equivalent mass $m_p = h\nu/c^2$ (from $m_p c^2 = h\nu$), than the mass m_e of the electron. Similarly, a bouncing basketball retains most of its energy but changes its direction each time it collides with the much more massive earth.

The above requirement on the equivalent mass of the photon is equivalent to saying that its energy $h\nu$ is much less than the electron rest energy $m_e c^2$, namely that $h\nu \ll m_e c^2$ where h is the Planck constant, ν the frequency of the photon, m_e the electron mass, and c is the speed of light. The rest energy of an electron is $m_e c^2 = 0.51$ MeV. The photon energies must be less than this but more than the few electron volts that bind the outer electrons to an atom or molecule because we require the photon energy in Thomson scattering to be substantially greater than the binding energy.

In the case of Thomson scattering, the probability for scatter turns out to be independent of frequency. The probability of an interaction is often described as a "cross section for interaction" σ which is an equivalent area or bull's-eye of the target object. (The cross section is further described in Section 4 of this chapter.) The constant cross section for Thomson scattering turns out to be

$$\sigma_T = \frac{8\pi}{3}r_e^2 = 6.65 \times 10^{-29} \, \text{m}^2 \quad \text{(Thomson cross section)} \tag{10.1}$$

where

$$r_e = \frac{e^2}{4\pi\epsilon_0 m_e c^2} = 2.82 \times 10^{-15} \, \text{m} \quad \begin{array}{l}\text{(Classical radius} \\ \text{of the electron)}\end{array} \tag{10.2}$$

is the *classical radius of the electron*. This is the name given to a combination of fundamental physical quantities ($e^2/m_e c^2$ in cgs units) that has the dimension of length. Further, we see from (1) that r_e is indeed very close, within a factor of $(8/3)^{1/2}$ of behaving exactly like the radius of the bull's-eye target the photon must strike if it is to be scattered by this process.

X-ray scattering from atmospheric molecules in a room near an x-ray generating machine is an example of Thomson scattering. Most of the electrons involved are the outer electrons of air molecules, and their binding energies are much smaller than the incident x-ray energy.

The white diffuse nebulosity surrounding the sun during a *total solar eclipse* is another example of Thomson scattering. Sunlight is Thomson-scattered by the free electrons in the ionized gas of the million-degree solar corona. Since the scattering probability is independent of photon frequency, the scattered light is white like the source of the light (unlike the Rayleigh scattering off air molecules).

Thomson scattering is also important in accreting x-ray sources where a plasma of protons and electrons is falling onto a compact neutron star. The rate of descent of this plasma may be slowed or even stopped by outgoing x rays that interact with the electrons in the plasma via Thomson scattering. Since it is the infall energy of the plasma that gives rise to the x rays, this leads to a maximum x-ray luminosity, the *Eddington luminosity*.

Compton scattering

If the photon energy is very high (in this context), approaching the rest-mass energy of an electron, $h\nu \approx m_e c^2 = 511$ keV, the probability of scattering from free electrons begins to decrease rapidly with frequency. This high-energy extension of Thomson scattering is called *Compton scattering*. In this regime, the photon will transfer a considerable portion of its momentum (and energy) to the electron, and the balance is carried off by the scattered photon. The cross section gradually decreases from its low-energy value, σ_T, toward zero. (Look ahead to Fig. 6c.) The energy of the scattered photon varies with the scattering angle, just as a baseball does when it hits a bat. This variation can be calculated by conserving energy and momentum in a relativistic calculation.

The threshold for this process suggests that it is applicable only to gamma-ray astronomy. However, *inverse Compton scattering* is important in many branches of astronomy. In this process, high-energy electrons interact with lower-energy photons, boosting the photons to higher energies.

For example, the energetic electrons in the hot plasmas known to exist in clusters of galaxies (from the observed x-ray emission) will scatter photons of the *cosmic microwave background* (CMB) up to higher energies and into different directions. The CMB radiation photons passing through the cluster en route to an observer will, with a low probability, be so scattered. Low-energy photons will be removed from the direction of the observer while photons initially traveling in some other direction will be scattered into it with increased energies.

An astronomer observing the CMB in the direction of a cluster of galaxies will thus detect a slightly reduced intensity of low-energy photons and a slightly increased intensity of high-energy photons compared to view directions with no cluster in the line of sight. In other words, the blackbody spectrum of the CMB (Fig. 11.9b) is slightly distorted. It is not an apparent increase in the temperature of the CMB because increasing the temperature of blackbody radiation raises the intensity at *all* frequencies (Fig. 8b). This distortion of the CMB spectrum in the direction of clusters of galaxies is known as the *Sunyaev–Zeldovich effect*.

Photon absorption in the CMB

Gamma rays at extremely high energy ($>10^{14}$ eV) will interact with the cosmic microwave background (CMB) radiation and thereby cease to exist. The interaction converts two photons into an electron–positron pair, sometimes simply called an "electron pair",

$$\gamma_{\text{gamma}} + \gamma_{\text{cmb}} = e^- + e^+ \qquad \text{(Pair production on microwave} \qquad (10.3)$$
$$\text{background)}$$

The CMB pervades the entire universe and has a thermal temperature of only 2.73 K; it is presumed to originate in a hot early phase of the universe and has since cooled to this frigid temperature. All the stars and planets are embedded in a sea of photons, each with a tiny energy, on average, of $E_{v,\text{av}} = 6 \times 10^{-4}$ eV. (For blackbody radiation, $E_{v,\text{av}} = 2.7\,kT$; see (2.13)). A high-energy gamma ray will interact with a CMB photon to create an electron pair only if there is sufficient energy ($2m_e c^2$) in the center of mass to create the particles. This requires energies in the "laboratory" frame of reference approaching 10^{15} eV (1 PeV).

Consider a gamma ray above the 1-PeV energy threshold. The cross section for a pair-production interaction is so large that it has only an $e^{-1} = 0.37$ probability of surviving a journey of 26 000 LY. Since the center of the Galaxy is about 25 000 LY distant, even galactic astronomy is somewhat compromised, and extragalactic astronomy is out of the question.

This discussion might make it appear that our universe is quite transparent below the threshold, e.g., at $\sim$1 TeV $= 10^{12}$ eV. Unfortunately, this is not true because the expansion of the universe causes photons traveling within it to be gradually reddened. Thus the gamma rays *and* the CMB photons were all of higher energies in times past. Gamma rays originating at great distances from the sun started their journey at these earlier epochs. Thus they could well have found themselves above the threshold for pair production. It turns out that a 1-TeV gamma ray detected at the earth is not likely to have originated more distant than redshift $z \approx 8$; see (9.31) for definition of z. Emission at $z \approx 8$ would have taken place when the universe was only $1/(1 + z) = 0.11$ of its current "size". Earlier phases of the universe are thus not accessible to 1-TeV astronomy.

At this writing there is some speculation that 1-TeV gamma rays might have a substantially shorter range because of a diffuse *extragalactic infrared background* that could be due to bursts of star formation in the early formation of galaxies (*starbursts*). It is difficult to measure such a background directly. The study of gamma-ray spectra from gamma-ray-emitting galaxies is a probe of such a background.

Photon–atom interactions

Photoelectric absorption and absorption lines

An important example of the attenuation of a photon beam in astrophysics is absorption by the hydrogen atoms in interstellar space. In such an interaction, the photon is completely absorbed by the atom and ceases to exist. The converted photon energy can excite the atom to a higher energy state if the photon has exactly the correct energy for the transition. The energy levels of hydrogen and some of the transitions are shown in Fig. 1. This is called a *bound–bound* transition. Such absorption will give rise to *absorption lines* in the spectrum of the detected radiation.

A photon without the correct energy to effect a bound–bound transition will not interact and will proceed unimpeded unless it has sufficient energy to ionize the atom. In this event, it completely ejects the electron from the atom. This process is called a *bound–free transition*, or *photoelectric absorption.* Part of the photon energy goes into overcoming the potential (ionization) energy, 13.6 eV for ground-state hydrogen, and part goes into the kinetic energy of the ejected electron.

We first encountered photoelectric absorption in Section 6.2 in the context of the photomultiplier tube and proportional counter. It is discussed further in Section 5 below. We will encounter atom–photon interactions again in Section 11.3 in the context of continuum spectra. Finally we note that a molecular system can similarly absorb photons via ionization and excitation as well as by being split apart, *dissociation.*

Emission nebulae

When a photon is absorbed by an electron or atom (photoelectric absorption), a portion of the energy may be re-emitted by the atom in the form of spectral lines that can be measured from the earth. Dramatic examples of this are the *H II regions* in the vicinity of newly formed and very hot stars. Energetic ultraviolet photons from these stars can ionize all of the hydrogen in a large region surrounding the star (a *Strömgren sphere* or *H II region*).

The designation H II ("hydrogen two") refers to the ionized state of the hydrogen atom. The neutral atom (1 proton and 1 electron) is called H I ("hydrogen one"). This nomenclature is widely used in astronomy. In the case of carbon, C I is the neutral atom, C II has 1 electron removed, C III has 2 electrons removed, . . . and

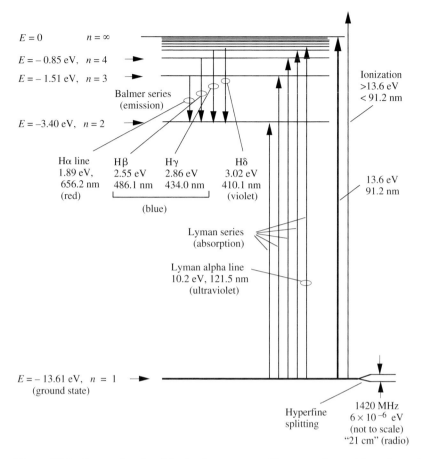

Figure 10.1. Energy levels of the hydrogen atom. Downward transitions (e.g., the Balmer transitions shown) give rise to emitted photons that can be observed by astronomers. Upward transitions between levels (e.g., the Lyman series shown) are caused by absorption of photons of the appropriate energy $E = h\nu$. The atom can be ionized from the ground state by absorption of a photon of energy $E > 13.6\,eV$ (wavelength $\lambda < 91\,nm$). The minuscule hyperfine splitting of the ground state is shown with greatly exaggerated spacing; downward transitions between these levels give rise to the 1420-MHz (21-cm) photons observed by radio astronomers.

C VII has all 6 electrons removed. An H II region consists of ionized hydrogen, that is, lots of unbound electrons and protons, and smaller amounts of heavier elements. The latter will be partially ionized to a degree that depends upon the density and temperature of the gas. (See Section 9.4.)

In a region of ionized hydrogen, a proton and an electron occasionally pass close to one another and recombine to become a neutral hydrogen atom most likely in an excited state. The decay to the ground state leads to the emission of the radiation in the form of emission lines at the frequencies characteristic of hydrogen. For example, the $n = 3$ to $n = 2$ Balmer transition yields a "red" optical photon at wavelength $\lambda = 656.2$ nm (transition shown in Fig. 1). It is known as the Hα *line* ("H alpha line"). The characteristic red radiation seen in color photos of H II regions, e.g., the Trifid nebula and Orion nebulas (Figs. 1.5 and 1.6 respectively, in grayscale) is this Balmer radiation.

Pair production near a nucleus

Photons of energies $\gtrsim 10$ MeV (gamma rays) that pass close to an atomic nucleus will interact with the electric field and spontaneously convert to an electron–positron pair. This is known as *pair production*. The process is fundamentally the same process as that discussed above (3); the electric field of the nucleus consists of virtual target photons.

The created $e^- e^+$ pair will propagate in the forward direction with most of the original energy; see the sketch of the EGRET gamma-ray detector, Fig. 6.5. The probability that pair production will occur in a given interaction increases rapidly with energy. Nevertheless, the numbers of nucleons are sufficiently small in interstellar space so that gamma-ray astronomy of the Galaxy and other distant galaxies is still viable. It is only at extremely high energies of $\gtrsim 10^{15}$ eV that it becomes opaque due to the interactions with the CMB (see above).

10.3 Extinction of starlight

Optical photons traveling in the plane of the Galaxy are heavily scattered by clouds of dust (*interstellar grains*) which limit the range of visibility to, very roughly, 5000 LY, a distance that varies considerably for different viewing directions. (The Galaxy is $\sim 100\,000$ LY in diameter.) Very distant galaxies can be seen if one looks away from the plane of the Milky Way where the dust along the line of sight is minimal.

The grains reduce, by scattering, the amount of light coming to us directly from a given star. The blue photons are preferentially scattered so the residual light is also noticeably redder. This diminution of brightness and reddening is called *interstellar extinction, interstellar reddening* or *interstellar absorption*. The latter phrase is rather misleading because the photons are largely scattered, not absorbed (See Fig. 2a). This is a large and important effect that optical astronomers must take into account for almost all of their observations. It also is a good diagnostic of the dust and magnetic field content of the *interstellar medium (ISM)*.

Grains in the interstellar medium

The nature of the scattering grains can be deduced from the overall (broadband) observed frequency dependence of the extinction and from specific narrow spectral

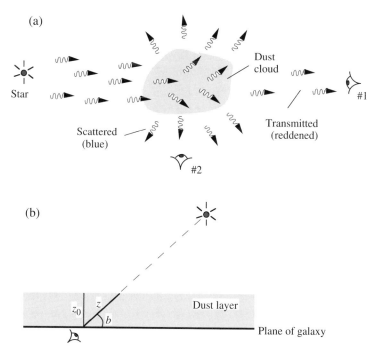

Figure 10.2. (a) Starlight passing through interstellar grains toward observer #1 is fainter and redder because the blue light is preferentially scattered by the grains toward, for example, observer #2. (b) Simple model for absorption of optical light by interstellar grains in the Galaxy. The grains are assumed to occupy a horizontal slab that varies in density only in the vertical direction.

features they generate in stellar spectra. The conclusions are that grains must be solid with sizes of order 200 nm, somewhat smaller than the wavelength of visible light, and that they consist of graphite, ice (water), silicates, etc.

Among the many wavelengths in the impinging light, the smaller (blue) wavelengths are closest to the grain size and hence experience the most scattering. Thus the blue light is preferentially scattered out of the beam en route to the earth. Stated differently, the red light penetrates the dust clouds more efficiently. Infrared radiation is even more penetrating.

Interstellar grains are a significant fraction of the interstellar medium, $\gtrsim 1\%$ by mass. The interstellar medium in turn makes up roughly 10% of the $\sim 6 \times 10^{10} \, M_{\odot}$ of normal (baryonic) matter in the disk of the Galaxy. The dust grains in the disk thus have a total mass in excess of $10^7 \, M_{\odot}$. That is a lot of mass tied up in grains.

A dust grain is very much larger and more massive than a hydrogen atom. Thus even though the mass in grains constitutes a full $\sim 1\%$ of the interstellar *mass* density, their *number* density is manyfold times less than for hydrogen. Typical numbers for the Galaxy are $\sim 500\,000$ hydrogen atoms/m^3 and only ~ 100 dust grains per cubic *kilometer*! The small number densities of grains interfere with starlight much more than the numerous hydrogen atoms because of their extremely high efficiency for scattering optical photons. It is this large effect that makes us so aware of their existence. The interstellar medium is very clumpy; extinction per unit path length along various lines of sight can vary by factors up to ~10.

Interstellar grains are believed to originate from the huge stellar winds of ionized gas (*plasma*) flowing out from *M-giant* stars, and also from mass ejections by other types of stars: carbon stars, planetary nebulae and probably novae and supernovae. These plasmas contain some heavy elements which then condense to grains of iron and silicates. Subsequently atoms of lighter elements strike the grain and are accreted onto it, possibly forming an ice coating.

The light removed by grains from starlight should appear as diffuse blue radiation from the sky, and one might expect it to appear as a diffuse halo surrounding individual stars. However, for most ordinary stars, this light is too faint to be detectable. Reflection nebulae such as the Pleiades are a special case. They are regions of recent star formation wherein the stars are especially bright and the quantities of dust especially high. The bluish scattered light is highly visible and dramatic in the regions between stars in both black and white and color photos of the Pleiades (Fig 1.7). Most stars in the Galaxy are older and in less dense regions of the interstellar medium.

The information about the grains provided by a single star is an average of the conditions all along the path traveled by the photons en route to the earth. Studies of numerous stars that lie at different distances and directions from us provide more detailed information about the distribution of the dust. Its clumpiness leads to the term *dust clouds*.

A very important effect of the interstellar grains is that they slightly polarize the starlight passing through them. This is an indicator of *interstellar magnetic fields* that tend to align the (non-spherical) grains. It is this alignment and the dependence of the scattering probability on grain size (in the direction of the electric field vector of the electromagnetic wave) that gives rise to the polarization. Maps of the sky showing polarization direction as a function of celestial position trace the magnetic field directions.

Extinction parameters

Extinction coefficient

The *extinction* A_V is the number of magnitudes by which light in the visual V band (Section 8.3) is dimmed by the intervening dust. The average or typical amount of extinction of visible light (V band) in the plane of the Galaxy (within ~500 LY of the plane) is about 0.6 magnitudes for each 1000 LY of distance along the line of sight. For galactic latitudes b $\lesssim 2°$,

$$A_V \approx 0.6 \frac{r}{1000 \text{ LY}} \text{ mag} \qquad \text{(Extinction; } b \lesssim 2°; \qquad (10.4)$$
$$V \text{ magnitudes)}$$

where r is the distance to a source given in LY. One must use this expression only as a rough approximation because of the clumpiness of the grains.

The extinction A_V is measured in magnitudes; it is the observed magnitude m_V less the magnitude $m_{V,0}$ that would be measured if the intervening dust were not present.

➡ $A_V \equiv m_V - m_{V,0}$ (Extinction definition) (10.5)

Note that A_V is positive since always $m_V > m_{V,0}$.

The 0.6-mag decrease in light in 1000 LY of travel in the galactic plane (4) amounts to a 58% decrease of the V-band light. This percentage follows from (8.14) where $m_2 - m_1 = 0.6$. Thus, in optical light, one sees only the closer portions of the Galaxy; the galactic center, 25 000 LY distant, can not be seen. The expected brightening of the Milky Way in the direction of the galactic center (in Sagittarius) is much diminished for the observer who sees, for the most part, only the local region of several thousand light years. (There are lines of sight between dust clouds that permit longer views.) If the extinction were significantly greater, the Milky Way would not even be apparent. Only the very closest stars would be visible, and they are isotropically distributed about the sun.

Extragalactic sources

It is fortunate that this extinction is sufficiently low to let us see out of the Galaxy in directions away from the galactic plane. In this way we can study and learn about other galaxies. In the direction of the galactic pole, perpendicular to the plane of the Galaxy, the visual-band extinction is about 0.2 magnitudes. It is possible to construct a simple model that will predict the amount of extinction at other galactic latitudes, but excluding the galactic plane.

Consider the geometry of Fig. 2b. The dust is assumed to occupy a horizontal slab of uniform density and height z_0. The line-of-sight distance at angle b through the dust is:

$$z(b) = z_0/(\sin b) = z_0 \csc b \qquad \text{(m; path length in slab)} \qquad (10.6)$$

What is the relation between path length z and extinction A_V? The extinction A_V (in magnitudes) describes how many factors of 2.51 the radiation flux is reduced; an extinction of $A_V = 1.0$ mag corresponds to a reduction in flux by a factor of

$10^{0.4} = 2.51$. Assume that a path length $z = 1.0\,z_1$ causes such a reduction. Now if the path is doubled to $z = 2.0\,z_1$, the radiation that survived in the first case will be further reduced by a factor of 2.51, or an additional 1.0 mag, for a total of $A_V = 2.0$. Similarly, a third unit of path so $z = 3.0\,z_1$ would result in a total of $A_V = 3.0$. It is apparent from these examples that z and A_V are proportional to one another,

$$A_V \propto z \qquad \text{(Extinction vs. distance)} \qquad (10.7)$$

for the uniform-density model. This relationship was assumed implicitly in (4) and is demonstrated more formally below; see (35).

Now introduce (6) into (7),

$$A_V \propto \csc b \qquad\qquad (10.8)$$

The proportionality factor is obtained by direct measurement of A_V at the galactic pole, $b = 90°$, because $\csc 90° = 1.0$. One obtains ~ 0.18 mag. Thus the approximate visual extinction A_V along a line of sight at galactic latitude b is

➡ $$A_V \approx \frac{0.18}{\sin b} = 0.18\ \csc b\ \text{mag} \qquad \begin{array}{l}\text{(Extinction through} \\ \text{Galaxy; } b \gtrsim 10°)\end{array} \qquad (10.9)$$

This is sometimes referred to as the *cosecant law*. It is applicable for observations of sources *outside* the (MW) Galaxy, i.e., of other galaxies.

The expression (9) is approximately correct over a wide range of galactic latitudes away from the galactic equator, approximately $10° < |b| < 90°$. At lower latitudes the extinction becomes very large and uncertain.

Measurements of extragalactic objects at numerous galactic latitudes can be used to confirm that the flat-slab model of the dust is approximately correct. The $\csc b$ dependence of the path length in our model remains valid even if the dust density decreases with height from the galactic plane as long as it does not vary in the lateral direction. This is approximately true in the vicinity of the sun from where we do our observing.

Color excess (reddening)

The *reddening* of the starlight due to scattering by grains may be described as a *color excess*, E_{B-V}, where

$$E_{B-V} \equiv A_B - A_V \qquad \text{(Color excess for } B \text{ and } V \text{ bands)} \qquad (10.10)$$

Recall that the difference in magnitudes, $m_B - m_V \equiv B - V$, is a color to the human eye because the difference in magnitudes translates to a ratio of the B and V fluxes. With the introduction of dust grains (in a thought experiment), both the blue light (*B* filter) and the yellow light (*V* filter) will become fainter. However, the blue light will decrease more than the yellow light because of the frequency dependence of the scattering. Thus both the *B* and *V* magnitudes increase, but *B* increases more than *V*. This represents the change in the perceived color. The *difference* in the two colors $(B - V) - (B - V)_0$ is described as a *color excess* E_{B-V}. It may be written as $(B - B_0) - (V - V_0) = A_B - A_V$, from (5), which justifies (10).

The extinction A_V and the reddening or color excess, E_{B-V}, go hand in hand; if the one increases because of more dust grains along the line of sight, the other also increases. From measurements, one finds

$$A_V \approx 3E_{B-V} \tag{10.11}$$

Thus a visual extinction of $A_V = 0.6$ mag per 1000 LY in the galactic plane (4) corresponds to a color excess of $E_{B-V} \approx 0.6/3 = 0.2$ mag per 1000 LY. That is, for a star at 1000 LY, the *V* magnitude would be increased 0.6 mag by extinction, but the *B* magnitude would be increased an additional 0.2 mag for a total of 0.8 mag. This latter result may be obtained from the expression

$$A_B \approx 4 \, E_{B-V} \tag{10.12}$$

which follows from (10) and (11).

Frequency dependence

The measured dependence of the extinction $A(\nu)$ upon frequency over a wide range of frequency is given in Fig. 3. The extinction is due to the integrated effect of grains of many different sizes, each with its own efficiency for scattering the light. The ordinate is the ratio of $A(\nu)/A_V$. The effective frequencies of the *V* and *B* bands are shown. The curve passes through unity at the *V* band as it should. In the optical region, the rapid increase of extinction with frequency is apparent.

This increase and the broad peak in the ultraviolet at $\lambda \approx 220$ nm provide information about the sizes and constituents of the grains. For example, the 220-nm peak could be due to small uncoated grains of graphite. The (approximate) straight-line character of the data in the visible region demonstrates that $A(\nu) \propto \sim \nu^1$. This is roughly in accord with the increase of A_B over A_V discussed above.

Astronomers must take care to properly account for extinction when deducing the spectral properties of an object, because the frequency-dependent extinction will modify the spectrum. The measured extinction of an object in the Galaxy, through its spectrum or colors, can be used to deduce the distance to the object if one also measures the extinction of other stars in the same direction in order to build up a model of extinction vs. distance. In this manner one takes into account the clumpiness of the absorbing medium.

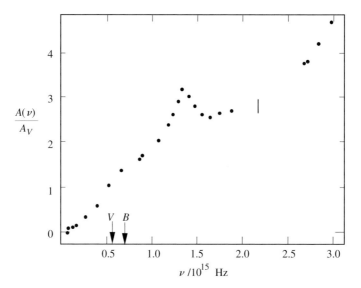

Figure 10.3. Measured extinctions in magnitudes versus photon frequency. The ordinate is the extinction coefficient $A(\nu)$ at frequency ν normalized to the extinction at the visual band, A_V. The effective frequencies of the V and B bands are indicated. The absorption increases approximately linearly with frequency in the optical region. The large bump in the ultraviolet at $\lambda \approx 0.22\,\mu m$ (1.4×10^{15} Hz) may be due to enhanced scattering by small graphite particles. [Adapted from A. Savage and B. Mathis, *Ann. Rev. Astron. Astrophys.* **17**, 75 (1979), with permission]

Dust–hydrogen association

The grains in the Galaxy are quite closely clumped with the hydrogen in the interstellar medium. We now describe how this was determined.

Studies from satellites of individual hot stars at ultraviolet frequencies provide information about the interstellar hydrogen. The ultraviolet photons en route to the earth from the star being studied can excite neutral hydrogen and also molecular hydrogen and hence be absorbed. Thus, the stellar spectra of bright hot stars show absorption lines from these constituents of the interstellar medium. The strength of these features permits one to estimate the amount of hydrogen summed over the distance from the star to the earth.

The data yield a *column density* (atoms/m^2) of neutral hydrogen atoms (H I), including those in H$_2$ molecules, along the line of sight to the star at distance r,

$$N_{\rm H} \equiv \int_0^r n_{\rm H}\,{\rm d}x \xrightarrow[n_{\rm H}\,=\,{\rm const.}]{} n_{\rm H} r \qquad \text{(atom/m}^2\text{; column} \qquad (10.13)$$
$$\text{density)}$$

This is the total number of hydrogen atoms in a column 1 m^2 in cross sectional area and of length r (see Fig. 8.9). It would be more precise to call $N_{\rm H}$ the *number column density*.

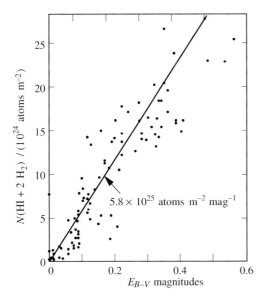

Figure 10.4. Correlation between hydrogen column density (atoms/m^2) and color excess E_{B-V} due to interstellar grains (dust) along the line of sight to ~100 stars. Each point represents a given star. The reddening (abscissa) due to interstellar grains was obtained from studies of the reddening of the same stars in optical light. It is expressed as the color excess E_{B-V}. The hydrogen values (ordinate) are from studies with the Copernicus satellite of ultraviolet absorption lines in the light from hot stars. The clustering of points along a line indicates that hydrogen and dust cluster together in the clumpy interstellar medium. [Adapted from A. Savage and B. Mathis, *Ann. Rev. Astron. Astrophys.* **17**, 86 (1979), with permission]

The dust content along the same line of sight may be obtained from the degree of reddening of optical light from the star. This is possible because the star can be spectroscopically classified, and the intrinsic spectral distribution (or color) of each type of star is presumed to be well known. This and the measured spectral distribution (or color) yield the amount of extinction, or equivalently the color excess E_{B-V}, for the star in question.

The color excess and hydrogen column densities are obtained for a number of independent stars over a large range of distances and directions. The values for each star may be plotted as points on an N_H vs. E_{B-V} plot (Fig. 4). The two quantities are found to be reasonably well correlated; the data points tend to follow more or less a straight line. The best fit to the points in the figure gives the correlation:

$$N(\text{H I} + 2\text{H}_2)/E_{B-V} = 6 \times 10^{25} \text{ atoms m}^{-2} \text{ mag}^{-1} \tag{10.14}$$

where $N(\text{H I} + 2\text{H}_2)$ is the column density of hydrogen *atoms* (one for each neutral atom and two for each molecule). If the color excess E_{B-V} equals 1.0, the column density will be about 6×10^{25} hydrogen atoms/m^2 according to (14).

This correlation suggests that the diffuse interstellar clouds that contain the dust also contain the hydrogen in approximately proportional amounts. Directions that happen to have lots of dust also have lots of hydrogen. The ratio of mass in grains relative to that in hydrogen quoted above, $\gtrsim 1\%$, derives from these measurements. Keep in mind that this result applies to our region of the Galaxy, in the solar neighborhood; it could differ elsewhere.

The ultraviolet absorption lines used for these measurements can be measured only for the brighter (and hence closer) stars. Radio astronomers, in contrast, detect emission from low-lying states of neutral hydrogen and can derive values of N_H out to very large distances. X-ray astronomers can derive values of N_H from x-ray spectra of x-ray emitting compact stars (neutron stars and black holes) throughout the Galaxy (see below). Molecular hydrogen, a very important component of the interstellar medium, unfortunately does not emit a radio signal, but as we have seen, it is detected in absorption by ultraviolet astronomers.

X-ray astronomers often make use of the correlation of Fig. 4. If they derive a value of N_H from their x-ray data for a particular source, they can find E_{B-V} from Fig. 4. This gives the extinction A_V, from (11), expected in the optical counterpart star of the x-ray source. Sometimes the source is in the galactic plane behind so much dust that it is too faint to be detected at optical wavelengths. In such cases, the source can sometimes be located in the infrared for which the extinction is less.

Heretofore we have noted the correlation between N_H and E_{B-V} but we have not commented on the linearity of the correlation, namely the straight-line character on the linear-linear plot of Fig. 4. We will demonstrate below that the definition of the color excess E_{B-V} implies that it is proportional to the grain column density, $N_g \propto E_{B-V}$. Thus, if the grain and hydrogen number densities are everywhere in a constant ratio, we would expect to find the linear relation, $N_H \propto E_{B-V}$, as we do.

10.4 Cross sections

The amount of extinction may be described on the atomic level in terms of the concept of cross section. This is a measure of the probability that a given atom (or ion or dust grain) will absorb or scatter a photon. It typically varies with the frequency of the radiation.

Cross section as a target

The concept of cross section is illustrated in Fig. 5a. The atom is visualized as having a target area, or "bull's-eye", which a photon may or may not strike. This target area is called the *cross section* σ with units of m^2. The value of the cross section for a given atom is tiny, e.g., 10^{-24} m^2. In this simplistic picture, the photon is taken

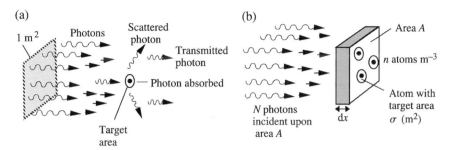

Figure 10.5. (a) Photon beam encountering a small target (atom). Individual photons may (hypothetically) "strike the target" and be absorbed or scattered, or they
may proceed on undisturbed if they "miss" the target. (b) Beam of N photons impinging on targets of cross section σ in a thin slab of area A and thickness dx. The
fractional change in the number of atoms, dN/N, is obtained from this geometry. This expression can be integrated to obtain the exponential attenuation of the
number of atoms in the beam as it passes through matter.

to be a point-like object (with no physical size). Whenever a photon strikes one of
the targets, the photon is taken to be absorbed or scattered by the target atom. If
it misses, it is taken to continue on without an interaction. A cross section, like a
probability, is meaningful only if a well defined process is specified, e.g., "the cross
section for the absorption of a photon of frequency ν by a hydrogen atom through
the ejection of an electron in the $n = 1$ shell".

The size of the target area (cross section) reflects the probability for the absorption
(or scattering) to occur. If the process is quite likely to occur, the cross section is
relatively large; conversely, unlikely processes are described with a small cross
section. The absorption process is actually a quantum-mechanical probabilistic
effect. A photon passing very close to the atom may have a high probability of
interacting and being absorbed while one passing farther away has a low probability.

However, all that one measures is how many photons of a beam are absorbed, not
the details of each interaction. The simplistic all-or-nothing target interpretation is
a way to visualize the meaning of cross section, and it can be used in quantitative
expressions. The cross section for Thomson scattering quoted earlier (1) is an
example of this target-area interpretation.

A conceptual measurement that would determine the cross section is the following. Place one atom in a photon beam of transverse area 1 m^2 (Fig. 5a). The
cross section is the number of scattered (or absorbed) photons divided by the total
number flux in the beam (Fig. 5a),

$$\sigma \, (\mathrm{m}^2) = \frac{\text{Number of scatters per second } (\mathrm{s}^{-1})}{\text{Number of photons in 1 m}^2 \text{ beam per second } (\mathrm{m}^{-2} \, \mathrm{s}^{-1})}$$

$$(10.15)$$

or equivalently

➡ $$\sigma\,(\text{m}^2) \equiv \frac{\text{Rate of scattering (s}^{-1})}{\text{Flux (m}^{-2}\,\text{s}^{-1})} \qquad \text{(Cross section defined)} \qquad (10.16)$$

Consider a beam with 100 photons traversing $1\,\text{m}^2$ in one second (flux $= 100\,\text{m}^{-2}$ s^{-1}) that yields 2 scatters each second (rate $= 2\,\text{s}^{-1}$) from a single target. We would infer that the cross section of the target must be $0.02\,\text{m}^2$, in accord with (16).

Mean propagation distance

The relation between the cross section and the mean (average) distance that the photon will travel is derived here. It takes on several forms.

Exponential absorption

A beam of photons impinges on a thin slab of material (Fig. 5b). The slab has thickness $\mathrm{d}x$ and area A (m^2). It contains many scattering atoms, each with cross section σ (m^2) and a random position in the slab. The slab is taken to be sufficiently thin so that the sum of all the individual cross sections in area A is much less than area A. In the limit of an infinitesimal slice, there are no overlapping target areas.

Let the number of target atoms per unit volume be n (particles/m^3). The number of targets in our slab is therefore $nA\,\mathrm{d}x$, and the total area blocked by all of the target areas is $\sigma n A\,\mathrm{d}x$. Divide this by the slab area A to obtain the *fractional* area blocked, $\sigma n\,\mathrm{d}x$. Now a uniform beam of N photons impinges on the slab in some fixed time. The fraction of these photons absorbed by the slab, $\mathrm{d}N/N$, will be equal to the fractional area blocked, $\sigma n\,\mathrm{d}x$. Thus,

$$\frac{\mathrm{d}N}{N} = -\sigma n\,\mathrm{d}x \qquad (10.17)$$

where the minus sign represents the fact that the number N decreases as the beam passes through the slab in the positive x direction; $\mathrm{d}x > 0$ and $\mathrm{d}N < 0$.

The absorption in material of substantial thickness x, is obtained by adding the contributions of many thin slabs, by integration,

$$\int_{N_i}^{N(x)} \frac{\mathrm{d}N}{N} = -\int_0^x \sigma n\,\mathrm{d}x \qquad (10.18)$$

where N_i and $N(x)$ are the initial and final (at position x) number of photons in the beam, and x is the thickness traversed by the beam. For constant σ and n along the path, integration yields

$$\ln\frac{N(x)}{N_i} = -\sigma n x \qquad (\sigma \text{ and } n \text{ constant}) \qquad (10.19)$$

➡ $$N(x) = N_i\,\exp(-\sigma n x) \qquad \text{(Exponential attenuation)} \qquad (10.20)$$

Thus one finds that, for a uniform medium (fixed σ and n), the beam intensity $N(x)$ decreases exponentially as a function of the distance traversed, x. If the cross section σ or the density of atoms n is increased, the number of surviving photons at a given x decreases, as expected. At distance $x = (\sigma n)^{-1}$, the number has been reduced to $1/e$ of its initial value.

Mean free path

A calculation of the average propagation distance, known as the *mean free path* x_m, turns out, fortuitously, to be equal to the $1/e$ distance. Thus,

$$x_m = (\sigma n)^{-1} \qquad \text{(m; mean free path)} \qquad (10.21)$$

and (20) may be written,

$$N(x) = N_i \, \exp(-x/x_m) \qquad (10.22)$$

The intensity $N(x)$ decreases to N_i/e after traversing the thickness $x = x_m$. At this distance, $63\% \, (1 - e^{-1})$ of the photons are absorbed. One can think of the mean free path as a typical absorption distance. Note that (20) and (22) are identical except that different parameters are used to describe the characteristics of the material being traversed.

Mass units and opacity

Yet another form of (20) makes use of the matter density ρ (kg/m³) of the absorbing material. Multiply and divide the exponent in (20) by ρ,

$$N(x) = N_i \exp\left(-\frac{\sigma n}{\rho}\rho x\right) \qquad (10.23)$$

Introduce the definition of *opacity*,

$$\kappa \equiv \frac{\sigma n}{\rho} \qquad \text{(m}^2\text{/kg; opacity)} \qquad (10.24)$$

and the *column mass density* ξ,

$$\xi \equiv \rho x \qquad \text{(kg/m}^2\text{; column mass density)} \qquad (10.25)$$

and rewrite (23)

$$N(x) = N_i \, \exp(-\kappa \rho x)$$
$$N(\xi) = N_i \exp(-\kappa \xi) \qquad (10.26)$$

The opacity κ(m²/kg) is another commonly used version of the cross section; note from (24) it may be written as $\kappa = \sigma/m$ where m is the mass of the target particle. Its units (m²/kg) indicate that it is the blocked area (sum of all the atomic cross sections) per kilogram of material.

The column mass density $\xi (\text{kg/m}^2)$ describes the distance traversed by the photons in terms of the amount of matter encountered along the way. Consider a column of matter 1 m^2 in cross section; for a given density of matter, its length can be expressed as the number of kilograms it contains, in units of kg/m^2.

In the upper expression of (26), κ and ρ are assumed to be constant. In the lower, changes in density ρ are subsumed into $\xi (\text{kg/m}^2)$ which tells us how much material has been traversed, independent of its distribution along the path. In this case, though, the exponential relation is valid only if κ is constant along the path, as it would be if the mix of types of atoms along the path were fixed so that the cross section per kilogram (κ) does not change.

These expressions lead to other versions of the mean free path. From the last term of (26), the flux is reduced to $1/e$ its initial value in the distance $\xi = 1/\kappa$, giving the mean free path,

$$\xi_m = 1/\kappa \qquad\qquad (\text{kg/m}^2) \qquad (10.27)$$

in units of the column mass density, kg/m^2. Also, the central term of (26) yields

$$x_m = 1/\kappa\rho \qquad\qquad (\text{m; mean free path}) \qquad (10.28)$$

which is the relation between opacity κ and mean free path x_m.

The use of the mass units ξ (kg/m^2) to measure thickness, as in (26), is quite sensible since most absorption processes depend on the number of atoms or electrons encountered by the photon in its travels, and the number encountered is roughly proportional to the mass traversed (kg/m^2). A path through 1 kg/m^2 of lead contains approximately the same number of electrons and nucleons (protons and neutrons) as does 1 kg/m^2 of cotton. Thus one might expect equal absorption in the two materials. In fact, they can be quite different because cross sections depend on the atomic number, the ionization state, and the molecular structure of the target atoms. Mass units thus highlight the physical character of the interactions themselves.

The matter thickness to the center of the Galaxy is $\sim$0.2 kg/m^2, the thickness of 1 m of water is 1000 kg/m^2, and the thickness of the standard earth atmosphere is $\sim$10 300 kg/m^2. The atmosphere is equivalent to 10 m of water or $\sim$0.9 m of lead; the density of lead is 11 300 kg m^{-3}. It is this material that protects us from the energetic cosmic rays that permeate interstellar space and have energies sufficient to not be deflected by the earth's magnetic field.

Optical depth

Astronomers use yet another form of notation to describe absorption, the *optical depth*, $\tau \equiv x/x_m = \kappa\rho x$, which is simply the thickness in units of the mean free path, a dimensionless quantity. The formal definition, allowing for variation of κ

Table 10.1. *Absorption parameters*[a]

Thickness traversed	Exponential attenuation	Mean free path	Condition
x (m)	$N(x) = N_i \exp(-\sigma n x)$	$x_m = 1/\sigma n$	Uniform medium
	$N(x) = N_i \exp(-\kappa \rho x)$	$x_m = 1/\kappa \rho$	Uniform medium
ξ (kg/m^2)	$N(\xi) = N_i \exp(-\kappa \xi)$	$\xi_m = 1/\kappa$	κ constant along path
τ (–)	$N(\tau) = N_i \exp(-\tau)$	unity	

[a] x: thickness (m); σ: cross section (m^2); ρ: mass density (kg/m^3); ξ: thickness in mass units (kg/m^2); n: number density (m^{-3}); N_i: initial number of photons in beam; τ: optical depth (–); κ: opacity (m^2/kg); $N(x)$: final number after traversing thickness x.

and ρ along the path, is ρ is

$$\tau \equiv \int_0^x \kappa(x) \rho(x) \, dx \xrightarrow[\text{const.} \kappa, \rho]{} \kappa \rho x \qquad \text{(Optical depth, dimensionless)} \qquad (10.29)$$

which yields, from (26),

$$N(\tau) = N_i e^{-\tau} \qquad (\tau = \text{optical depth}) \qquad (10.30)$$

This form preserves the exponential dependence even if κ and ρ vary with x, which is often the case. An optical depth of 1.0 means that the photon beam is attenuated by a factor of e^{-1}. An optical depth much less than unity means negligible absorption, and a large value means substantial absorption.

These several ways of describing distances traveled, mean free path, and the exponential absorption are summarized in Table 1.

Cross section and extinction coefficient

Since a cross section σ_g gives rise to extinction of starlight by interstellar grains, one can develop a relation between cross section σ_g and the extinction coefficient A_V. It turns out that the two quantities are proportional to one another as we now demonstrate.

Consider starlight traversing a thickness r of uniform grain density n_g. The photon flux $\mathscr{F}_p(r)$ (m^{-2} s^{-1}) decreases with distance similarly to $N(x)$ in (20). Take the natural log of both sides of (20) and express the numbers N in terms of photon fluxes,

$$\ln \frac{\mathscr{F}_p(r)}{\mathscr{F}_{p,0}} = -\sigma_g n_g r \qquad (10.31)$$

where $\mathscr{F}_{p,0}$ is the flux one would measure with no intervening absorbing material. Convert to the base 10 logarithm,

$$\log \frac{\mathscr{F}_p(r)}{\mathscr{F}_{p,0}} = -\sigma_g n_g r \log e \tag{10.32}$$

This ratio of fluxes can be expressed in terms of the magnitude difference, from (8.12),

$$m - m_0 = -2.5 \log \frac{\mathscr{F}_p}{\mathscr{F}_{p,0}} \tag{10.33}$$

Substitute (32) into (33),

$$m - m_0 = +2.5 \, (\log e) \, \sigma_g n_g r \tag{10.34}$$

and recall the definition of the extinction coefficient (5),

$$A_V \equiv m - m_0 = +2.5 \, (\log e) \, \sigma_g n_g r \propto \sigma_g n_g r \tag{10.35}$$

The extinction coefficient is thus found to be proportional to the cross section σ_g as well as to the density n_g and distance r. The dependence on r is a restatement of (4) and (7). If the matter were not uniformly distributed, one would replace $\sigma n r$ with $\kappa \xi$, where the opacity κ (m^2 kg^{-1}) is another version of the cross section. One could choose to substitute the grain column density N_g, after (13), for $n_g r$, to obtain

➡ $$A_V \propto \sigma_g N_g \tag{10.36}$$

which exhibits the linearity between A_V and σ_g that we set out to show.

The expressions (35) and (36) take us from the microscopic physics of the individual interactions to the macroscopic attenuation of starlight by large column densities of interstellar grains. For example, the extinctions in Fig. 3 for various frequencies are a direct indicator of the variation with frequency ν of the cross section $\sigma_{av}(\nu)$ for photon–grain interactions, averaged over the grains along the lines of sight to a number of stars.

Since $A_V \propto E_{B-V}$ (11), we have, from (36), $E_{B-V} \propto N_g$ as we examine many different stars. If further the hydrogen and grains are clumped with a constant number abundance ratio everywhere, the two column densities along the lines of sight to stars at various locations in the Galaxy will be proportional to one another, $N_g \propto N_H$ from star to star. Hence we have finally

$$N_H \propto E_{B-V} \tag{10.37}$$

which demonstrates the expected linearity of the correlation in Fig. 4.

10.5 Photoelectric absorption in the interstellar medium

The photoelectric effect is an important mode of photon absorption in the Galaxy. It effectively prevents photons with energies $h\nu$ from 13.6 eV to about 2 keV (ultra-violet to "soft" x-ray) from traveling galactic distances in the interstellar medium. At higher photon energies, the cross sections for this process have decreased sufficiently for the interstellar medium to become transparent for distances comparable to the radius of the Galaxy. This allows one to carry out x-ray observations ($\gtrsim 2$ keV) of sources near the galactic center.

Photoelectric effect

The photoelectric effect occurs when a photon ejects an electron from an atom; the photon ceases to exist. The most abundant atom in interstellar space, hydrogen, can be ionized by photons with energies above 13.6 eV ($\lambda = 91.2$ nm), i.e., by ultraviolet or x-ray photons. For pure hydrogen, the cross section for the photoelectric effect is effectively zero below 13.6 eV; at 13.6 eV, it jumps abruptly up to a high value. This transition is called the *K edge* (Fig. 6a). This alludes to the inner orbital K shell of electrons ($n = 1$) from which the electron is ejected.

At photon energies greater than 13.6 eV, the neutral hydrogen atoms can still absorb the photons. The excess photon energy appears as kinetic energy of the ejected electron. As the photon energy increases, beyond 13.6 eV, the cross section decreases roughly as the inverse 8/3 power of the photon energy or frequency,

$$\sigma \propto \nu^{-8/3} \qquad \text{(Cross section for photoelectric absorption)} \qquad (10.38)$$

This takes the form of a straight line of slope $-8/3$ on a log σ − log ν plot (Fig. 6a).

Heavier elements, e.g., oxygen, have electrons in higher energy levels, e.g., $n = 2$ (L shell) and higher, as well as in the K shell. The cross sections for these elements will show not only a K edge but also "L" edges at the energies where the $n = 2$ electrons can be ejected, an energy lower than that of the K edge (Figs. 1 and 6b). The L edges are actually a triplet because the $n = 2$ state has coupling between the electron spins and orbits. The *L* edges are somewhat less pronounced than the K edge. There are also M edges, even less pronounced, from the $n = 3$ level, etc.

The cross sections for heavier elements such as carbon and oxygen are very large at and just above their K edges which are in the x-ray range. Therefore they play an important role in x-ray absorption. The global view of cross section variation with photon energy for the several processes mentioned herein is shown in Fig. 6c.

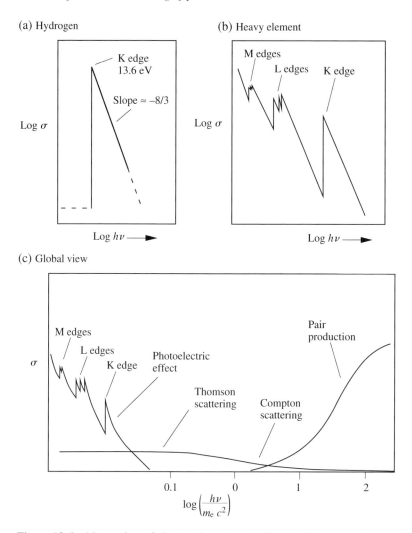

Figure 10.6. Absorption of photons in matter. (a) Sketch of cross section vs. photon energy for photoelectric absorption by hydrogen, near the K edge at $h\nu = 13.6$ eV. (b) A similar sketch for a heavier element. (c) Sketch showing global view of the several processes that can lead to photon absorption. The vertical scale is nominally linear, but the relative amplitudes of the several processes are not to scale. The abscissa is the photon energy in units of the rest mass of an electron, $m_e c^2 = 0.5$ MeV. The Thomson/Compton scattering and pair-production cross sections are sufficiently small so these processes are generally not important in the interstellar medium. Thus x rays and gamma rays can usually traverse the Galaxy without significant absorption. [(c) is adapted in part from E. Fermi, *et al.*, *Nuclear Physics*, University of Chicago Press, 1950, p. 50]

Table 10.2. *Solar-system abundances*[a]

Element (Z)		Abundance by number	Element (Z)		Abundance by number
H	1	1.00	Mg	12	3.85×10^{-5}
He	2	9.75×10^{-2}	Si	14	3.58×10^{-5}
C	6	3.62×10^{-4}	S	16	1.85×10^{-5}
N	7	1.12×10^{-4}	Ar	18	3.62×10^{-6}
O	8	8.53×10^{-4}	Fe	26	3.23×10^{-5}
Ne	10	1.23×10^{-4}			

[a] Abundances from meteorites; adapted from N. Grevesse and E. Anders, in *Cosmic Abundances of Matter*, ed. C. J. Waddington, AIP, New York, 1989, p. 1. The hydrogen abundance is set to 1.00. The elements presented are those with abundances greater than 3.6×10^{-6}.

Cosmic abundances

The net effect of all the atoms in the interstellar medium must be taken into account if the true transparency or attenuation in the Galaxy is to be obtained; different elements dominate the absorption at different frequencies. The photoelectric cross sections for hydrogen and other elements that make up the interstellar medium are well known from earth-bound laboratory experiments and theoretical considerations.

Abundances by number

The relative abundances of the most abundant elements in the solar system are presented in Table 2. They come from studies of meteorites, stony solar-system objects that have fallen to the earth's surface from their circumsolar orbits. Studies of the spectra of the sun's photosphere yield abundances that are in good agreement.

Since the solar system formed from the debris of previous episodes of star formation and evolution (e.g., stellar winds and supernova remnants), these relative abundances may be taken as characteristic of the interstellar medium elsewhere in the Galaxy at the time the solar system was formed, hence the terms *cosmic abundances* and *universal abundances* are sometimes used. Of course fluctuations from place to place in the Galaxy are expected.

Table 2 gives the abundances in terms of the relative *numbers* of atoms. The ratio of the numbers of helium atoms to hydrogen atoms in Table 2 is about one in ten. The other elements are much less abundant, one in 2500 or less.

Mass fractions

In contrast to relative *number* abundances, one can describe the relative abundances in terms of their contributions to the *mass* of the interstellar medium. The fraction of the presumed interstellar mass that resides in a given element can be characterized as a *mass fraction*. The several elements are often grouped to yield the mass fractions X, Y, and Z, for hydrogen, helium and all heavier elements respectively. Since these fractions presumably include all the mass (excluding dark matter), the sum of the fractions must be unity,

$$X + Y + Z = 1 \tag{10.39}$$

The values of these fractions may be calculated from the values in Table 2, giving each element a weighting proportional to its abundance and atomic weight. Let M_A be the total mass of atoms of atomic weight A in some volume of interstellar gas. Then, from the data of Table 2 (including omitted elements),

$$X \equiv \frac{M_{\mathrm{H}}}{\sum_A M_A} = 0.71 \pm 0.02 \qquad \text{(Hydrogen mass fraction)} \tag{10.40}$$

$$Y \equiv \frac{M_{\mathrm{He}}}{\sum_A M_A} = 0.27 \pm 0.02 \qquad \text{(Helium mass fraction)} \tag{10.41}$$

$$Z \equiv \frac{\sum\limits_{A>4} M_A}{\sum\limits_A M_A} = 0.019 \pm 0.002 \qquad \begin{array}{l}\text{("Heavy element"} \\ \text{mass fraction)}\end{array} \tag{10.42}$$

Equation (41) indicates that ~27% of the total mass (of all elements) is in helium, a value significantly larger than the ratio (0.10) of helium to hydrogen (only) by number. The difference arises because each helium atom has ~4 times the proton mass and because the ratios are relative to hydrogen alone in one case and to the total mass of all elements in the other. When quoting abundances one should take care to specify precisely the quantity being used. The two conventions given here are quite standard, namely numbers relative to hydrogen alone (Table 2) and masses relative to the total mass (40)–(42).

Propagation distances in the interstellar medium

The summed contributions of all the elements in the interstellar medium yield a net cross-section curve for photoelectric absorption that can be used to calculate propagation distances in the Galaxy.

Effective cross section

An effective cross section in the interstellar medium may be calculated from our knowledge of the relative abundances of the elements (Table 2). The result for the UV to x-ray region is plotted in Fig. 7a. (The optical region at ~2 eV is off the plot,

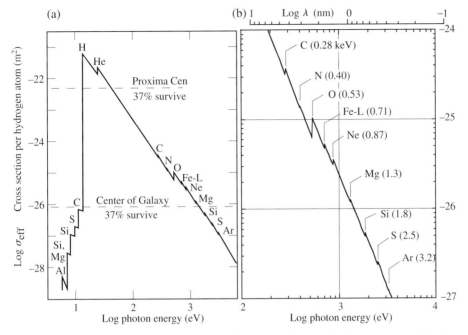

Figure 10.7. Effective photoelectric cross section *per hydrogen atom* of the interstellar medium versus photon energy, for (a) the entire UV to x-ray range and (b) the x-ray range enlarged. The ordinate is the cross section of a single hydrogen atom enhanced to include its share of the cross sections due to heavier atoms. The cross section at any given photon energy is dominated by the element whose K edge is next below the photon energy. Horizontal dashed lines indicate the cross sections that yield attenuation to $e^{-1} = 0.37$ for Proxima Cen and the galactic center, for a typical particle density in the interstellar medium, $n_H = 0.5 \times 10^6$ m^{-3}. Note that the scales of the ordinates are compressed relative to the abscissas. [<13.6 eV from Cruddace *et al.*, *ApJ* **187**, 497 (1974); 13.6–280 eV from H. Marshall personal communication; >280 eV after Morrison and McCammon, *ApJ* **270**, 119 (1983)]

immediately to the left.) This plot is a summation of photoelectric curves like those of Fig. 6, for all different elements and taking into account the relative numbers of each species in the interstellar medium, and normalizing to the hydrogen content as described below.

The discontinuities represent ionization energies of the several elements. At the lowest photon energies (in the ultraviolet), the cross section is due to ejection of electrons from the higher levels of the heavier elements, i.e., Al, Si, Mg, S, C. At the K edge of hydrogen, 13.6 eV, the cross section jumps upward by 5 decades! At this energy, the interstellar medium becomes very opaque. The region above this energy is known as the *extreme ultraviolet*.

Thereafter the cross section decreases approximately according to the −8/3 law (38) except for small jumps due to K edges of the heavy elements. This region is magnified in Fig. 7b with the K-edge photon energies indicated in keV. At cross sections below about 10^{-26} m², the interstellar medium of the Galaxy is quite transparent to radiation (horizontal dashed line for galactic center). Thus, as noted earlier, one can do galactic astronomy in the ultraviolet up to 13.6 eV ($\lambda = 91.2$ nm) and in the x ray beyond ~2 keV.

Most of the effective cross section in the x-ray region (beyond the K edge of carbon) is due to the heavy elements indicated. Their smaller numbers are easily offset by their much greater cross sections (per atom) in the regions near their K edges. These same heavy elements in interstellar grains give rise to optical extinction.

The ordinate in Fig. 7, σ_{eff}, is the *effective cross section per hydrogen atom*. Each H atom is assigned its share of the cross section that is due to other elements. For example, we have seen that there is 1 He atom for every ~10 hydrogen atoms in the interstellar medium. The effective cross section per H atom therefore includes the cross section of a single H atom *and* an additional ~1/10 the cross section of a single He atom. The trace amount of each heavier element is included in the same manner. There is 1 carbon nucleus for each ~2800 hydrogen atoms; thus 1/2800 of the carbon atom cross section is included, and so on. Thus,

$$\sigma_{\text{eff}} = \sigma_{\text{H}} + \sum_i \sigma_i f_i \qquad \text{(Cross section per H atom)} \qquad (10.43)$$

where σ_i is the cross section of element i and f_i is the number abundance relative to hydrogen.

This way of presenting the total cross section, "per H atom", is often quite convenient. Radio astronomers can determine the amount of hydrogen in different parts of the Galaxy by detecting its emission at 1420 MHz and measuring intensities and Doppler shifts. They are thus able to determine densities and locations of hydrogen clouds. The other elements are more difficult to measure and generally are not well known for most regions of the Galaxy. If one assumes that the local (solar system) cosmic abundances are approximately those of the rest of the Galaxy, one can use the cross section σ_{eff} of Fig. 7 in conjunction with the radio measurements of N_{H} to estimate the total absorption for photons from various places in the Galaxy.

Survival distances

The depletion of a photon beam of a given frequency may now be calculated from (20), $N(x) = N_i \exp(-\sigma nx)$, for a uniform medium of hydrogen number density n_{H} and effective cross section σ_{eff}. In this case, $\sigma nx \rightarrow \sigma_{\text{eff}} n_{\text{H}} r = \sigma_{\text{eff}} N_{\text{H}}$ where r is the distance to the source of the photons and N_{H} is the column density of hydrogen atoms (m^{-2}) from the observer to the source. Thus, in terms of photon fluxes, the attenuation at frequency v is

$$\frac{\mathscr{F}_\mathrm{p}(v)}{\mathscr{F}_{\mathrm{p},0}(v)} = \exp(-\sigma_\mathrm{eff}(v)N_\mathrm{H}), \qquad \text{(Attenuation in the ISM} \qquad (10.44)$$
$$\text{at frequency } v)$$

where $\sigma_\mathrm{eff}(v)$ is obtained from Fig. 7, and N_H may be inferred from the radio data.

Knowledge of σ_eff and N_H yields the ratio of photon fluxes. For an astronomical source, the two fluxes in the ratio may be considered to be those one would measure at the earth, the one in the numerator with the absorbing material in place and the one in the denominator without it, hypothetically. If the flux at the earth is measured, one thus finds the intrinsic (without absorption) flux.

Consider 1.0-keV x rays traveling in the plane of the Galaxy, e.g., from the galactic center to the sun, a distance of $\sim$25 000 LY. A nominal value for the interstellar number density of neutral hydrogen n in the galactic plane is $n \approx 0.5 \times 10^6$ m^{-3}, but one must keep in mind that the actual value from place to place in the Galaxy can vary by many orders of magnitude, from $\sim 10^3$ to $\sim 10^{10}$ m^{-3}. The curve of Fig. 7b, for an energy of 1.0 keV, yields $\sigma_\mathrm{eff} = 2.4 \times 10^{-26}$ m^2. Thus, the values,

$$n_\mathrm{H} = 0.5 \times 10^6 \text{ H atoms m}^{-3} \qquad \text{(Galactic center to sun)}$$
$$r = 25\,000 \text{ LY} = 2.4 \times 10^{20} \text{ m} \qquad (10.45)$$
$$\sigma_\mathrm{eff} = 2.4 \times 10^{-26} \text{ m}^2 \text{ (1 keV x rays)}$$

may be used in (44), recalling that $N_\mathrm{H} = n_\mathrm{H}\, r$,

$$\frac{\mathscr{F}_\mathrm{p}}{\mathscr{F}_{\mathrm{p},0}} = \mathrm{e}^{-2.9} = 0.055 \qquad \text{(1.0 keV)} \qquad (10.46)$$

This tells us that 5.5% of the 1-keV radiation from the center of the Galaxy would reach the earth if our assumptions are correct. At higher photon energies, the cross section is less, and the Galaxy quickly becomes quite transparent; at lower energies, the absorption rapidly becomes greater.

The problem can be turned around to find the photon energy at which the photon flux will be attenuated to $\mathrm{e}^{-1} = 0.368$ of its initial value as it travels from the galactic center to the sun. Adopt the values of n_H and r in (45), and $\mathscr{F}_\mathrm{p}/\mathscr{F}_{\mathrm{p},0} = 0.368$, and then solve (44) for σ_eff. Then enter Fig. 7b to find the photon energy corresponding to σ_eff. The result is $E \approx 1.5$ keV. At this photon energy, 37% of the photons would survive the trip from the galactic center to the earth.

This calculation serves to fix one point on the 37% curve of Fig. 8 which gives propagation distances (in terms of the column density N_H) that can be reached at different photon energies. In other words, the 37% curve tells us that, at $E = 1.5$keV,

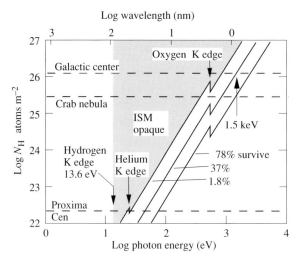

Figure 10.8. Distances of photon travel in the interstellar medium as dictated by the photoelectric effect. (The earth's atmosphere is not included nor is extinction by dust.) This plot can be derived (approximately) from the cross section data of Fig. 7, but here we omit most of the K edges. The hydrogen column density $N_H = n_H r$ is plotted against photon energy. The shaded region is excluded from observation. The solid lines indicate, for a given energy photon, the "distance" (actually column density N_H) that a photon beam has traveled when it has been reduced to $e^{-0.25}$ (78%), e^{-1} (37%), and e^{-4} (1.8%) of its initial numbers. The values of N_H to three celestial objects are shown for an assumed uniform hydrogen number density of $n_H = 0.5 \times 10^6$ m^{-3}. [Adapted from M. Oda, in *Proceedings of 9th International Cosmic Ray Conference*, 1965; also in Rossi, *IAU Symposium #37*, p. 1]

photons can penetrate to $N_H = n_H r = 1.2 \times 10^{26}$ H atoms/m^2, approximately the galactic center distance, with a ~37% chance of survival.

Astronomy through gases

Similar calculations provide the entire 37% curve as well as the $e^{-0.25} = 0.78 = 78\%$ and $e^{-4} = 1.8\%$ curves (Fig. 8). At a given N_H, higher x-ray energies yield greater survival fractions because the cross sections are lower, while lower photon energies yield smaller survival fractions. The entire *shaded* region to the left of the 1.8% curve is quite inaccessible to photon astronomy. The photon energy where the opacity sets in is a function of N_H, or roughly the distance to the source, as given in the figure.

Below 13.6 eV, the interstellar gases in the Galaxy are quite transparent because the photons have insufficient energy to eject a ground-state electron from the atom. (This ignores the large contribution of interstellar dust to the absorption of optical radiation.) In principle such photons could eject electrons from the higher $n = 2$,

3 states, which would require less energy. But, since very few hydrogen atoms in cold interstellar space are in these higher states, this makes little contribution to the absorption.

Photoionization can also take place in the higher atomic levels of heavier elements, such as Si, but the net absorption is small; see the edges to the left of the hydrogen K edge (Fig. 7a). The Hubble satellite carries out ultraviolet observations to 10.8 eV, just beyond the 10.21 eV Lyman α line (Fig. 1) and not quite to the 13.61 eV hydrogen K edge.

On the far side of the cross-section peak (Fig. 7), distant objects again become visible. The distances of Proxima Cen, the Crab nebula and the Galactic center (4.3 LY, 6 000 LY and 25 000 LY respectively) are indicated on Fig. 8, still under the assumption of a uniform hydrogen density $n = 0.5 \times 10^6$ H atoms/m^3. The value of $n_\mathrm{H}r$ for the line of sight to the Crab nebula would be, from (13),

$$N_\mathrm{H} \approx n_\mathrm{H}r = 0.5 \times 10^6 \times 6000 \times 0.95 \times 10^{16} = 2.9 \times 10^{25} \text{ H atoms/m}^2$$
$$(10.47)$$

The Crab, being closer than the galactic center, becomes visible at a somewhat lower energy, $\sim$0.8 keV for 37% survival (Fig. 8), compared to the 1.5 keV calculated above for the galactic center. Proxima Cen is even closer; it can be reached at x-ray–ultraviolet energies as low as 50 eV (37%) or 25 eV (2%).

This discussion has been based in part on an interstellar medium that is a uniform gas of neutral atoms. In fact, interstellar space also contains large molecular clouds, interstellar grains, and regions of low-density, highly ionized gas. Our sun appears to be in one of the latter regions. The clumpiness of the interstellar medium causes the visibility to be much better (in terms of LY) in some directions than in others.

Observations of nearby objects have been successfully carried out in the seemingly excluded *extreme ultraviolet* region, above $\sim$20 eV with the ROSAT and EUVE satellites. Numerous nearby late-type stars and white dwarfs have been detected with these systems. Infrared astronomy is relatively immune to interstellar grains so, like x-ray astronomy, it is a prime tool for studies deep into the Galaxy.

Extragalactic astronomy is usually carried out with lines of sight directed toward the pole of the Galaxy to minimize absorption. This requires that the Galaxy be transparent to the radiation of interest for relatively short distances, several hundreds of light years, rather than the 25 000 LY to the galactic center.

Our discussion pertains to absorption in *interstellar space*; the radiation may or may not be able to penetrate the earth's atmosphere (Fig 2.2). Space observatories on rockets, balloons, or satellites are required for much of the frequency band shown in Fig. 8.

Problems

10.2 Photon interactions

Problem 10.21. Locate color photos of the Pleiades star cluster and the two H II regions, the Lagoon and the Trifid nebulae. Contemplate the dust grains and hydrogen atoms from whence come, respectively, the blue and red light. Do you find any dust in the H II regions? Any hydrogen in the Pleiades? (Hint: look for posters around the astronomy department, in any general astronomy textbook, or at astronomy sites on the internet, e.g. Astronomy Picture of the Day (APOD).

10.3 Extinction of starlight

Problem 10.31. (a) Calculate the mass m_g of a typical interstellar grain. Adopt the density of water ($\rho_G = 1000$ kg/m^3) for the grain density and a spherical grain size (diameter) of 0.2 μm. Compare to the mass and size of a hydrogen atom. (b) Find the average *number* density (#/m^3) of grains in the interstellar medium (ISM) of the Galaxy if they constitute ~1% of the *mass* of the interstellar hydrogen. The number density of hydrogen atoms in the ISM is $n_H = 500\,000$ m^{-3}. Do not confuse the mass density of a single grain ρ_G with the mass density ρ_g of grains per unit volume of the ISM. (c) How many grains would occupy a volume comparable to that of the Empire State Building in New York City? Assume it fills a small city block (~70 m square) and is ~100 stories high. [Ans. ~10^{-18} kg; ~10^{-6} m^{-3}; a few]

Problem 10.32. (a) Roughly how much absorbing dust in the Galaxy would be required to make the stars in the sky appear to be isotropically distributed, i.e., so the bright Milky-Way band across the sky would not be visible in a survey that records all stars brighter than $V = 15$ mag and none fainter? Express your answer in terms of the extinction A_V of visible light per 1000 LY. Consider that the dust and stars extend with uniform densities up to a height of ±500 LY from the galactic plane and no further, and that the observer is located exactly at the galactic plane. Assume that all stars in the Galaxy are identical to the sun, with absolute magnitude $M_V = +4.8$. (b) The eye can see to about $m_V = 6$ mag on a clear dark night. How much extinction is required in this case? (c) What is the effect in part (a) of having stars with a distribution of luminosities? (d) To what distance would sun-like stars be visible (i.e., $V < 15$ mag) with the typical galactic plane extinction of 0.6 per kLY given in (4)? [Ans. ~10 mag/kLY; —; —; ~2 kLY]

Problem 10.33. Show that the csc b dependence of the extinction A in (9) remains valid even if the density of grains varies with height (distance from the plane) as long as it does not vary in the lateral directions (see Fig. 2b).

Problem 10.34. The bright (B3V) star at the end of the handle of the Big Dipper is η U Ma ($m_V = 1.9$; $M_V = -1.6$); it lies in a direction close to the much more distant galaxy M51 at galactic latitude $b = +70$. (a) Estimate the extinction A_V of the light from this star. Assume that the star is embedded within a uniform galactic dust layer with the extinction given in (4). Hint: first find an estimated distance to η U Ma with no extinction correction, then iterate to improve the distance estimate. (b) Estimate from (9) the extinction A_V for light from the galaxy M51. Compare the two answers. [Ans. $\sim$0.1 mag; $\sim$0.2 mag]

10.4 Cross sections

Problem 10.41. Consider gamma rays interacting with the cosmic microwave background (CMB) to produce electron pairs via the reaction (3), $\gamma_{\text{gamma}} + \gamma_{\text{cmb}} = e^+ + e^-$. (a) Confirm the statement in Section 2 that the required gamma-ray energy E in eV for this reaction, $\gamma_{\text{gamma}} + \gamma_{\text{cmb}} = e^+ + e^-$, to occur is $\gtrsim 10^{15}$ eV. Use the relativistic expression, $E = (m_e c^2)^2/(h\nu)$, for the threshold gamma-ray energy for such an interaction (for a two-photon head-on collision), where $m_e c^2 = 0.51$ MeV is the electron rest energy and $h\nu = 6 \times 10^{-4}$ eV is the average photon energy in the CMB. (Optional: you may derive this threshold expression if you know some special relativity.) (b) How far (in light years) will a gamma ray of energy $E = 1$ PeV $(= 10^{15}$ eV) travel with a 37% chance of not interacting with the CMB. Repeat for a 1% chance. At this energy, the cross section for the interaction with a given photon of the CMB is $\sigma \approx 1 \times 10^{-29}$ m^2. The number of CMB photons per unit volume is $n = 2.0 \times 10^7 T^3$ m^{-3} where the temperature of the CMB is $T = 2.735$ K. Assume the universe is static. How do your answers compare to the size of the Galaxy? [Ans. —; $\sim$10^4 LY, $\sim$10^5 LY]

Problem 10.42. What is the distance in mass units ξ (kg/m^2) to the galactic center if the physical distance is $x = 25\,000$ LY and the average density of interstellar gas is $n = 0.5 \times 10^6$ hydrogen atoms per m^3? Neglect the presence of heavier atoms. [Ans. $\sim$0.2 kg m^{-2}]

Problem 10.43. Find x_{av}, the average distance traversed by photons in a beam being attenuated as $N(x) = N_i \exp(-\sigma n x)$. Carry out the integration required by the formal definition of an average taking into account all path lengths from 0 to ∞. Hint: the probability of an interaction occurring at the distance x is proportional to the number of photons remaining in the beam at that x. What is the relation of your answer to the quantity designated x_{m} in the reading?

10.5 Photoelectric absorption in the interstellar medium

Problem 10.51. (a) Use the plot of effective cross section (Fig. 7a) to find the photon energy where 1/10 of the photons survive a journey in the ISM, through a hydrogen column density of $N_{\text{H}} = 10^{24}$ m^{-2}. Plot your result as a point on Fig. 8 to illustrate how the curves of Fig. 8 are constructed. (b) In what distance will 1.0 keV photons suffer 50% absorption in the ISM if the neutral hydrogen gas is of density 5×10^5 atoms m^{-3}. Again use Fig. 7. [Ans. $\sim$100 eV; $\sim$8 kLY]

Problem 10.52. Consider that a cross section varies with frequency ν as a power law, $\sigma = k\nu^{\alpha}$, where k and α are constants. Demonstrate that a plot of $\log \sigma$ vs. $\log \nu$ will yield a straight line. How is the slope of the straight line related to the given parameters? What is the slope of the long straight-line segment of Fig. 7a? [Ans. slope $\approx -8/3$]

Problem 10.53. Derive the *mass* fractions as defined in (40), (41), and (42) from the *number* abundances in Table 2. Assume that essentially all the mass is in the listed elements. You will have to look up the atomic weights elsewhere, e.g. on the physics.nist.gov website. Compare your results to the values quoted in the text.

11

Spectra of electromagnetic radiation

What we learn in this chapter

The distribution with frequency of radiation from a source is called a **spectrum**. It can be plotted as an energy spectrum or as a number spectrum and as a function of either frequency or wavelength. **Conversions** from one to another are possible and useful. **Continuum spectra** are without spectral lines though spectral lines may be superposed upon them. They can arise from **interactions of atoms and free electrons**, for example in the solar atmosphere. Three kinds of such spectra encountered in astronomy are **thermal bremsstrahlung** from an optically thin gas, **blackbody radiation** from an optically thick gas in thermal equilibrium, and **synchrotron radiation** from a gas of extremely energetic electrons in the presence of magnetic fields. **Antenna temperatures** used by radio astronomers are a measure of specific intensity. The total power radiated by unit area of a blackbody, σT^4, allows one to relate approximately the **radius of a star** to its **luminosity** and **temperature**.

Spectral lines arise from atomic transitions in emitting or absorbing gases. They provide powerful **diagnostics** of the regions that form the lines. **Stars** exhibit mostly **absorption lines** while **gaseous nebulae** exhibit **emission lines**. Some of the latter are **forbidden lines** which occur only at the extremely low gas densities found in space. The **shapes of spectral lines** reveal the presence of **turbulent motions** and the effects of **collisions**, the latter providing the **local density**. The **curve of growth** of a spectral line relates the **equivalent width** to the **atomic column density** along the line of sight. The **formation of spectral lines** is quantified with the **radiation transfer equation**. Its **solution** for different conditions gives insight to the formation of both absorption and emission lines.

11.1 Introduction

The spectral distribution of electromagnetic radiation from a celestial object can reveal much about the physical processes taking place at the object. The *spectral*

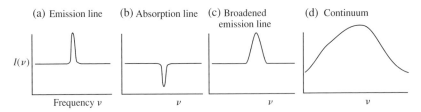

Figure 11.1. Idealized spectra of radiation: sketches of (a) line emission, (b) line absorption, (c) a broadened emission line, and (d) continuum emission. The broadening could arise from differing Doppler shifts due to differing velocities of different portions of the cloud, e.g.,from rotation or turbulence.

distribution is the variation of intensity with frequency. One aspect of such studies is the study of *spectral lines*. This is an excess or deficiency of radiation at a specific frequency relative to nearby frequencies (Figs. 1a,b,c). These are called *emission lines* and *absorption lines* respectively. These lines can be quite narrow or they can be substantially broadened due, for example, to Doppler shifts arising from thermal and turbulent gas motions.

Another kind of spectral distribution is the *continuum spectrum* (Fig. 1d). This is a spectrum that varies smoothly with frequency. An emission line may or may not be superposed upon a continuum spectrum. By its very nature, an absorption line must lie upon a continuum background (Fig. 1b).

The intent of this chapter is to familiarize the reader with plotting conventions and with three commonly encountered continuum spectral shapes (optically thin thermal bremsstrahlung, synchrotron radiation and blackbody radiation). Spectral lines are formed as radiation passes through a medium of atoms. We develop the radiative transfer equation which describes this process. Its solution for various limits provides insight into the formation of spectral lines. Finally, we examine the significance of line strengths and shapes.

11.2 Plots of spectra

The manner in which astronomical spectra are presented (graphically and mathematically) is closely associated with the way we think about them. Here we discuss variants of the measurable quantities presented in Sections 8.2 and 8.4 as well as some practicalities of their use.

Energy and number spectra

The (*energy*) specific intensity has previously been defined (8.26) as follows:

$$I(\nu): (\text{W m}^{-2} \text{ Hz}^{-1} \text{ sr}^{-1}) \qquad (\textit{Energy} \text{ specific} \qquad (11.1)$$
$$\text{intensity)}$$

This quantity is appropriate for a diffuse (resolved) celestial source, one larger in angular size than the telescope resolution. It is often rewritten in terms of other combinations of units. For example, one could choose,

$$I_p(\nu): \text{(photons s}^{-1}\text{ m}^{-2}\text{ eV}^{-1}\text{ sr}^{-1}) \qquad \textit{(Number} \text{ specific} \qquad (11.2)$$
$$\text{intensity)}$$

which illustrates the two major variants: (*i*) the intensity is expressed as number of photons per second rather than power (energy per second), and (*ii*) the band is expressed in energy units (eV or J) rather than in frequency or wavelength units. A spectrum plotted with photons rather than energy in the numerator is often called a *photon spectrum* or a *number spectrum* in contrast to an *energy spectrum*. Note that the quantity "photons" is a dimensionless number,

$$\text{(photons s}^{-1}\text{ m}^{-2}\text{ Hz}^{-1}\text{ sr}^{-1}) \equiv (\text{s}^{-1}\text{ m}^{-2}\text{ Hz}^{-1}\text{ sr}^{-1}) \qquad (11.3)$$

The choice between energy and photon spectra is quite arbitrary. The information content is the same in either case because at any frequency ν, the energy E of a photon is known to be

$$E = h\nu \qquad \qquad \text{(J; photon energy)} \qquad (11.4)$$

where $h = 6.63 \times 10^{-34}$ J s. The units of Hz are inverse seconds (s^{-1}) so the units balance.

The *number* specific intensity is converted to *energy* specific intensity simply by multiplying the number of photons in a narrow energy band $I_p\,d\nu$ by the energy, $E = h\nu$, of each photon in that band,

$$I(\nu)\,d\nu = h\nu I_p(\nu)\,d\nu$$
$$I = h\nu I_p \qquad \qquad (11.5)$$

In this conversion, one must use self-consistent units, for example $h\nu$ in joules and I and I_p in the SI units of (1) and (3).

If the data represent the spectrum of a *point* (unresolved) source, the *spectral flux density S* describes the radiation. In energy units,

$$S(\nu): \text{(W m}^{-2}\text{ Hz}^{-1}) \qquad \qquad \text{(Energy spectral flux} \qquad (11.6)$$
$$\text{density)}$$

or in terms of photons,

$$S_p(\nu): (\text{s}^{-1}\text{ m}^{-2}\text{ Hz}^{-1}) \qquad \qquad \text{(Number spectral flux} \qquad (11.7)$$
$$\text{density)}$$

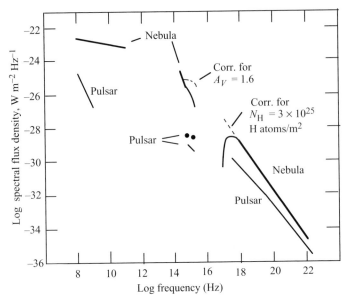

Figure 11.2. Spectrum of the Crab nebula and Crab pulsar from radio through gamma-ray frequencies. The log of the energy spectral flux density is plotted vs. log frequency. The straight-line segments are power-law spectra typical of the synchrotron emission process. The dashed lines show the spectrum that would be observed in the absence of interstellar dust and atoms. The time-average pulsar fluxes are shown as lines; typical peak fluxes of pulses are shown at two frequencies in the optical (dots). [Courtesy G. Fazio]

An example is the spectrum of the Crab nebula (Fig. 2). The Crab is not really a point source; it is a nebula several arcminutes in angular extent. However, if the instrument beam encompasses the entire source, it appears as an unresolved point source. Stated otherwise, the flux plotted is the specific intensity integrated over the angular extent of the nebula.

Astronomers tend to choose units that yield numbers close to unity because it is easier to say them. Since radio sources are many orders of magnitude less intense than the SI unit of spectral flux density (W m^{-2} Hz^{-1}), radio astronomers found it convenient to define a *jansky*, abbreviated as Jy,

$$1 \text{ Jy} \equiv 10^{-26} \text{ W m}^{-2} \text{ Hz}^{-1} \tag{11.8}$$

The jansky was named after Karl Jansky, the discoverer of celestial radio waves (in 1932). This unit is used in Table 8.2.

Spectral reference band

Radio astronomers usually work in frequency units; optical and infrared astronomers think and work in terms of wavelength; while x-ray and gamma-ray astronomers typically use energy units of keV or MeV. Therefore one often sees the intensity or flux quoted with respect to (per unit) wavelength λ or energy E rather than frequency ν.

Frequency and wavelength

The conversion of the specific intensity between the frequency and wavelength units was quoted in (8.43). We now derive this conversion, but we do it for the flux density S (W m^{-2} Hz^{-1}; frequency units). The flux density in wavelength units S_λ is power per unit area per unit wavelength interval:

$$S_\lambda: (\text{W m}^{-2}\ \Delta\lambda^{-1}) \text{ or } (\text{W m}^{-2}\ \text{m}^{-1}) \text{ or simply (W/m}^3) \tag{11.9}$$

The spectral flux density S in our usual frequency units may be written with subscript ν to distinguish it from S_λ,

$$S(\nu) \equiv S_\nu \qquad\qquad (\text{W m}^{-2}\ \text{Hz}^{-1}) \tag{11.10}$$

The energy flux $\mathscr{F}$ (W m^{-2}) in a given band ν_1 to ν_2 (or the corresponding λ_1 to λ_2) must be independent of the manner in which it is expressed, giving

$$\mathscr{F} = \int_{\nu_1}^{\nu_2} S_\nu \, d\nu = -\int_{\lambda_1}^{\lambda_2} S_\lambda \, d\lambda \qquad\qquad (\text{W/m}^2) \tag{11.11}$$

The minus sign is necessary if both S_ν and S_λ are to be positive quantities. (Convince yourself of this; note that $\lambda_2 < \lambda_1$ if $\nu_2 > \nu_1$.) This equality (11) is valid over any arbitrary band pass, that is, for any arbitrary interval ν_1, ν_2 (and the corresponding interval λ_1, λ_2). The only way this could be true is if the equality also holds for the integrands in (11),

$$S_\nu \, d\nu = -S_\lambda \, d\lambda \tag{11.12}$$

The frequency and wavelength in a vacuum are related as

$$\lambda = c/\nu \tag{11.13}$$

which yields the relation between the differential intervals $d\lambda$ and $d\nu$, by differentiation,

$$d\lambda/d\nu = -c/\nu^2 \tag{11.14}$$

Substitute into (12),

$$S_\nu = -S_\lambda(d\lambda/d\nu) = S_\lambda(c/\nu^2) = S_\lambda(\lambda^2/c) \tag{11.15}$$

where (13) was invoked again.

This is the desired conversion from S_λ to S_ν. It may be restated with our convention, $S \equiv S_\nu$, and with the functional arguments,

$$\blacktriangleright \qquad S(\nu) = S_\lambda(\lambda)\frac{\lambda^2}{c} \qquad\qquad (11.16)$$

The expression (16) properly transforms the bandwidths. The specific intensity I can be converted with the same ratio λ^2/c according to the same line of reasoning.

Frequency and energy

Energy units in the reference band in use by high-energy astronomers are usually keV or MeV rather than joules. Thus one might see S (erg cm^{-2} s^{-1} keV^{-1}). The conversion of this expression to our standard SI units S (W m^{-2} Hz^{-1}) requires only numerical multiplicative factors. The conversion of erg cm^{-2} to J m^{-2} is made with the factors 10^{-7} J/erg and 10^4 cm^2/m^2. The conversion of keV^{-1} to Hz^{-1} requires the conversion factor $1/(1.6 \times 10^{-16})$ keV/J and then the expression $E = h\nu$ which provides the conversion factor $h = 6.63 \times 10^{-34}$ J/Hz (Section 2.3). The conversion requires multiplication by the product of these four factors, or 4.14×10^{-21}.

X-ray astronomers sometimes use units of I(keV s^{-1}cm^{-2} keV^{-1} sr^{-1}). The two keVs may seem confusing, but remember that the energy term in the numerator refers to the accumulated energy and that in the denominator to the width of the energy band in which the data are being accumulated. This awareness makes straightforward the various conversions between these units.

Spectral bin widths

Data may be taken with various frequency resolutions. A high resolution (narrow bandwidth) allows one to detect and study narrow spectral lines. Since a number must be stored or telemetered for each of the many narrow spectral bands, a high resolution can require a lot of data storage space (or telemetry from a space vehicle) if the overall bandwidth is large. If, in addition, high time resolution is required, the spectral numbers must be stored for every time interval. The required data storage can be huge.

The observation time required for a spectral measurement must be such that sufficient signal or statistics are obtained in each narrow spectral or time band (often called *channel* or *bin*). For example, $N = 100$ counts (photons) must be accumulated in each channel if each is to yield a 10% uncertainty in the measured value. (See Section 6.5; we assume here negligible background.) If the 2.0 to 12.0 keV band is divided into 10 channels of 1-keV width and if the accumulation rate is uniform across the channels, a total of 1000 counts must be accumulated. Improved energy resolution of 0.01 keV, again with 10% precision, requires 1000 channels accumulating a total of 10^5 counts. The observation time would be 100 times longer than that for the 10-channel case.

Thus, measurements with high spectral resolution carry a high price and must be used only when needed, such as for the detailed measurement of a specific spectral line. If only the broad spectral shape is desired, as in Fig. 2, then broad frequency bins will suffice.

11.3 Continuum spectra

Important examples of continuum spectra are *blackbody radiation, synchrotron radiation*, and *optically thin thermal radiation*. Each has a characteristic shape on a spectral plot and each reflects particular physical conditions in the source. For example, if the specific intensity or the spectral flux density of a radio source increases with frequency, the mechanism that gives rise to the emission is most likely blackbody radiation, and if it decreases with frequency it is most likely to be synchrotron radiation. The Crab nebula spectrum is an example of a continuum source with a spectrum that extends from radio to gamma-ray energies (Fig. 2). The radiation is mostly due to synchrotron radiation.

We present an observer's view of these emission mechanisms here. We derive the spectral shapes from basic physical principles in *Astrophysics Processes* (see Preface).

Free–bound, bound–free and free–free transitions

On the atomic level, continuum spectra can arise from *free–bound* or from *bound–free* transitions of the atom. The former occurs when an electron makes a transition from an unbound (free) state to a bound state, thereby emitting a photon with the energy of the transition. Since the unbound states are a continuum of energy levels, the energies, or frequencies, of the emitted photons are not restricted to well-separated discrete levels. Thus a continuum of photon frequencies (a continuum spectrum) is observed in emission.

The *bound–free* transitions are the reverse process wherein an energetic photon is absorbed and an electron is ejected into the continuum of free states. These are the photoelectric reactions discussed in Sections 10.2 and 10.5. Since photons at any energy above the ionization energy can be so absorbed, a continuum of absorption in the photon spectrum results.

Free–free transitions occur when an electron has a near collision with an atom but never becomes bound. If the electron loses energy, a photon is emitted; if it gains energy a photon is absorbed. Again, since the energy levels are not discrete, the emitted or absorbed photons exhibit a continuum emission or absorption spectrum.

An example of the free–bound and bound–free transitions occurs in the photosphere of the sun where a *second* electron can attach itself to the hydrogen atom to form a negative *hydrogen ion* H⁻. As might be expected, the second electron is very weakly bound to the atom; it takes only 0.75 eV to separate it from its parent atom. This is to be compared to the 13.6-eV binding energy of the ground state. These (second) electrons can easily be detached by visible photons, which have energies well in excess of 0.75 eV (Fig. 2.1).

Most of the light from the sun arises from these ions which are constantly being formed (free–bound) and dissociated (bound–free) in the hot photosphere. The result is a continuum spectrum, except for the absorption lines that arise from the temperature gradient in the solar photosphere (see below).

Optically thin thermal bremsstrahlung

A common mechanism for the emission of photons is *thermal bremsstrahlung* from an *optically thin plasma*. A plasma is a cloud of ionized atoms or molecules. Astrophysical plasmas will typically have a mix of elements, e.g., the solar system abundances (Table 10.2). The dominant component, hydrogen, will be ionized, for the most part, into its constituent protons and electrons. The ions and electrons are typically assumed to be in approximate *thermal equilibrium*; their speeds follow the Maxwell–Boltzmann distribution. A thermal plasma must be sufficiently hot for its components to remain ionized at least in part. The term "optically thin" means that any emitted photon is very likely to escape the cloud without being absorbed by another atom; the optical depth of the cloud is small, $\tau \ll 1$; see (10.30).

Radiation from a hot plasma

In a plasma cloud, the radiation arises when electrons are accelerated in near collisions with ions and thereby emit photons. The term "bremsstrahlung" means "braking radiation" in German. In an electron–ion near collision, the less-massive electron undergoes a large acceleration due to the mutual Coulomb force. An accelerated electric charge will always radiate away some of its energy. The radiated photons constitute the observed radiation. Bremsstrahlung gives rise to the x rays in your dentist's x-ray tube. High-energy electrons beamed into a metal target are rapidly braked to a stop by Coulomb interactions. The deceleration results in an intense beam of x rays.

The dependence of the volume emissivity j (W m⁻³ Hz⁻¹) of emitted radiation upon temperature T and frequency ν may be approximately obtained in a semi-classical derivation. This derivation (not worked out here) takes into account only the free–free collisions of electrons and ions. It thus yields only a continuum spectrum. A complete solution taking into account free–bound and bound–free transitions would yield strong emission lines.

The resultant continuum spectrum is

$$j(\nu,T) \;\propto\; g(\nu,T,Z)Z^2 n_e n_i T^{-1/2} e^{-h\nu/kT} \qquad \text{(W m}^{-3}\text{ Hz}^{-1}\text{)} \quad (11.17)$$

where h and k are the Planck and Boltzmann constants respectively ($h = 6.63 \times 10^{-34}$ J s; $k = 1.38 \times 10^{-23}$ J K^{-1}), n_e and n_i are the number densities (m^{-3}) of the electrons and ions respectively, and Z is the atomic number (charge number) of the ions. The *Gaunt* factor $g(\nu,T,Z)$ is a slowly varying term (decreasing with frequency) that takes into account the quantum mechanical effects in the free–free collisions, such as *screening* by nearby electrons. In the approximation, $g = $ constant, the dependence of j on frequency at a fixed temperature is a simple exponential. The exponential reflects the thermal Maxwell–Boltzmann distribution of electron speeds.

Plasma parameters determined

The quantity that one can measure from afar is the specific intensity I which is related to j according to (8.50),

$$I = \frac{1}{4\pi} \int_0^\Lambda j(r)\, dr \qquad \text{(Specific intensity; hydrogen plasma;} \quad (11.18)$$
$$\text{W m}^{-2}\text{ Hz}^{-1}\text{ sr}^{-1})$$

$$\propto \frac{1}{4\pi} \int_0^\Lambda n^2 T^{-1/2} \exp(-h\nu/kT)\, dr$$

where we assume $g = $ constant and adopt a pure hydrogen plasma, so $Z = 1$ and $n_i = n_e \equiv n$. The integration is along the line of sight through the entire depth Λ of the cloud; recall that the cloud is optically thin. If one further assumes that the densities and temperature are constant through the depth of the cloud,

$$I \xrightarrow[j\,=\,\text{const.}]{} \frac{j}{4\pi}\Lambda \qquad \text{(Constant } n,T) \quad (11.19)$$

$$\propto n^2 \Lambda\, T^{-1/2} \exp(-h\nu/kT)$$

This function is sketched in Fig. 3 for two temperatures, $T_2 > T_1$, on linear, semilog and log-log axes. The unspecified proportionality constants in (17) are taken to be equal in the two cases; thus Fig. 3 compares the spectra of two hydrogen plasmas that have different temperatures, but which are otherwise identical.

The linear plot (Fig. 3a) shows the typical exponential decrease with increasing frequency. The higher temperature for the T_2 curve is reflected in two ways. It has less amplitude at $\nu = 0$, as a consequence of the $T^{-1/2}$ dependence, and it falls to $1/e$ of its $\nu = 0$ amplitude at a greater frequency than does the T_1 curve. The higher temperature plasma emits more high-energy photons as might be expected.

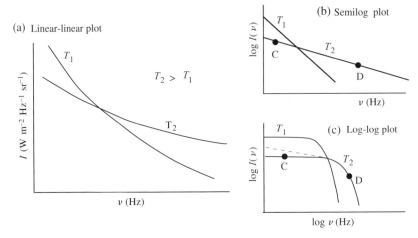

Figure 11.3. Sketches of exponential spectra on linear-linear, semilog, and log-log plots for two sources with differing temperatures, T_2 and T_1. In each case, most of the energy is emitted at the higher frequencies. Measurements of the flux at points C and D permit one to solve for the temperature T of the plasma as well as the factor $n_e^2 \Lambda$ where n_e is the electron density and Λ is the line-of-sight thickness of the cloud. The dashed line in (c) illustrates the effect of the Gaunt factor for the T_2 curve in a realistic thermal bremsstrahlung spectrum. These plots do not show the prolific emission lines typical of a real plasma.

The semilog plot (Fig. 3b) shows the same features with the typical straight-line character of exponential functions.

The log-log plot (Fig. 3c) is interesting; it shows a flat spectrum out to a certain region where it starts to decrease rapidly. (We show the effect of a hypothetical Gaunt factor g with a dashed line.) The turnover occurs near the frequency given by $h\nu \approx kT$. At this frequency, the kinetic energy of the emitting electrons, $\sim kT$, is about equal to the energy of the individual emitted photons $h\nu$. Most of the power is emitted in this frequency region. At lower frequencies, $h\nu \ll kT$, the exponential function in (17) is approximately unity; recall that an exponential with a small argument ($x \ll 1$) may be approximated as $e^x = 1 + x \ldots \approx 1.0$.

On a logarithmic abscissa, the low-frequency portion of the function near $h\nu \ll kT$ may be stretched out indefinitely toward $\nu = 0$, to 0.1 Hz, to 0.01 Hz, to 0.001 Hz, etc.; this is why it is flat. Relatively little power is included in the low-frequency regions because the bandwidths are so small. For example, the band 0.1 to 1 Hz has a bandwidth of 0.9 Hz while the band 0.001–0.01 Hz has a bandwidth of only 0.009 Hz. The power in each decade is proportional to the product of I and the bandwidth. At low frequencies, I is constant but bandwidth becomes negligibly small. This explains qualitatively our statement that most of the emitted power is at frequencies near the cutoff at $h\nu/kT \approx 1$.

Measurement of the specific intensity curve $I(v)$ provides direct information about both the product $n^2\Lambda$ and the temperature T of the cloud. Consider the approximate expression for I(v) (19), for a hydrogen cloud of constant temperature and density along the line of sight with a fixed Gaunt factor. The temperature T is obtained directly from the value of v where I is at e^{-1} of the maximum, namely at point D in Fig. 3c; at this frequency, $hv/kT = 1$. The value of I at low frequencies (point C) yields the product $n^2\Lambda T^{-1/2}$. Since T is known, we obtain $n^2\Lambda$. The unspecified proportionality constant in (19) is a combination of well known physical and numerical constants, so absolute values of T and $n^2\Lambda$ are obtained.

Shocks in supernova remnants, stellar coronae, H II regions

The actual spectral form for thermal bremsstrahlung is not a pure exponential. The Gaunt factor causes the flat portion of the log-log plot to decrease slowly with increasing frequency. Also the several atomic elements in cosmic plasmas lead to strong emission lines superposed on the quasi exponential continuum. A theoretical calculation (Fig. 4) of the expected radiation from a plasma of temperature 10^7 K containing "cosmic" abundances of the elements (e.g. Table 10.2) shows both of these effects.

On the semilog plot of Fig. 4, the Gaunt factor appears at low photon energies hv as an excess above the extrapolated straight-line continuum seen at higher frequencies. This continuum decreases by a factor of about e $= 2.7$ for each increase of hv by a factor kT as expected from (17). Note that the emission lines typically exceed the continuum in intensity by two decades, a factor of ~ 100.

Such hot x-ray emitting plasmas are found in shock waves propagating outward from the sites of supernova explosions, for example in the supernova remnant Puppis A (Fig. 5a). Coronae in the vicinity of stars like our sun are hot plasmas of temperatures reaching 10^6 K and more. In some cases, e.g. Capella (Fig. 5b), the radiation is sufficiently hot and intense that the currently orbiting Chandra observatory can resolve its spectral lines. In both cases, the emission lines are plotted on a linear scale which dramatically illustrates how the spectral line intensities greatly exceed the continuum, in accord with Fig. 4.

Thermal bremsstrahlung emission is found in radio emitting low-temperature plasmas of emission nebulae such as H II regions of the Orion nebula (Fig. 6b). The ideal spectrum for such a source is sketched in Fig. 6a. The flat ($\alpha \approx 0$) portion is the low-frequency end of an optically thin thermal bremsstrahlung spectrum with $g =$ constant (compare to Fig. 3c).

At very, very low frequencies, plasmas become optically thick; the nebula becomes opaque to its own radiation because the number of *phase space* states the photons can occupy are limited at the low frequencies. (Phase space is six-dimensional

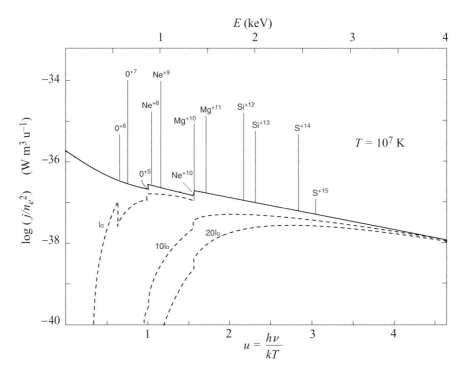

Figure 11.4. Theoretical x-ray spectrum of an optically thin plasma at $T = 10^7$ K on a semilog plot. The ordinate is the log of the volume emissivity j (17) divided by electron density squared. The abscissa is unity at the frequency where the exponential term in (17) equals e^{-1}. The various atomic levels are properly calculated, and strong emission lines are the result. The dashed lines show the effect of photoelectric absorption of x-rays by interstellar gas. [From Tucker and Gould, *ApJ* **144**, 244 (1966)]

with 3 momentum and 3 position coordinates.) The spectrum thus descends like the low-frequency part of a blackbody spectrum which is similarly constrained by available phase space states. This portion of the spectrum has the ν^2 Rayleigh–Jeans dependence (see below).

Synchrotron radiation

A mechanism that can give rise to a spectrum of very different shape is *synchrotron radiation*. This occurs when very high energy (relativistic) electrons spiral around magnetic field lines due to the $q\boldsymbol{v} \times \boldsymbol{B}$ force on an electric charge q moving with velocity $\boldsymbol{v}$ in a magnetic field $\boldsymbol{B}$. The electrons emit electromagnetic radiation because the spiraling motion constitutes an acceleration, and accelerating charges emit photons. Because the charges are relativistic, the radiation is particularly intense, and it can reach to extremely high energies, even to x rays and gamma rays. It is also

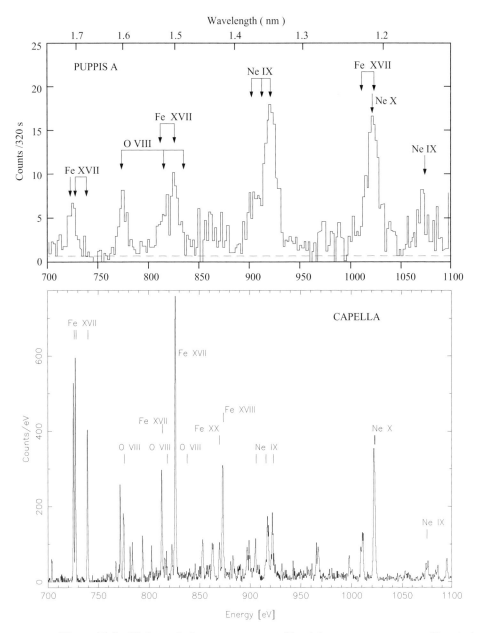

Figure 11.5. High resolution x-ray spectra of (top) the supernova remnant Puppis A from the historic Crystal Spectrometer on the Einstein observatory (launched 1978) and (bottom) the bright star Capella from the High Energy Grating Spectrometer on the Chandra observatory (launched 1999), for the same part of the spectrum. The resolution of the latter is much improved over the former. The locations of some spectral lines expected at x-ray temperatures are indicated. The relative strengths of the observed lines indicate temperatures, densities and compositions of the plasmas (a) in the shock wave that originated in the supernova explosion of Puppis A and (b) in the plasma of the active corona of Capella. [Top: adapted from Winkler *et al.*, *ApJ Lett.* **246**, L27 (1981); bottom: from C. Baluta and K. Flanagan from archival calibration data; also see Canizares *et al.*, *ApJL* **539**, L41 (2000)]

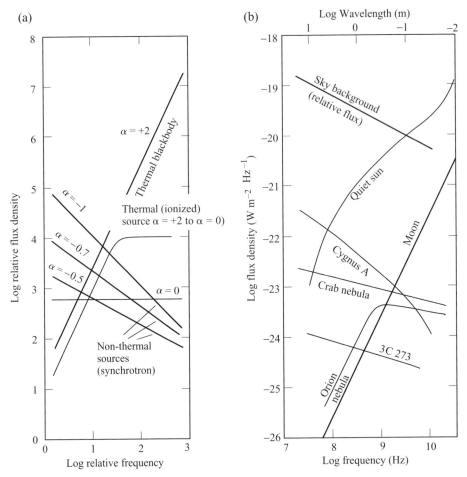

Figure 11.6. (a) Sketches of ideal radio spectra that are (*i*) rising (optically thick thermal, typically blackbody), (*ii*) falling (non-thermal, typically synchrotron radiation), and (*iii*) flat (optically thin thermal radiation). (b) Sketches of real radio spectra. Note that, at very low frequencies, the flat optically thin spectra become optically thick and take on the ν^2 power-law character of blackbody intensity for $h\nu \ll kT$. The sky background was measured over an unspecified solid angle. [Adapted from J. D. Kraus, *Radio Astronomy*, 2nd ed. Fig. 8-9, Cyg-Quasar Books, 1986, with permission]

beamed in the direction of travel of the radiating electron, like the headlight of an automobile. It can also exhibit polarization that reflects the large scale coherence of magnetic field directions in the source region.

The observed spectrum from a source radiating by the synchrotron process reflects primarily the distribution of energies of the radiating electrons. The energy spectrum of these electrons often has a characteristic shape, a *power law*, and this

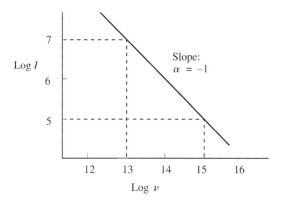

Figure 11.7. Schematic of a power-law spectrum, $I \propto \nu^{\alpha}$ where α = constant. On a log-log plot, as shown here, the spectrum is a straight line with slope α.

leads to a radiated photon spectrum that is also a power law. The spectrum of the radiation can thus be represented, of course, by a power-law function,

$$I = K\nu^{\alpha} \tag{11.20}$$

where I is the specific intensity, K is a constant, and α is a constant exponent. In nature, power-law spectra usually decrease with frequency, in which case α is negative. Because of this, a positive *spectral index* $\beta \equiv -\alpha$ is often defined so that $I = K\nu^{-\beta}$. It is important to specify which convention you use; we use the convention (20).

A power law is most conveniently plotted in a log-log format (Fig. 7). In this format, a power law is a straight line with a slope equal to the exponent, α. This is apparent if one takes the logarithm of (20),

$$\log I = \log K + \alpha \log \nu \tag{11.21}$$

The exponent α is thus the *logarithmic slope*. From it, one can deduce the logarithmic slope of the energy spectrum of the radiating electrons.

The slope is obtained from a spectrum by using logarithmic intervals. For instance, in Fig. 7,

$$\alpha = (5 - 7)/(15 - 13) = -1 \tag{11.22}$$

The slope can be measured as a ratio of distances if the ordinate and abscissa both have the same scale factors, that is, a decade is the same distance on both axes. It is often helpful to construct log-log plots with equal scales because the slope which equals α can readily be estimated by eye.

Blackbody radiation

Spectrum

An emitting body can be *optically thick*. The conditions are such that the photons scatter, or are absorbed and re-emitted, many times prior to being emitted from the surface. In this case one obtains a spectral shape known as the *blackbody spectrum*. The spectrum depends upon the temperature of the emitting body. Many ordinary objects emit radiation that approximates a blackbody spectrum. Objects at room temperature ($T \approx 300$ K) emit photons with a spectrum characteristic of $T \approx 300$ K (infrared radiation), and a piece of iron heated until it glows a yellow color emits at $T \approx 1100$ K. At ~5000 K, the color is "white hot"; which is typical of the sun. These spectra are also called optically thick thermal spectra in contrast to the optically thin thermal bremsstrahlung discussed above.

Blackbody radiation is a well defined physical quantity. It is the radiation one would find inside an evacuated cavity of a given temperature with "black" walls that absorb and re-emit radiation freely such that they are in thermal equilibrium with the radiation in the cavity. At a given frequency and temperature, the specific intensity $I(v,T)$ of the radiation inside the cavity is a well determined function that can be calculated from basic physics principles. It could not be explained with classical physics, but in the early years of the last century, Max Planck was able to explain it in terms of quantum mechanics.

The specific intensity for blackbody radiation, the *Planck radiation law*, in terms of the frequency v and temperature T, is

$$\Rightarrow \quad I(v,T) = \frac{2hv^3}{c^2} \frac{1}{e^{hv/kT} - 1} \qquad \text{(Planck radiation law;} \qquad (11.23)$$
$$\text{W m}^{-2}\text{ Hz}^{-1}\text{ sr}^{-1})$$

where h and k are the Planck and Boltzmann constants and c the speed of light.

Again this is a continuum spectrum; the ideal blackbody spectrum has no spectral lines. Also note that the amplitude at a given temperature and frequency is completely specified. There are no other parameters such as the density or charge of the atoms in the source. At a given frequency, there is no such thing as fainter or brighter blackbody radiation at a specified temperature T. The intensity is fixed by T; blackbody radiation *is* blackbody radiation, period.

The expression (23) is plotted for two temperatures in Fig. 8 with linear axes and with log-log axes. The exponential temperature dependence causes the amplitude to change dramatically even for a modest change in temperature, a factor of 2 in the figure. For this reason, it is often convenient to use logarithmic axes.

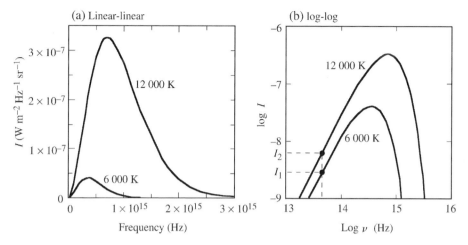

Figure 11.8. Sketches of blackbody spectra for two temperatures on both linear and log-log plots, according to (23). The temperatures differ by a factor of 2. Note the following features: the power-law behavior at low frequencies, $I \propto \nu^2 T$, the rapid decrease at high frequencies caused by the exponential term, the frequency of the maximum intensity increasing with temperature, and the rapid growth as a function of temperature. At a specified low frequency, the temperature T may be used as a shorthand for the specific intensity I, since $I \propto T$.

Radio spectra and antenna temperature

At frequencies significantly less than that of the peak frequency, the blackbody function varies linearly with the temperature and quadratically with frequency:

$$\Rightarrow \qquad I(\nu, T) \approx \frac{2\nu^2 kT}{c^2} \; \propto \; \nu^2 T \qquad \text{(Rayleigh–Jeans law)} \qquad (11.24)$$

This expression follows from an expansion of the exponential, $e^x = 1 + x + \cdots$ for $x \ll 1$. It is called the *Rayleigh–Jeans law* (R-J law) or more appropriately, the *Rayleigh–Jeans approximation*. Note that it is a power law in frequency, $I \propto \nu^2$, so that it becomes a straight line with slope 2 on the low-frequency end of a log-log plot (Fig. 8b). It is also directly proportional to temperature T; a factor of 2 increase in T leads to a factor of 2 increase in the specific intensity in this low-frequency portion of the spectrum.

The lowest temperature encountered in space is $T = 2.73$ K (the cosmic background radiation), and it has a peak at $\nu = 160$ GHz or at wavelength $\lambda = 1.9$ mm. This is at the extreme upper end of the radio band. In the radio portion of the electromagnetic spectrum, therefore, the blackbody spectrum for all expected temperatures is rising. It is therefore usually appropriate to invoke the low-frequency approximation, the R-J law (24), for measurements in the radio region.

Since the specific intensity is directly proportional to the temperature of the radiation in the R-J approximation, radio astronomers often report intensities as equivalent temperature $T(K)$ at some specified frequency. The phrases *antenna temperature* and *brightness temperature* are used to describe this temperature. For example, the two data points (filled circles) in Fig. 8b may be described as either the values of I_1 and I_2 in SI units or as the temperatures 6000 K and 12 000 K, respectively. Note that the same antenna temperature T represents larger values of I at higher frequencies.

Even if the object does not have a blackbody spectrum, the antenna temperature may be used to represent a measurement of I at some frequency v. At another frequency, the antenna temperature could well have a different value. (It would be the same if the spectrum were precisely blackbody.) One converts from antenna temperature T to specific intensity I with the expression (24), or more precisely (23), where the frequency of the observation is used for v.

Radio astronomers traditionally characterize broadband spectra as rising, falling, or flat according to the behavior of the spectral flux density as frequency increases. Figure 6a shows a variety of theoretical spectral shapes. The rising spectra all have slope +2 consistent with (24) and are called *thermal spectra*. The falling spectra are called *non-thermal*. These latter spectra were so named because they do not rise as does a thermal spectrum. Later, it was realized that such spectra frequently indicate synchrotron radiation. The term "non-thermal" remains appropriate because the emitters are ultra-high-energy electrons that are *not* in thermal equilibrium with their surroundings.

Real examples of radio spectra are shown in Fig. 6b. The quiet sun and the moon are largely thermal whereas the radio galaxy Cygnus A, the quasar 3C273, the Crab supernova remnant, and the sky background are all non-thermal (synchrotron radiation). The Orion nebula (an H II region of ionized plasma) exhibits an approximately flat spectrum that becomes optically thick at low frequencies.

Cosmic microwave background

A beautiful example of blackbody radiation is the remnant radiation from the early hot, dense universe. The cosmic background radiation (CMB) would have had a blackbody spectrum of temperature $T \approx 4000$ K when the cooling and expanding universe first became optically thin to this radiation. This occurred when protons and electrons combined to become neutral hydrogen. By now the radiation has cooled to $T \approx 3$ K, and is expected to still have the spectral shape of blackbody radiation.

Such radiation was discovered in 1964 by Penzias and Wilson at Bell Laboratories in New Jersey; a Nobel-prize winning discovery. Another earlier indication, in 1941, of this radiation was derived from optical spectral-line observations of starlight absorbed by cyanogen (CN) molecules in interstellar space, but this was before the predictions of such radiation so the result received little notice. (The method of the CN detection was discussed in Section 9.4.) The 1964 discovery was a great surprise because the scientific community did not generally recognize at that time that such radiation would be a consequence of the big bang theory. In fact, its existence had been suggested in the 1940s and its temperature predicted to be ~5 K in a 1950 paper.

Early measurements of the CMB are shown in Fig. 9a where they are compared to a blackbody spectrum. Note that the detections at several different frequencies by several different groups verified that the spectrum rises $\propto \nu^2$ as expected for blackbody radiation, but the high frequency fall off could not be confirmed with ground-based experiments.

The galactic background (non-thermal synchrotron radiation) is also shown in Fig. 9a. One reason the thermal radiation was not discovered earlier is that the galactic background dominated the radio fluxes at lower frequencies where most measurements were made. It is only at the higher frequencies that the thermal fireball radiation exceeds the non-thermal galactic background.

The spectral shape of the CMB has been determined with great precision by the COBE satellite (Cosmic Background Explorer; launched 1989). The published spectrum (Fig. 9b) shows dramatically the agreement with a blackbody spectrum on both sides of the peak. The temperature is found, also with high precision, to be $T = 2.725 \pm 0.002$ K.

This precise agreement with a blackbody spectrum strongly supports the model of a hot early universe (big bang). The radiation is highly isotropic; it arrives from all directions with nearly the same intensity. This too is consistent with expectations; the entire universe is immersed in this radiation. Ongoing studies of minuscule fluctuations of CMB intensity (temperature) from point to point on the sky with satellites and balloons further probe the nature of the early universe.

Stars

The overall shape of the spectra of many stars approximates that of a blackbody, and blackbody formulae are often adopted to describe their energy output. Nevertheless the spectrum is substantially distorted by absorption lines and by the effect of the temperature variation with depth in the stellar atmosphere. In the spectrum of Canopus (Fig. 10a), the absorption lines are obvious deviations.

There is also severe *continuum absorption* rightward of 365 nm (3.40 eV) due to electrons in the $n = 2$ state of hydrogen being ejected into the continuum. Photons more energetic than 3.40 eV can be absorbed by the transition. In hot stellar atmospheres, sufficient numbers of atoms are found in the $n = 2$ level to make this absorption substantial.

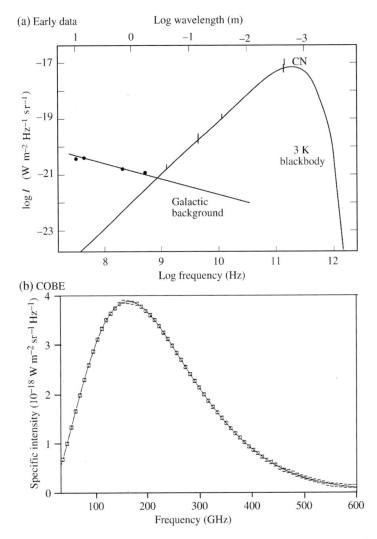

Figure 11.9. (a) Early (∼1969) measurements of the $T = 3$ K cosmic microwave background (CMB) radiation, a remnant of the early hot universe. The measurements near the peak are from observations of the molecule cyanogen (CN) in the interstellar gas. The non-thermal background at low frequencies is synchrotron radiation due to electrons spiraling in the magnetic fields of the Galaxy. (b) Spectrum measured with the COBE satellite in the 1990s. The squares indicate measurements and the solid line is the best-fit blackbody fit to them. The fit is incredibly good over a large range of frequencies. Note that this is a linear-linear plot. [(a) B. Partridge in *Am. Scientist* (1969); (b) courtesy NASA/COBE Science Team]

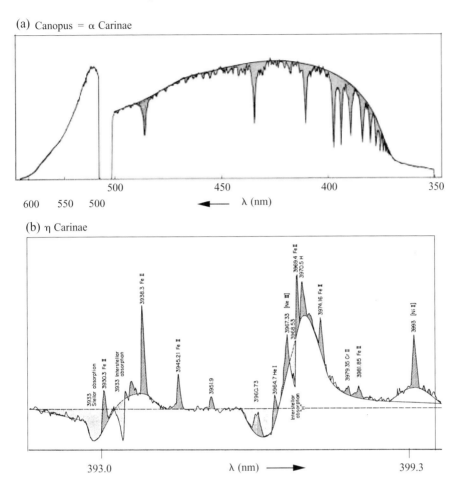

Figure 11.10. Examples of absorption and emission lines in sketches by the spectroscopist, Lawrence Aller. (a) Spectrum of star α Carinae (Canopus), a star of Type F0, not unlike our sun (Type G2), showing absorption lines. The vertical and horizontal scales are changed leftward of 500 nm. (b) Partial spectrum (390–400 nm) of η Carinae, a star of type known as a luminous blue variable (LBV) which underwent huge mass ejections in the 1840s. This spectrum shows many narrow emission lines (He I, Cr II, Fe II, Ne II, Ne III, etc.) which are emitted by fluorescence of ejected gas. Very broad absorption and emission features are due to ionized calcium. The latter are emitted by fluorescence of ejected gas. The narrow absorption lines are due to interstellar gas between us and η Carinae. [(a) Adapted from L. Aller, *Atoms, Stars & Nebulae*, 3rd Ed., 1991, p. 60, Cambridge University Press; (b) *ibid.*, p. 285; reprinted with permission of Cambridge University Press]

Consider 1 m^2 of the stellar surface and let it radiate as a perfect blackbody into all directions in the upper hemisphere. Integrate the spectrum over all frequencies and all directions, taking into account the projected area, $\cos \theta$, at angle θ to obtain the total power radiated by the 1 m^2. It turns out to be

➡ $$\mathscr{F} = \sigma T^4$$ (W m^{-2}; blackbody radiation) (11.25)

where σ is the *Stefan–Boltzmann constant*:

$$\sigma = \frac{2\pi^5 k^4}{15\, c^2\, h^3} = 5.670 \times 10^{-8} \text{ Wm}^{-2}\text{ K}^{-4}$$ (Stefan–Boltzmann (11.26) constant)

The calculated flux (25) is that which passes in one direction through a surface immersed in a blackbody cavity. The total flux (in both directions) through the surface would be zero. The flux (25) increases rapidly with temperature. A doubling of the temperature yields a power greater by a factor of 16.

Since the spectrum from a normal star approximates that of a blackbody, one can use (25) to estimate its luminosity in terms of its surface temperature T and radius R. The surface area of the spherical star is $4\pi R^2$ and its luminosity is $L \approx 4\pi R^2 \sigma T^4$. The approximate equality indicates that the star does not emit as a perfect blackbody. In Section 9.4, we defined an *effective temperature* T_{eff} (9.13) to yield the exact relation,

➡ $$L = 4\pi R^2 \sigma T_{\text{eff}}^4$$ (W; luminosity of (11.27) spherical object)

The total power L radiated by a star thus varies as the fourth power of T_{eff} and the second power of its radius R.

Models of normal stars tell us that the more massive stars are both larger and hotter. In a simple model, the luminosity increases extremely fast with temperature, approximately as $L \propto T_{\text{eff}}{}^5$ for lower mass stars where the dominant fusion process is initiated by a proton–proton interaction, the *p–p process*. For the massive hotter stars where fusion interactions involving carbon, nitrogen and oxygen take place, the model indicates $L \propto T^{13}$. A small rise of surface temperature signifies a huge increase in the output from the thermonuclear reactions that power the star.

11.4 Spectral lines

Spectral lines provide powerful diagnostics of the conditions in the emitting region of a celestial source. Normal stars exhibit absorption lines due to decreasing temperature (with altitude) in the photosphere (Fig. 10a) while ejected gas near a star or an active corona can result in emission lines (Fig. 10b). Here we discuss the several types of spectral lines and their measurable characteristics. In the following section, we present the physics of radiation propagation that creates the lines.

Absorption and emission lines

Spectral lines arise from atoms or molecules undergoing transitions between two energy states differing in energy by ΔE. Such transitions in the hydrogen atom are shown as arrows in Fig. 10.1. If the atom is going from a high (excited) energy state to a lower energy state, the excess energy is emitted as a photon of energy $\Delta E = h\nu$. If many atoms do this, many photons with the same energy are emitted giving rise to an *emission line* (Fig. 1a), provided the photons can emerge without further scatters.

On the other hand, if these atoms are being excited to a higher energy state through the absorption of photons, only those photons of the correct energy, $\Delta E = h\nu$ will be absorbed. If the absorbing atoms are between us and the source of the original photons (say, a hot star), a deficiency of photons at that frequency, an *absorption line*, will be observed.

Each type of atom or molecule emits or absorbs radiation at frequencies characteristic of that atom; the observed frequencies therefore indicate the type of atom involved. The sodium doublet ($\lambda \sim 589$ nm) is one example, and the Hα line of hydrogen at $\lambda = 656.2$ nm is another.

The emission and absorption processes are a function not only of the kinds of atoms that are present but also of the conditions of temperature and pressure in which the atoms find themselves. For instance, the hotter stars do not show hydrogen absorption because the hydrogen is entirely ionized. Thus the conditions in the stellar atmosphere are directly indicated by the presence or absence of certain lines. See for example our discussion of the Saha equation in Section 9.4.

Origin of spectral lines

Figure 11 shows how the emission and absorption lines arise. A hot incandescent lamp emitting a continuum spectrum illuminates a cool cloud containing sodium (Na) atoms. Three observers analyze the light with a prism; each has a different perspective, and each sees a different spectrum. Each can choose to observe the light emerging from the prism directly by eye (observers A,B) or with the aid of a lens and piece of film (observer C).

Observer A studies the light coming directly from the lamp and sees the continuum spectrum. Observer B studies the light from the cloud and observes the Na doublet in emission. The emission-line spectrum arises from the re-emission of the radiation initially absorbed by the Na atoms. If the gas is sufficiently hot, collisions of the gas atoms will also excite the atoms to produce the lines of interest;

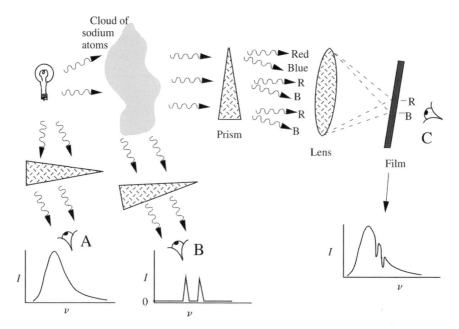

Figure 11.11. Origins of the spectral lines. A continuum source of light illuminates a cloud of sodium (Na) atoms. Three observers, A, B, and C see respectively, the continuum spectrum, emission lines, and the continuum with absorption lines.

the illuminating source need not be present, though it could be the source of the heating. In either case, the characteristic photons are emitted into all directions. Observer C studies the light coming from the direction of the lamp, such that the light has passed through the cloud. The continuum spectrum of the lamp is observed but the sodium atoms have removed many of the photons at the two frequencies of the Na doublet. Thus the doublet is observed in absorption.

In the laboratory there are two classic cases that produce lines. Emission lines are observed from a heated gas such as a Bunsen burner flame, and absorption lines are observed when a cool gas is placed in front of a hot source. The former case is a variant of observer B's situation. The latter is the case of observer C. It is also characteristic of stellar atmospheres which typically exhibit absorption lines.

For the case of observer C, both absorption *and* emission processes along the path from the lamp to observer C must be taken into account. In each differential layer of gas, radiation from the lamp is absorbed at the particular frequencies characteristic of the gas atoms and is diminished as it does so. The atoms in the layer are also emitting radiation characteristic of its cooler temperature. The net spectrum seen by observer C is obtained by integrating these effects layer by layer through the cloud, for each frequency element of the spectrum. The formalism for this *radiative transfer* calculation is presented in the next section.

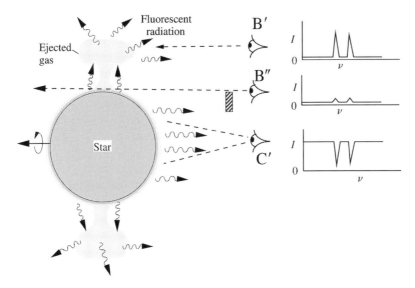

Figure 11.12. Light from a star exhibiting absorption (observer C′) and emission lines (B′, B″). The view of the stellar disk is blocked for B″ as in a solar eclipse because the chromospheric emission lines seen by B″ are weak compared to the light from the disk. A normal star viewed with no such aid usually exhibits only absorption lines formed in the photosphere, unless large amounts of ejected hot or fluorescing gas are in its vicinity.

Stars and nebulae

Radiation from stars exhibits both absorption and emission lines (Fig. 12). When the star is viewed directly, the decreasing temperature with increasing radius in the photosphere results in the production of absorption lines (observer C′). These are known as *Fraunhofer lines* after the discoverer of such lines in the solar spectrum.

Consider now that a large volume of gas is ejected from the star, possibly because the star is rotating so rapidly (see arrow) that centrifugal force ejects part of the atmosphere in the equatorial regions. Photons emitted from the ejected cloud yield emission lines (observer B′). Of course, the star is so distant that an observer on the earth can not distinguish the different parts of the cloud; all the light (and the spectral lines) appear to come from the same place in the sky. Thus, the emission lines must compete with the continuum in the detected spectrum.

Most stars do not have large quantities of ejected gas and thus exhibit only absorption lines (Fig. 10a). A tangential line of sight through the outer layers of a star's atmosphere (chromosphere) does give rise to emission lines, but they are not strong enough to overcome the light from the stellar disk with its absorption lines at the same frequencies. In this case, the emission lines serve to fill in the absorption lines only slightly. Emission lines from the transparent outer layers of the sun (chromosphere) may be observed by blocking the light from the solar disk, either artificially or with the moon during a solar eclipse (Fig. 12, observer B″).

Stars that do in fact exhibit emission lines must have large amounts of gas that are strongly illuminated by the star or are hot in their own right. Examples are centrifugally ejected gas, an *accretion disk*, an intense *stellar wind*, and a very *active corona*. Massive and hence highly luminous stars, called *luminous blue variables*, can produce major ejections of gas that then exhibit emission lines to distant observers. A portion of the spectrum from such a star, η Carinae, is in Fig. 10b.

Another important example of *emission lines* in optical astronomy is the Balmer radiation emitted from H II regions (Section 10.2). As noted, the prominent hydrogen Hα emission line gives rise to the red regions in photographs of these nebulae.

Permitted and forbidden lines

Emission lines that arise from *allowed transitions* are called *permitted lines*. The selection rules of quantum mechanics allow these transitions to occur rapidly. The emitted radiation is *electric-dipole radiation*. If the atom is in an upper state of a permitted transition, the transition will occur after a very short time, $\sim 10^{-8}$ s.

In contrast, the emission of electric-dipole radiation between certain states of the atom is forbidden by the selection rules of quantum mechanics (angular momentum and parity). These transitions in fact can take place, but with a much lower probability of occurrence. The observed lines are called *forbidden lines*. In such cases, the atom will remain in the upper state for a mean lifetime of order 1 s. The actual mean lifetime depends upon the particular atomic states involved; it is a value between 0.01 s and 100 s for most forbidden transitions. Such a long-lived upper state is called a *metastable state*.

In laboratory situations, even in an excellent vacuum, collision times between atoms will be much shorter than the $\sim$1-s decay time. Thus the metastable states will be *collisionally de-excited* long before they get around to freely radiating a photon. Thus one rarely sees forbidden spectral lines in earth laboratories.

In contrast, the densities within a celestial emission nebula are exceptionally low, $\sim 10^{8}$ m^{-3}, compared to an excellent laboratory vacuum of $\sim 10^{14}$ m^{-3} (10^{-9} torr), and collisions occur infrequently. Hence, the atoms in space will often decay via forbidden transitions. (They are excited to the upper states by infrequent collisions with electrons.) In some regions such as the Orion nebula, Fig. 1.7, green light is prominently seen; this is due to emission from two forbidden lines of doubly ionized oxygen, O III, at $\lambda = 496$ nm and $\lambda = 501$ nm. Forbidden lines are often indicated with brackets, e.g., the latter line would be designated "[O III] 500.7 nm". The most prominent lines from the Orion nebula are listed in Table 1.

Table 11.1. *Prominent emission lines in
Orion nebula*

Wavelength (nm)[a]	Name/color
[O II] 372.61, 372.86	Two ultraviolet lines
H I 486.1	Hβ line, blue
[O III] 495.9, 500.7	Green pair
H I 656.2	Hα line, red
[S III] 953.2	Infrared line

[a] Brackets indicate forbidden lines

Spectral lines at non-optical frequencies

Spectral lines are studied in all bands from the radio through gamma-ray. In radio astronomy, the study of line emission and the Doppler shifts in frequency of these lines has provided valuable information about the existence of molecules in space (Fig. 13) which are the building blocks of life. The Doppler shifts of "21-cm" spectral lines from the hyperfine (spin flip) transition of hydrogen (Fig. 10.1) reveal the revolution of stars and gas about the center of the Galaxy.

In x-ray astronomy, as noted in the previous section, high-resolution spectra of supernova remnants reveal the heavy elements ejected in the explosion (Fig. 5a). Observations of stellar coronae similarly reveal the elements therein (Fig, 5b). In general, x-ray line studies of Fe (iron) and other elements provide diagnostics of temperatures and densities of the hot ($T \approx 10^7$ K) plasmas in the source regions.

Line strengths and shapes

Equivalent width

A typical spectral line will have a *profile* that may be more or less Gaussian in shape and which can be severely distorted under certain conditions (Fig. 14). The total (integrated) area of an absorption or emission line in the observed spectrum is the measure of its strength or total power. The irregular shape of some lines makes it useful to define a quantity that is a measure of this integrated power relative to the other light from the star.

This quantity is called the *equivalent width* (EW). The EW is defined as the width (in wavelength or frequency) of the nearby continuum flux that contains the *same power* (same area on the plot) as the real line, that is, the dark shaded areas in Fig. 14. A large EW means the line is quite pronounced compared to the continuum flux. In

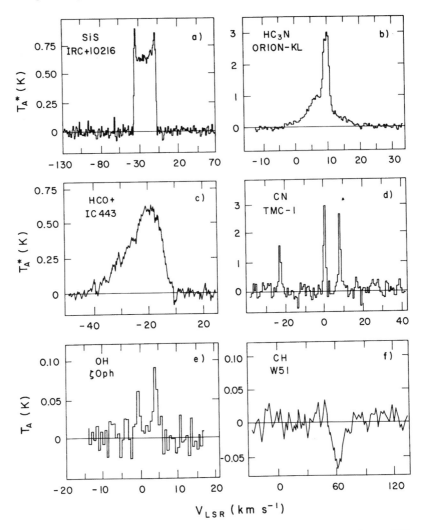

Figure 11.13. Sample radio spectra showing molecular lines from different sources. The objects are (a) late-type carbon star, (b) active molecular cloud core region, (c) a supernova remnant, (d) dark cloud core, (e) diffuse cloud, (f) giant molecular cloud. The observed lines are identified as being specific lines in SiS, CN, etc., so their rest frequencies are known. The abscissa indicates the radial Doppler velocity that is associated with a given observed frequency for the specified line. These plots thus give directly the radial velocities of the emitting clouds. [From B. Turner & M. Ziurys, in *Galactic and Extragalactic Radio Astronomy*, eds. G. Verschuur and K. Kellermann, 2nd Ed. Springer, 1988, p. 210]

the case of absorption (Fig. 14a,c), this hypothetical (dark-shaded) line has total absorption over the equivalent width. For emission (Fig. 14b,d), the equivalent width can be extremely large if the line is intense and the continuum is very small; it will be infinite if there is no continuum flux!

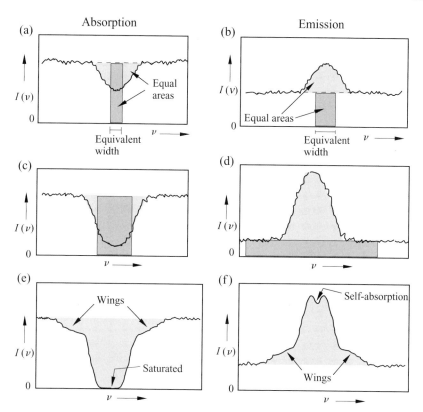

Figure 11.14. Rough sketches of hypothetical spectra for weak lines (a,b), moderately strong lines (c,d), and intense lines (e,f). The concept of equivalent width EW is illustrated in (a–d). The weak wings due to pressure broadening become prominent for the more intense lines. The emitting cloud can begin to absorb its own radiation if it is sufficiently dense (f).

Damping and thermal profiles

The profile of a spectral line from a sample of unperturbed gas has a shape governed primarily by two factors. First, the atoms in the gas will have a thermal spectrum governed by the Maxwell–Boltzmann (M-B) velocity distribution. The atoms are receding from and approaching the observer; radiation emitted or absorbed by them will be shifted in frequency due to the Doppler shift. This broadens the lines because both red and blue shifts are present. The Doppler shift of the frequency is proportional to the velocity component along the line of sight. The M-B distribution gives the number of atoms at each speed. Thus the line shape is governed by the M-B distribution; this is called *thermal Doppler broadening*, or simply, *thermal broadening*.

The M-B distribution is a Gaussian (or normal) distribution of speeds (6.3). The speeds translate directly to a Gaussian distribution for the line shape, that is, the line amplitude $\kappa_1(\nu)$ vs. frequency ν. For an emission line,

$$\frac{\kappa_1}{\kappa_{1,0}} = \exp\left[-\frac{(\nu - \nu_0)^2}{2\sigma^2}\right]$$

(Emission line shape;   (11.28)
Doppler broadening)

where $\kappa_{1,0}$ is the amplitude at $\nu = \nu_0$, where ν_0 is the central frequency of the line, and where σ is the standard deviation (characteristic width) of the Gaussian (6.3). The parameter κ_1 represents the frequency dependent opacity (m^2/kg). We explore the connection between spectral lines and opacity in the next section.

The second factor that broadens the spectral line is known as the *damping profile*. Classically, this is the damping term of the classical oscillator. Friction or radiation shortens the life of the oscillator and broadens the resonance curve (amplitude vs. frequency). In the quantum-mechanical interpretation, the limited lifetime Δt of the initial state leads to an uncertainty ΔE in the energy of that state, according to the *Heisenberg uncertainty principle*,

$$\Delta E\,\Delta t \gtrsim \frac{h}{2\pi} \equiv \hbar$$

(Heisenberg uncertainty   (11.29)
principle)

where Δt is the time the atom is in the upper state and h is the Planck constant. The longer the atom is in the state, the more precisely its energy can be measured. A large *transition probability* leads to a short life in the state and a large energy uncertainty. The emitted (or absorbed) photons thus yield a broadened line shape.

The expected damping line profile obeys the classical formula,

$$\frac{\kappa_2}{\kappa_{2,0}} = \frac{\gamma}{(\nu - \nu_0)^2 + (\gamma/2)^2}$$

(Damping curve)   (11.30)

where γ is the full width at half the maximum height of the line profile and $\kappa_{2,0}$ is an adjustable amplitude factor. It turns out that γ is proportional to the transition probability (not shown here). A large transition probability, i.e., a shorter lifetime, thus corresponds to a wider profile, or energy uncertainty, as expected.

The line shapes for the two effects (damping and thermal broadening) are shown in Fig. 15 for the case where the width at half maximum of the damping curve is half that of the thermal curve. In stellar atmospheres the central portion of the damping profile is quite narrow. Thus one might expect that the damping would not be important. But the exponential in (28) drives the thermal term strongly toward zero as one moves off center, and the damping curve dominates in the wings.

The combined thermal/damping line profile can be visualized in the following manner. Consider that each moving atom emits photons with the narrow damping profile centered on the Doppler-shifted line frequency ν' of that particular atom,

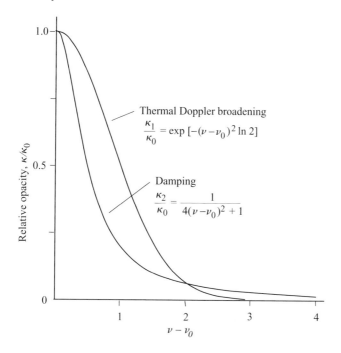

Figure 11.15. Line profiles due separately to the thermal and the damping terms, from (28) and (30) respectively. The opacity k is plotted against frequency offset from the line center ($v - v_0$). Both curves are normalized to unity at $v = v_0$, and the widths at half maximum are set to 1.0 and 0.5 for the thermal and damping curves, respectively. Note how the wings of the damping curve dominate at large offsets from the line center.

$$\kappa_2'(v) \propto \frac{\gamma}{(v - v')^2 + (\gamma/2)^2} \tag{11.31}$$

Summation over all atoms (that is, over all Doppler velocities) is a summation of many such profiles centered on the various frequencies v' of the thermal distribution. This smears the two distributions together.

Formally, the damping function $\kappa_2'(v)$ is weighted with the amplitude $\kappa_1(v')$ of the thermal distribution at frequency v' and then integrated over all v'. The net profile $f(v)$ is thus

$$f(v) \propto \int_{-\infty}^{+\infty} \left[\frac{\gamma}{(v - v')^2 + (\gamma/2)^2} \right. \qquad \text{(Combined profile)} \quad (11.32)$$
$$\left. \times \exp\left(-\frac{(v' - v_0)^2}{2\sigma^2} \right) \right] dv'$$

which is called the *convolution* of the two functions. When the central part of the damping profile is narrow, the convolution simply maps out the central portion of the thermal curve, while the damping curve is mapped at off-center positions large compared to the thermal width. If the damping width is down by a factor of 100 or 1000, as it sometimes is, the wings will be at such a low level they will not affect the line shape.

Turbulent motions and collisional broadening

Lines can also be Doppler broadened by rapid motions of clouds of emitting or absorbing atoms; this is known as *bulk turbulent motion*. The velocities of the several clouds are not necessarily thermally distributed so the shape of the line may differ from that quoted above. Bulk motion will significantly affect the line shapes when the bulk velocities approach or exceed the thermal velocities.

Broadened lines can also arise from collisions between the gas atoms which will de-excite the atoms before they would decay naturally. The atom is in its initial state for even a shorter time than in its unperturbed state, and this further broadens the (quantum-mechanical) damping profile according to (29). The damping profile is thus a measure of the number of collisions and hence of the density of the gas. (The number of collisions varies as particle density squared n^2.)

In turn, the density n is related to the pressure in the region where the line is being formed, as $P = nkT$ where k is the Boltzmann constant. This relation follows directly from the ideal gas law $PV = \mu RT$ where μ is the number of moles in a sample of volume V and where R ($= N_0 k$) is the universal gas constant (N_0 is Avogrado's number). Thus, for gases of about the same temperature, measurement of collisional broadening yields the relative pressures. This turns out to be useful for stellar classification when comparing different stars of the same photospheric temperatures. Thus

$$\text{Collisional broadening} \leftrightarrow \text{Pressure in photosphere } (T \approx \text{constant}) \quad (11.33)$$

Collisional broadening is sometimes called *pressure broadening*.

Saturation and the curve of growth

The *curve of growth* (Fig. 16) of a spectral line describes the measured strength (equivalent width, EW) as a function of the number N of absorbing (or emitting) atoms along the line of sight (atoms/m^2). Consider the absorption case (Figs. 14a,c,e). When the strength of the line is weak (Fig. 14a), the atoms along the line of sight are so few they block only a small portion of the beam. An increase in the number of atoms removes a proportional amount of radiation so that the curve grows linearly with N. As more atoms are added, eventually there are sufficient numbers of low velocity atoms to completely absorb the photons near the central frequency. At this point, the specified line is *saturated*, and the addition of more atoms has only a small effect; the EW is increases only very slowly with N.

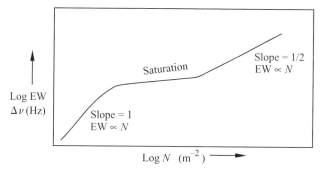

Figure 11.16. The curve of growth or the variation of the equivalent width (EW) with the column density of absorbing atoms N (atoms/m^2), for the absorption case. The flattening at intermediate N is due to the saturation of the central part of the absorption line (Fig. 14e). On this log-log plot, the linear dependence for weak lines and the (approximate) square root dependence for intense lines appear as slopes of 1 and 1/2 respectively.

As more and more atoms are added, the weak wings due to collisional broadening finally become important (Fig. 14e). At this stage the EW begins to increase again, approximately as $N^{1/2}$. This is a slower rate of increase in EW than for the unsaturated state. The $N^{1/2}$ dependence is not derived here. The growth of an emission line (Fig. 14 b,d,f) is largely similar to the absorption case.

The shape of the curve of growth will vary depending on the relative widths of the thermal and damping terms. Knowledge of the curve of growth enables one to determine column densities of different elements in stellar atmospheres and hence the chemical compositions.

11.5 Formation of spectral lines (radiative transfer)

Radiation propagating through a gas is transformed by emission and absorption processes. The result is the observed spectrum including spectral lines. Here we set up the differential equation for an elementary case of *radiative transfer* and solve it for several different conditions. This allows us to understand the formation of spectral lines in terms of the frequency dependence of the optical depth.

Radiative transfer equation (RTE)

The differential equation that governs the absorption and emission in a layer of gas follows from the geometry of Fig. 17. A uniform cloud ("source") of temperature T_s, depth Λ, and optical depth τ_Λ lies between the observer and a background source at some other temperature T_0.

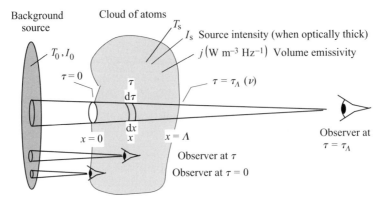

Figure 11.17. Geometry for the radiative transfer equation. The background surface emits with specific intensity I_0 and the intervening gas cloud emits thermal radiation with specific intensity I_s when it is optically thick. An observer in the cloud at position x, or optical depth τ viewing leftward will detect radiation from the cloud atoms at lesser τ and from the background source to the extent it is not absorbed by the cloud.

For the immediate discussion, we refer to radiated intensities at some single frequency (in a differential band) without regard to the entire spectrum. In fact, the overall spectral shape may be inconsistent with a single temperature. Hence we discuss intensities without necessarily defining a temperature. Nevertheless, in stellar atmospheres, temperatures can be defined in regions of *local thermodynamic equilibrium (LTE)*. In this case, a higher intensity at a given frequency does represent a higher temperature.

In the absence of the cloud, the observer in Fig. 17 would detect a specific intensity I_0 (W m^{-2} Hz^{-1} sr^{-1}) from the background source (T_0) at the frequency in question. We refer to I_0 as the *background intensity*. If the intervening cloud is in place, it will absorb some of the radiation from the background source. In addition, the cloud will emit its own thermal radiation in the direction of the observer. If the cloud is optically thick, the emerging radiation would exhibit the specific intensity I_s characteristic of blackbody radiation at T_s.

Intensity differentials

Consider a beam of photons moving in the direction of an observer at some location in the cloud. The differential equation that describes absorption of the photons in a differential path length dx is, from (10.17), d$N/N = -\sigma n\,$dx, where dN/N is the fractional change in the number of photons in the beam, σ (m^2) is the cross section per atom, and n (m^{-3}) is the number density of atoms. The fractional change of the photon number will be equal to the fractional change in the specific intensity, giving,

$$\frac{dI_1}{I} = -\sigma n \, dx \qquad \text{(Absorption in layer } dx) \qquad (11.34)$$

where dI_1 is one of two contributions to the total change dI.

The cloud also contributes photons to the beam. The thermal emission originating in the layer at x in dx of the cloud can be described with the volume emissivity j (W m^{-3} Hz^{-1}). This gives rise to an element of specific intensity from the layer in question which is, from (8.48),

$$dI_2 = \frac{j \, dx}{4\pi} \qquad \text{(Thermal emission from gas)} \qquad (11.35)$$

The sum of these two effects yields the net change in intensity I of the beam,

$$dI = -I\sigma n \, dx + \frac{j \, dx}{4\pi} \qquad \text{(Net change in } I \text{ in } dx \text{ at } x) \qquad (11.36)$$

This is the differential equation that allows us to find, by integration, the variation of beam intensity as it traverses the material on its way to the observer.

Intensity variation with optical depth

Rewrite (36) to be a function of optical depth τ. Recall the definition of the opacity, $\kappa \equiv \sigma n/\rho$ (10.24), where ρ (kg/m^3) is the mass density. Opacity is the cross section per kilogram of material (m^2/kg). Substitute $\kappa\rho$ for σn into (36) and rearrange,

$$\frac{1}{\kappa\rho} \frac{dI}{dx} = -I + \frac{j}{4\pi\kappa\rho} \qquad (11.37)$$

The product $\kappa\rho$ or σn is simply the inverse of the mean free path x_m with units of (m^{-1}); see Table 10.1. Thus the product $\kappa\rho x$ is the *number of mean free paths* in the distance x for fixed κ and ρ. In other words it is the *optical depth* $\tau = \kappa\rho x$, a dimensionless quantity previously defined (10.29). The denominator $\kappa\rho \, dx$ of the left side of (37) is thus equal to $d\tau$ since κ and ρ do not change appreciably in an incremental distance dx.

The equality (37) demands that the rightmost term have units of specific intensity. Since j is the volume emissivity of our cloud, we define this term to be the cloud intensity, or the *source intensity I_s*,

$$I_s \equiv \frac{j \, (\text{W m}^{-3} \, \text{Hz}^{-1})}{4\pi(\text{sr}) \, \kappa\rho \, (\text{m}^{-1})} \qquad \begin{array}{l} \text{(Source intensity defined:} \\ \text{W m}^{-2} \, \text{Hz}^{-1} \, \text{sr}^{-1}) \end{array} \qquad (11.38)$$

This expression has the form of (8.53), the relation between j and I for an optically thin plasma of thickness Λ, namely $I = j_{av}\Lambda/4\pi$. Here, the mean free path $(\kappa\rho)^{-1}$ plays the role of the cloud thickness Λ. In our optically thick case, an observer can "see" only a depth of about one mean free path into the cloud. The source intensity (38) is thus the intensity an observer embedded in the cloud would measure if her view were limited by optically thick conditions ($\tau \gg 1$).

It follows from the above considerations that the differential equation (37) may now be written as

➡ $$\frac{dI(\tau)}{d\tau} = -I(\tau) + I_s \qquad\qquad \text{(Equation of radiative} \qquad (11.39)$$
$$\text{transfer)}$$

where we express I as a function of τ, the optical depth of the cloud in the observer's line of sight (Fig. 17). This is the differential *radiative transfer equation* (RTE) which may be solved for the unknown quantity $I(\tau)$, the specific intensity at optical depth τ for our chosen frequency.

In (39), $I(\tau = 1)$ is the specific intensity measured by an observer within the cloud at the depth of one mean free path into the gas. Note that depth is measured from the left edge of the cloud. At $\tau = 0.1$ or $\tau = 3$, the function $I(\tau)$ is the specific intensity at depths of 0.1 and 3 mean free paths respectively. If the entire cloud has optical depth τ_Λ (corresponding to thickness Λ), the function $I(\tau_\Lambda)$ is the specific intensity measured by the observer outside the cloud.

The quantity $I(\tau)$ is distinct from I_s. It includes the radiation from the background source I_0 as modified by absorption and emission in the cloud. The background radiation is the "initial condition" we apply to the differential equation (39). The source function reflects the volume emissivity of the cloud itself.

The quantities τ, $I(\tau)$ and I_s in (39) are all functions of frequency; namely $\tau(\nu)$, $I(\nu)$, and $I_s(\nu)$. We continue to consider one frequency only and suppress the argument ν. The function j, and hence I_s, can vary with position in the cloud, i.e., both can be functions of τ. This is the case in stellar atmospheres where the temperature varies continuously with altitude. In the following, we consider I_s to be a constant throughout the cloud; the important consequences of (39) are well illustrated in this case.

Local thermodynamic equilibrium

If the gas of the cloud were in *complete thermodynamic equilibrium*, the radiation and matter would all be in thermal equilibrium at some temperature T; the specific intensity $I(\tau)$ would not vary throughout the cloud. In this case, the derivative in (39) equals zero, $dI/d\tau = 0$, and the observed intensity $I(\tau)$ is given by

$$I(\tau) = I_s \qquad\qquad\qquad \text{(Perfect thermal} \qquad (11.40)$$
$$\text{equilibrium)}$$

which is independent of τ. Since $I(\tau)$ is the specific intensity for complete thermo-dynamic equilibrium, its spectrum must be the Planck (blackbody) function (23). In turn, the source intensity I_s must also have a blackbody spectrum.

The solutions we seek are, in general, not for complete thermodynamic equilib-rium because they involve a gas at one temperature and incoming photons repre-sentative of a slightly different temperature. Also, the limited extent of the cloud implies that radiation is leaving the volume of the cloud, so that complete equilib-rium can not exist near the surface. Nevertheless, in solving the RTE one can make the assumption of *local thermodynamic equilibrium* (LTE).

Under LTE, the matter (e.g., protons and electrons) in a local region is in equi-librium with itself, but not necessarily with the radiation. That is, the matter obeys strictly the Boltzmann–Saha–Maxwell statistics, i.e., (9.14) and (9.15), for the lo-cal temperature, but the photon distribution is allowed to deviate slightly from it. Nevertheless, the radiation emitted from the local region follows the frequency de-pendence of the blackbody function for the local temperature, according to (40). The source function I_s for radiation emitted in a local region is therefore the blackbody function (23) for the (matter) temperature of the local region.

One can show that in the solar photosphere, the number density of particles is $\sim 10^5$ times that of the photons. Since every such photon or particle has about the same energy, $\sim kT$, in thermal equilibrium, the energy content is overwhelmingly contained in the particles. They can thus maintain their own temperature and radiate at that temperature in their local region even if photons from a lower and slightly hotter region diffuse up into their region and minimally distort the overall photon spectrum.

Solution of the RTE

Insight into the behavior of $I(\tau)$ according to the radiative transfer equation (39) can be gained simply from knowledge of the relative magnitudes of $I(\tau)$ and I_s. If $I(\tau) < I_s$ at some depth τ, the derivative in (39) is positive which tells us that $I(\tau)$ increases with optical depth. This is shown as the heavy line in Fig. 18a; note that it lies below the horizontal dashed line for I_s. Recall that in our case we hold I_s constant throughout the cloud. If, on the other hand, $I(\tau) > I_s$, then $I(\tau)$ decreases with depth (heavy line in Fig. 18b). In each case, $I(\tau)$ moves toward I_s and asymptotically approaches it at large optical depth.

At zero optical depth, $I(0)$ is equal to the background intensity I_0 because only the background source, and no part of the cloud, is in the observer's line of sight as is clear from Fig. 17. We also see this in both panels of Fig. 18. This obvious result also follows from the formal solution of the RTE to which we now proceed. We will find that the solution naturally provides for absorption and emission lines.

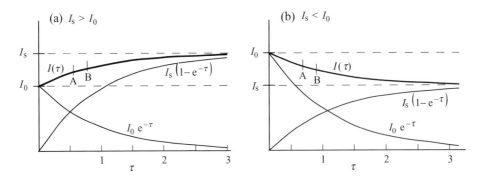

Figure 11.18. Plots of intensity $I(\tau)$ vs. optical depth τ from (44) for two cases: (a) source (cloud) intensity greater than the background intensity, $I_\text{s} > I_0$, and (b) $I_\text{s} < I_0$. As frequency is varied, the optical depth becomes higher at a resonance. If depth "A" is off resonance and depth "B" is centered at the resonance, case (a) yields an emission line and case (b) an absorption line.

The RTE (39) can be solved for $I(\tau)$ by integration as follows. Multiply (39) by e^τ,

$$\frac{dI}{d\tau} e^\tau + I e^\tau = I_\text{s}\, e^\tau, \tag{11.41}$$

rewrite the left side as $d(I e^\tau)/d\tau$, and integrate from 0 to τ,

$$\int_0^\tau d(I e^\tau) = \int_0^\tau I_\text{s}\, e^\tau\, d\tau \tag{11.42}$$

For our cloud with I_s independent of optical depth τ,

$$I(\tau) e^\tau|_0^\tau = I_\text{s}\, e^\tau|_0^\tau \tag{11.43}$$

Insert the limits and divide through by e^τ,

➡ $$I(\tau) = I_0\, e^{-\tau} + I_\text{s}(1 - e^{-\tau}) \qquad \text{(Solution of radiative} \qquad (11.44)$$
$$\text{transfer equation)}$$

This is the solution of the RTE. The first term on the right shows the decreasing effect of the background radiation I_0 as the optical depth increases, while the second term shows the increasing effect of the source (cloud) emission. These two terms and their sum are plotted in Fig. 18. These plots illustrate the variation of intensity with τ for a single chosen frequency.

Limiting cases

The solution (44) readily illustrates the formation of spectral lines if we consider the variation of τ (and also I_0 and I_s) with frequency. There are four cases to consider, one of which has two possibilities:

$I_0 = 0$: there is no background radiation illuminating the cloud
    (i) $\tau \ll 1$: the gas is optically thin
    (ii) $\tau \gg 1$: the gas is optically thick
$I_0 > 0$: background radiation illuminates the back of the cloud
    (iii) $\tau \ll 1$: the gas is optically thin (for $I_s > I_0$ and $I_s < I_0$)
    (iv) $\tau \gg 1$: the gas is optically thick

Case 1: $I_0 = 0$, $\tau \ll 1$

The condition $I_0 = 0$ means that $I(\tau)$ will be affected only by radiation from the cloud. The $\tau \ll 1$ condition allows us to expand the exponential, $e^{-\tau} \approx 1 - \tau$. The solution (44) then reduces to

$$I(\tau) = \tau I_s \qquad\qquad (I_0 = 0, \tau \ll 1) \qquad\qquad (11.45)$$

This tells us that the emission is proportional to the optical depth, for $\tau \ll 1$. This is reasonable because, for an observer located at $\tau \approx 0$ with leftward viewing detectors (Fig. 17), there are no atoms in view. The optical depth is zero and so is the detected intensity. As the observer moves to the right, toward increasing τ, the number of atoms in the line of sight increases linearly with τ. The cloud is optically thin so every layer $d\tau$ of the cloud that is in view contributes equally to the intensity (Fig. 8.8); hence $I \propto \tau$. Note that changes in mass density ρ and opacity κ along the line of sight are automatically incorporated into τ.

Now we address the frequency variation of the quantities in (45). Let the atoms in the cloud have an atomic transition or *resonance* at some frequency. At that frequency the cross section σ for absorption of incoming photons is high, and hence, so is the optical depth τ. In general, τ is a function of frequency and therefore so is the intensity I. We therefore rewrite (45) as

$$I(\nu) = \tau(\nu) I_s(\nu) \qquad\qquad (I_0 = 0, \tau \ll 1) \qquad (11.46)$$

Resonances at two distinct frequencies are hypothesized and illustrated in Fig. 19a (left panel) which is a plot of τ vs. ν for an observer at fixed position x. From (46), we see that high optical depths at these frequencies lead to high emerging fluxes $I(\nu)$ at these same frequencies, provided that I_s is a smooth function of frequency. A plot of I vs. ν for an arbitrarily chosen spectrum $I_s(\nu)$ (Fig. 19a, right panel) shows emission lines at the two resonance frequencies. Note that the spectrum lies well below the source spectrum I_s because $\tau \ll 1$, in accord with (46).

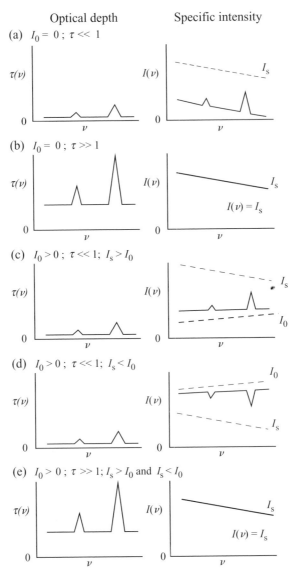

Figure 11.19. Frequency dependence of optical depth $\tau(\nu)$ (left) and specific intensity $I(\nu,\tau)$ (right) for an observer at some fixed position x in the cloud. I follows from $\tau(\nu)$ and the solution (44) of the radiative transfer differential equation. Arbitrary spectral shapes for the source (cloud) spectrum I_s and the background spectrum I_0 are shown as dashed lines. Emission lines are expected if the foreground gas is optically thin, $\tau \ll 1$, and hotter than the background source (c) or, in the limiting case, there is no background source (a). Absorption lines are expected if the background source is hotter than the foreground source, again if the cloud is optically thin (d). If the cloud is optically thick, the continuum spectrum of the cloud is observed (b,e).

At resonance frequencies, with their higher cross sections, the gas behaves as if it contains more atoms, or as if it were thicker. In this case, the observer would seem to "see" more emitting atoms, and hence greater intensity. At frequencies adjacent to a line, τ will be lower by definition and fewer atoms are seen. If τ is constant at these adjacent frequencies, as in the left panel of Fig. 19a, the output spectrum $I(\nu)$ will, according to (46), mimic the spectrum I_s away from the line as shown in the right panel.

Case 2: $I_0 = 0, \tau \gg 1$

Here the gas is very thick ($\tau \gg 1$), and the expression (44) reduces to,

$$I(\nu) = I_s(\nu) \qquad\qquad (I_0 = 0, \tau \gg 1) \qquad\qquad (11.47)$$

which we have previously deduced (40); here we use the variable ν rather than $\tau(\nu)$.

The output radiation given by (47) equals that of the continuum source specific intensity at all frequencies. If the source spectrum is the blackbody spectrum, the output spectrum at any depth $\tau \gg 1$ is also blackbody. Even though the resonances may exist (Fig. 19b), the intensity $I(\nu)$ has no dependence on τ, and hence no spectral lines will form.

In this case, the local observer sees the maximum possible number of emitting atoms at any frequency so the resonances are not apparent. It is like being immersed in a thick fog that is denser (more opaque) in some directions than others. Nevertheless, the appearance in all directions (frequencies) is uniform as long as the fog is totally impenetrable ($\tau \gg 1$) in all directions.

The observer in the fog sees only to a depth that yields enough water droplets to completely block the view. The same number of water droplets are thus seen in all directions, and the view appears uniform even though in some directions it penetrates less deeply than others. In our case, the increase of opacity at some frequency reduces the depth of view, but the observed intensity does not change.

The optically thick character of the gas allows the photons and particles to interact sufficiently to come into equilibrium thus giving rise to the continuum (blackbody) spectrum characteristic of the cloud.

Case 3: $I_0 > 0, \tau \ll 1$

In this case, there is a source behind the cloud. Since $\tau \ll 1$, we again use the Taylor expansion, $e^{-\tau} \approx 1 - \tau$, so that (44) becomes

$$I = I_0 + \tau(I_s - I_0) \qquad\qquad (I_0 > 0, \tau \ll 1) \qquad\qquad (11.48)$$

Consider two cases here, $I_s > I_0$ and $I_s < I_0$. In the former case, the output intensity is the background intensity I_0 plus another positive term. If the optical depth τ is higher at some frequency (i.e., greater opacity κ) than at surrounding frequencies, the emerging flux will be greater at that frequency. This yields an emission line (Fig. 19c).

In the case of $I_s < I_0$, the rightmost term is negative, and the emerging intensity is less than the background intensity. If again, the optical depth τ is especially large at some frequency, the emerging intensity is depressed even more at that frequency. This yields an absorption line (Fig. 19d).

These same conclusions extend to somewhat larger optical depths, $\tau \lesssim 2$, as illustrated in the plots of the function $I(\tau)$ vs. τ (Fig. 18). In the case of $I_s > I_0$ (Fig. 18a), the observed intensity I increases with optical depth τ. At a given frequency not at a resonance, the intensity I might be given by the value at point A in the plot. At the frequency of a resonance, the optical depth is higher (by definition), and the observed intensity is therefore higher (point B). Thus an emission line is observed. For $I_s < I_0$ (Fig. 18b), the increase in opacity again moves the observer from A to B, but in this case it yields a decrease in intensity, or an absorption line.

If the functions I_0 and I_s are each blackbody spectra, the one with the higher temperature will have the greater intensity at any frequency (Fig. 8). Thus we have $T_s > T_0$ for the $I_s > I_0$ case, and $T_s < T_0$ for the $I_s < I_0$ case. We conclude therefore that if the temperature of the foreground cloud T_s is *greater* than the background temperature T_0, a spectrum with *emission lines* will emerge, and that if the cloud is *cooler* than the background, a spectrum with *absorption lines* will emerge.

In most stellar atmospheres at the depth seen in visible light (the photosphere), the temperature decreases with altitude, i.e., toward the observer. The deeper hot layers are then the background radiation for the higher, cooler regions. Absorption lines are thus prevalent in stellar spectra at visible wavelengths.

In contrast, radiation from the sun at ultraviolet frequencies yields emission lines. The observed ultraviolet radiation comes from high in the solar atmosphere because the higher opacities in the ultraviolet limit the depth into which the observer can "see". In these higher chromospheric regions, the temperature is increasing with height (moving toward the 10^6-K corona). Thus the higher temperatures are in the foreground, and the spectra characteristically exhibit emission lines.

Case 4: $I_0 > 0$, $\tau \gg 1$

In this case, the gas is optically thick and (44) again yields

$$I(\nu) = I_s(\nu) \tag{11.49}$$

This is the same expression obtained when there was no background intensity (47). Since the gas is optically thick, the presence of the background source is immaterial (Fig. 19e). The radiation at any τ is simply the continuum source (blackbody) spectrum of the optically thick cloud. It does not matter whether $I_0 > I_s$ or $I_0 < I_s$.

Summary

This concludes our discussion of the limiting cases of the solution to the radiative transfer equation. In each case the result is a continuum spectrum that reflects one or both of the spectra I_0 and I_s with, in some cases, superimposed lines created by increases in the optical depth τ at certain frequencies. If the foreground cloud intensity (temperature) is greater than the background intensity (temperature), emission lines are formed. If the opposite is true, absorption lines are formed.

Problems

11.2 *Plots of spectra*

Problem 11.21. (a) The spectral flux density in wavelength units of some source varies as the inverse fourth power of the wavelength, $S_\lambda = K\lambda^{-4}$, where K is a constant. What is S_ν, expressed as a function of ν? See if you can do this from first principles without reference to the text. Give the units of S in both forms. (b) Develop an expression for the specific intensity in wavelength units, $I_\lambda = I(\lambda, T)$, for blackbody radiation. Start with the blackbody spectrum (23). Give the units of I_λ.

Problem 11.22. An x-ray astronomer measures the spectrum of the diffuse x-ray background over the range 2–60 keV and finds it to have an exponential shape. He reports the *energy* specific intensity to be

$$I_E = 3.6 \times 10^4 \exp\left(-\frac{h\nu}{23\text{ keV}}\right)\text{keV s}^{-1}\text{ m}^{-2}\text{ keV}^{-1}\text{sr}^{-1}$$

where $h\nu$ is given in keV. (a) Convert this to a *photon number* specific intensity $I_p(\nu)$ with units $\text{s}^{-1}\text{ m}^{-2}\text{ Hz}^{-1}\text{ sr}^{-1}$ and with the coefficient that gives the correct quantitative values. (b) This radiation is believed to come from the active galactic nuclei (AGN) of many distant galaxies, not from an isothermal optically thin plasma as might be inferred from its spectical shape; see Section 3. If it were such a plasma, what would be its temperature in kelvin? [Ans. $\propto I_E/\nu$; $\sim 10^8$ K]

11.3 *Continuum spectra*

Problem 11.31. Consider the sketches of thermal bremsstrahlung spectra on a log-log plot in Fig. 3c. The curves are for two identical plasmas, with constant identical Gaunt factors, except that their temperatures differ. Suppose that one is three times hotter than the other, $T_2 = 3T_1$. (a) At what photon energy $h\nu$ do the curves for T_1 and T_2 cross. Express your answer in terms of kT_1. (b) Make a quantitative log-log plot similar to Fig. 3c showing three thermal bremsstrahlung spectra (I vs. ν) for temperatures T, $2T$ and $3T$, drawn properly to scale, again for identical plasmas with identical constant Gaunt factors. [Ans. (a) $\sim 0.8\,kT_1$]

Problem 11.32. (a) Demonstrate that the Rayleigh–Jeans law (24) follows from the blackbody intensity (23) in the limit of low frequency. What is the condition on the frequency for this expression to be valid? (b) Find the temperature of the CMB radiation from the value of the fitted curve at 10 GHz in Fig. 9a. Would it have been easier to use Fig. 9b because the ordinate is a linear scale? Explain. [Ans. (b) $\sim$3 K]

Problem 11.33. (a) Write an expression for the radio portion of the spectrum $I(\nu)$ of the Crab nebula as presented in Fig. 2. Give your answer in the forms $S = K\nu^{\alpha}$ and $S = K'(\nu/\nu_0)^{\alpha}$ where ν_0 is some convenient frequency (e.g., some integer power of 10) in the region. Hint: find the latter form first. Include numerical values for α, K and K'. (b) Repeat for the x-ray/gamma-ray portion of the spectrum. [Ans. $\propto \sim \nu^{-0.25}$; $\propto \sim \nu^{-1.5}$]

Problem 11.34. (a) Integrate graphically under the curve for the spectral flux density S given in Fig. 2 for the Crab nebula to find the flux density $\mathscr{F}$ summed over all energy bands, from $\log \nu = 6.5$ to $\log \nu = 22.5$. Take small slices along the abscissa, (one decade of frequency) to minimize errors due to the logarithmic scale. Interpolate over regions with no data. (b) The Crab nebula is $\sim$6000 LY distant. What is its luminosity, from $10^{6.5}$ Hz to $10^{22.5}$ Hz? Compare to the luminosity of the sun. (c) Comment on the relative fluxes in the several frequency bands, radio, optical, etc. [Ans. $\sim 10^{31}$ W]

11.4 Spectral lines

Problem 11.41. (a) What is the approximate equivalent width (in units of nm) of the prominent absorption line shown at $\lambda \approx 485$ nm toward the left in Fig. 10a? (b) Estimate the equivalent width (in eV) of the Ne X emission line at $\sim$1022 eV in the Capella spectrum of Fig. 5. [Ans. $\sim$1 nm; $\sim$100 eV]

Problem 11.42. Calculate roughly the time for collisional de-excitation of a single oxygen (O III) atom in a metastable state if it resides in an emission nebula (H II region). The de-excitations take place because fast electrons collide with the relatively large and slow oxygen atom. The approximate time between collisions is the required answer. Let the density of electrons in the nebula be $n = 1 \times 10^8$ m^{-3}, the temperature of the electrons be $T \approx 7000$ K, and the size (diameter) of the oxygen atom be $d \approx 0.3$ nm. Compare your answer to the natural or spontaneous decay time of $\sim$100 s for the metastable states of Table 1 (assuming no collisions). Hints: (*i*) Consider the size of the electron to be negligible when estimating the cross section for the collision. (*ii*) The speed of the electron may be obtained from the relation $3kT/2 = m\nu^2/2$. (*iii*) The oxygen atom may be considered to be a stationary target (why?). [Ans. $\sim$3 days]

Problem 11.43. (a) For the damping profile (30), find the frequency where the maximum amplitude occurs, the value of $\kappa_1/\kappa_{1,0}$ at this frequency, and $(\nu - \nu_0)_{\text{HM}}$, the half width of the curve at one-half the maximum amplitude (HWHM). (b) Based on these results, make rough sketches of the damping curves for $\gamma = 0.5$, 1, and 2. (c) Repeat part (a) for the Doppler distribution (28). (d) Consider Fig. 15. What is the value of the parameter γ for the damping expression given there? Find an expression for the Doppler response κ_1 that has unit amplitude and HWHM twice that of the HWHM of the damping curve. Compare to the Doppler expression given in the figure. [Ans. ν_0, $4/\gamma$, $\gamma/2$; ___; ν_0, 1, σ $(2 \ln 2)^{1/2}$; 1, same]

11.5 Formation of spectral lines (radiative transfer)

Problem 11.51. Consider a stellar atmosphere where I_s varies with depth in the cloud as $I_s = a + b\tau$ where a is a positive constant and b is a constant that can be positive or negative. (In the text, we took I_s to be constant throughout the cloud.) Assume that conditions of local thermodynamic equilibrium are satisfied, and that the observer views the atmosphere head on, as in Fig. 17. (The variation in I_s arises through a variation the volume emissivity with position (38) which in turn is a consequence of temperature variation within the atmosphere. (a) Find the solution $I(\tau)$ of the equation of radiation transfer (39) for this situation. (b) Evaluate the solution for the case of no background source, $I_0 = 0$, with $\tau \ll 1$ and with $\tau \gg 1$. (c) Explain why spectral lines will or will not be formed in each of these two cases. If they are, what are the condition(s) on b that result in emission or absorption lines? In the $\tau \ll 1$ case, how would you constrain b so that only emission lines occur in the region $\tau < 0.1$, in the context of your approximations? [Ans. (b) $I(\tau \ll 1) \approx a\tau + (b - a)(\tau^2/2)$]

12

Astronomy beyond photons

What we learn in this chapter

Major new facilities that detect signals from the cosmos other than electromagnetic radiation are bringing new fields into the forefront of astronomy. **Neutrino observatories** study the energy-producing thermonuclear reactions at the center of the sun with detectors utilizing **chlorine**, **gallium**, and **pure water**, the latter making use of **Cerenkov radiation** from recoil electrons. The pioneering **Homestake mine experiment** and the huge **Super-Kamiokande experiment** are important examples. Neutrino astronomers detected a **flash of neutrinos** from the collapse of a star in the supernova **SN 1987A** and hope to see **extragalactic flashes** from **gamma-ray bursts**.

Cosmic ray observatories study highly energetic charged particles (mostly protons) entering the atmosphere from the Galaxy and probably extragalactic sources. The **element abundances** at energies $\lesssim 1$ GeV provide a **lifetime** ($\sim 10^7$ yr) for their **storage in the Galaxy**. The highest energy particles initiate **extensive air showers** (EAS) of particles in the earth's atmosphere, facilitating their study with detector arrays covering 10^3 km^2, such as the **HiRes Fly's Eye** and the **Auger project**. The most energetic such particles, ~ 10 to 300 EeV (10^{19} to 3×10^{20} eV) are probably **extragalactic in origin** and may arrive from the approximate directions of their origin. **Small EAS** initiated by TeV gamma rays high in the atmosphere produce **Cerenkov radiation** observed with ground based **mirror-PMT systems**, i.e., **TeV photon astronomy**.

Gravitational waves (G waves) are predicted by Einstein's **general theory of relativity** and searches for them have so far not reached the needed sensitivities. However, **binary radio pulsars** do exhibit **decaying orbits** that clearly indicate the loss of energy through the emission of gravitational waves. **Resonant bar** detectors were employed in the first searches and now huge (4 km) **laser interferometers** such as **LIGO** are reaching sensitivity levels that could detect **neutron star mergers** in the **Virgo cluster**, ~ 55 MLY distant. The **compact-star mergers** or **stellar collapses** that probably give rise to **gamma-ray bursts** could also generate detectable pulses of gravitational radiation. **Low frequency studies** could detect **binary compact star systems**, **mergers of supermassive black holes**, and background radiation from the **early universe**. The planned NASA/ESA **Laser Interferometer Space Antenna** (LISA) mission is intended to carry out such studies.

12.1 Introduction

Astronomers and physicists now observe the sky with instruments that are sensitive to signals that are not a form of Maxwell's electromagnetic waves or their equivalent, photons. These "telescopes" have been designed to detect (*i*) neutrinos from the sun, supernova explosions and nuclear interactions in the earth's atmosphere, (*ii*) cosmic ray protons and other heavier atomic nuclei that gain their high energies at acceleration sites in the Galaxy and possibly beyond, and (*iii*) gravitational waves which should be created by large rapidly accelerating masses, such as a binary system containing two neutron-stars that spiral into one another to form a black hole.

Neutrinos and cosmic ray fluxes are now being detected and studied. Gravitational waves have yet to be detected but there is much reason, both observational and theoretical, to believe they exist and will be found when experiments reach the required sensitivity.

These detection systems do not focus the particles from a distant point source to an image as do optical telescopes, and so are often called "experiments" or "observatories". Nevertheless they do have methods of determining arrival directions to within several degrees in some cases, and they do probe the secrets of the universe. Thus "telescope" in its broadest sense is certainly an appropriate term for these instruments.

12.2 Neutrino observatories

In this section we present the instruments of neutrino astronomy. Neutrinos are chargeless and massless (or almost so) and interact with matter with extremely low probability. Thus their detection requires detectors of large mass and hence volume. The field has had two major successes, (*i*) the discovery of neutrinos from the reactions ongoing within the sun and (*ii*) the spectacular discovery of neutrinos created in the collapse of a star in the relatively nearby Large Magellanic Cloud which resulted in supernova 1987A. Here we describe the solar work and describe several types of detectors, namely chlorine (the Homestake mine experiment), gallium (e.g., GALLEX), and water Cerenkov (e.g., Super-Kamiokande).

Neutrinos from the sun

Electron neutrinos v_e are emitted in the power-generating nuclear interactions that occur at the center of stars such as the sun. There are two other *flavors* of neutrinos, the *muon neutrino* v_μ and the *tau neutrino* v_τ but these are not emitted in the nuclear reactions within the sun. (Each of the three flavors has a counterpart antineutrino; there may also be a *sterile neutrino*.) Neutrinos are like photons in that they are chargeless particles with zero mass, or possibly with a very small mass ($mc^2 \lesssim$ few eV). They travel at, or very close to, the speed of light.

Unlike photons, neutrinos interact with matter only very rarely, i.e., with incredibly small cross sections (Section 10.4). For absorption on hydrogen the cross section is only $\sim 10^{-48}$ m^2. This corresponds to a mean free path of ~ 2 LY in material of density 10^5 kg/m^3, the density of the solar core. Since the sun is only a 2 light-seconds in radius, essentially all the neutrinos escape the sun without interacting. Large numbers must be raining copiously down onto the earth, roughly 10^{15} s^{-1}m^{-2}!

Detection of neutrinos from the sun is important because it would permit a good cross check on the nuclear physics that astrophysicists believe is taking place within it. However, detection is difficult because the same low cross sections that allow the neutrinos to escape the sun make difficult their detection. A very large detector mass is required to detect even a tiny proportion of the incident neutrinos. All others pass right on through the detector and nearly all pass right on through the earth!

The primary energy generation process in the sun is the conversion of hydrogen to helium. This occurs through several series of nuclear reactions that take four protons and convert them into one helium nucleus and two positive electrons (positrons). This conserves both the number of baryons (protons and neutrons) and charge. Additional products are gamma rays and electron neutrinos. The initial steps involve the interactions of two protons, so this is called the pp chain. There is also a cycle of nuclear interactions involving carbon, nitrogen and oxygen as catalysts (the CNO cycle) that has the same effect, namely the conversion of four protons to a helium nucleus, and two positrons, again with the emission of gamma rays and neutrinos.

The mass of the final products of one of these nuclear cycles is a bit less than that of the input products so a bit of excess energy is released. It is in the form of kinetic energy of the products of the reactions, including the neutrinos. The neutrinos freely escape the sun while the other particles contribute to the energy content of the sun. The standard solar model indicates that 98% of the energy is generated by the pp chain and the remaining 2% by the CNO process. The CNO cycle is more important in more massive stars which have higher central temperatures.

The spectrum of solar neutrinos from the pp process at the earth as expected from the standard model is shown in Fig. 1. The reactions or decays that give rise to each component are indicated. If the final state consists of only two particles, one of which is a neutrino, the neutrino will always have the same energy in the center of mass (CM) frame of reference. This follows from momentum conservation and the

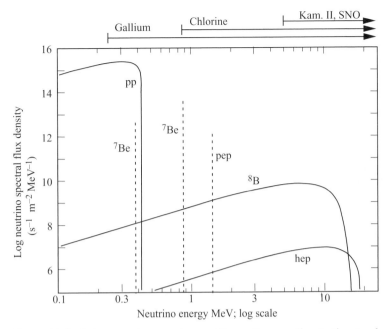

Figure 12.1. Solar neutrino spectrum at the earth according to the standard model of the sun for the pp chain sequences. The continuum spectra are in units of $s^{-1}\,m^{-2}\,MeV^{-1}$. The line fluxes (dashed lines) are the integrated fluxes (over energy) and are given in units of $s^{-1}\,m^{-2}$ on the same scale. The proton–proton (pp) and proton–electron–proton (pep) reactions each create a deuteron and a neutrino, and in the former case also a positron. The nuclide ^{7}Be and an electron interact to produce ^{7}Li and a ν_e while ^{8}B decays to ^{8}Be with the emission of a positron and ν_e. The ^{3}He interaction with a proton (hep) yields ^{4}He, a positron and a ν_e. Two-body final states yield a line while three-body final states yield a continuum spectrum. [Bahcall, Parsonneault & Basu, *ApJ* **555**, 990 (2001) astro-ph/0010346]

fact that the total available energy is fixed. In the solar interior with temperatures $T \approx 10^7$ K, thermal motions or energies correspond to keV energies, while the products of the reactions have MeV energies; the CM is essentially at rest. The neutrinos from a two-body final state will all therefore emerge with nearly the same energy. The spectrum of such neutrinos will be a line of very narrow width.

If, on the other hand, the final state consists of three particles, one of which is a neutrino, a spread of neutrino energies is obtained with a definite maximum neutrino energy. The maximum occurs when the two non-neutrino particles are emitted in the same direction. To conserve momentum, the neutrino must be emitted in the opposite direction with momentum equal to that of the other two. The associated energy is the most the neutrino can take from the reaction.

An ambitious effort has been underway since the late 1960s to detect and study the solar neutrinos at the earth. These neutrinos have indeed been detected in several different experiments. This is a major affirmation of our understanding of the nuclear physics in stellar interiors. However, the match with theory is not perfect; the numbers of detected neutrinos are deficient by a factor of 2 or 3 in the several experiments. This is known as the *solar neutrino problem*. It raises fundamental questions about the nature of the neutrino; e.g., it may not be massless.

Homestake mine experiment

The original experiment to detect solar neutrinos was carried out by Ray Davis with much encouragement from John Bahcall whose calculations showed the solar flux of neutrinos was indeed sufficient for detection by this experiment. He and others over the years continuously refined the theoretical model while experimentalists independently measured values of cross sections for the nuclear reactions, etc. All this auxiliary effort was required because there would have been no "solar neutrino problem" had there not been a well tuned (precise) solar model with which to compare the data.

The *Homestake mine experiment* was carried out in a gold mine in South Dakota, 1500 m below the surface. This reduced the background due to highly penetrating cosmic ray muons (Section 3). The detector was a huge (390 m^3) tank containing 630 Mg (= kkg = metric tons) of cleaning fluid (perchloroethylene; C_2Cl_4).

Neutrino–chlorine conversions

Solar neutrinos passing through the tank of C_2Cl_4 are occasionally absorbed in nuclear reactions with the nuclei of chlorine atoms. The reaction converts the chlorine nucleus into a radioactive argon nucleus. After some hundred days, the argon nuclei are removed from the tank by a filtering process and counted through the detection of their radioactive decays.

Specifically, in this conversion, an electron neutrino is absorbed in a chlorine-37 nucleus which becomes an argon nucleus, and an electron is ejected,

$$\nu_e + {}^{37}_{17}Cl \rightarrow {}^{37}_{18}Ar + e^- \qquad \text{(Neutrino absorption)} \qquad (12.1)$$

This reaction conserves electric charge, the number of baryons (p,n), and the number of leptons (ν,e), as it must. It can occur if the neutrino energy is greater than 0.81 MeV.

In this expression (1), the preceding subscript is the atomic number Z of the nucleus (i.e., the number of protons), and the superscript is the mass number A (neutrons plus protons). Chlorine can have various *isotopes* of different mass numbers, e.g., ${}^{35}_{17}Cl$, ${}^{36}_{17}Cl$, etc. The subscript "17" is redundant with the symbol Cl, but it is convenient for keeping track of the electric charges of the nuclei.

The atomic mass number A did not change in the reaction (1). In effect, a nuclear neutron was converted to a proton, and an electron was created and ejected to conserve charge. The electron is typically ejected with an energy much greater than atomic binding energies. Thus it is ejected completely from the atom. The argon atom, being one element up on the periodic table, is now short one electron. The ejected electron (or more likely a different nearby electron) soon becomes the extra orbital electron needed to neutralize the ^{37}Ar atom.

Argon decay

The incoming flux of neutrinos steadily produces ^{37}Ar atoms which accumulate in the tank. However, the ^{37}Ar is radioactive with a *half-life* of 35 days. In 35 days, $1/2$ of an isolated sample of atoms will have converted back to ^{37}Cl by the process called *electron capture*, or *K-shell capture*, wherein the nucleus absorbs an electron from the innermost (K or $n = 1$) shell with the concurrent emission of a neutrino,

$$^{37}_{18}\text{Ar} + \text{e}^- \rightarrow (^{37}_{17}\text{Cl})^{\#} + \nu_{\text{e}} \qquad \text{(Electron capture;} \qquad (12.2)$$
$$\tau_{1/2} = 35\,\text{d} \qquad \text{chlorine in excited (\#) state)}$$

The captured electron simply ceases to exist.

The resultant chlorine atom is neutral, but it is in an excited state because it is missing one of its innermost electrons, the one that was just absorbed by the nucleus. It therefore immediately decays to a lower energy state by filling the K shell, most probably with an electron from the $n = 2$ (L) shell. However, instead of giving up the energy as an emitted photon, it is likely to do so by ejecting another of the L-shell electrons from the atom. This is known as the *Auger effect*, and the ejected electron is known as an *Auger electron*. Its ejection leaves the Cl atom ionized.

$$(^{37}_{17}\text{Cl})^{\#} \rightarrow (^{37}_{17}\text{Cl})^{\#}_{\text{ion}} + \text{(Auger electron)} \quad \text{(Auger effect)} \qquad (12.3)$$

The Auger effect is favored for the lighter elements, and a radiative transition is favored for heavier elements.

The ^{37}Cl is now ionized and still is in an excited state with two vacancies in the L shell. It will immediately fill these with the emission of photons, and it will soon encounter and absorb a free electron to become neutral. The cycle is complete; the chlorine atom has returned to its original state before absorbing the neutrino.

It takes 35 days for $1/2$ of the created ^{37}Ar atoms in the tank to decay back to chlorine. After another 35 days only $1/2$ of the remainder will survive, i.e., $1/4$ of the original number. The record of the absorbed neutrinos would thus seem to be gradually lost. However, in practice, the steady influx of neutrinos is continuously creating excited ^{37}Ar nuclei. If an experimental run starts with no argon content in the tank, the creation process causes the number of ^{37}Ar to build up linearly, but as decays become important ($t \gtrsim 35$ d), the increase is slowed, and after several mean lives ($\sim$100 days), an equilibrium number of ^{37}Ar atoms is approached. This number is indicative of the intensity of the incident solar neutrino flux.

Sweeping for argon atoms

The number of ^{37}Ar atoms in the tank is determined as follows. Helium gas is pumped through the tank; it collects ("sweeps up") the individual ^{37}Ar atoms and thereby removes (purges) them from the liquid. This is done about every 100 days. The ^{37}Ar atoms are removed from the helium by trapping them in a charcoal filter at liquid nitrogen temperature. The atoms can then be detected because they are radioactive; they decay with a 35-d half-life and emit a neutrino (2) followed immediately by an Auger electron (3).

A proportional counter (Section 6.2) is used to detect the Auger electron; the gas sample is placed inside the proportional counter. The delicacy of this experiment can be appreciated from the fact that the number of ^{37}Ar atoms in the entire 390-m^3 volume of liquid at the time of the purging is expected to be only $\sim$50. Furthermore it takes 35 d for 1/2 of them to decay; thus less than 1 event per day is detected by the proportional counter! Great care to eliminate background events is required.

Solar neutrino problem

The low rates of neutrino detection are described with the "SNU" the *solar neutrino unit*.

$$1 \text{ SNU} = 10^{-36} \text{ captures s}^{-1} \text{ (target atom)}^{-1} \qquad (12.4)$$

The expected value from the sun depends upon assumptions about conditions in the sun, but for the standard solar model, it is about 8 SNU (7.7 ± 1.1). After more than 25 years of operation of the Homestake experiment, the measured rate was only 2.56 ± 0.23 SNUs or about 1/3 the expected value.

The model of the sun is unlikely to be significantly in error because it produces quite well the basic properties of the sun and describes well the evolution of stars like the sun. Furthermore, studies of the internal structure of the sun with acoustical waves that cause oscillations of the sun's size and shape (*helioseismology*) yield properties closely in accord with the model. Thus the discrepancy between the expected and observed neutrino rates appears to be real and may be revealing new physics to us.

The Homestake experiment was by far most sensitive to the ^{8}B-decay neutrino ($E_{max} = 14$ MeV; Fig. 1) because the cross section for capture by ^{37}Cl is greatest at high neutrino energies. These neutrinos are only a minuscule portion of the emitted neutrino flux. (Note that the ordinate scale of Fig. 1 is logarithmic.) The much more plentiful pp neutrinos ($E_{max} = 0.42$ MeV) can not be detected at all in the chlorine

(Homestake) experiment; the neutrinos are all below the 0.81 MeV threshold for the ^{37}Cl capture reaction.

Second generation experiments

Gallium detectors

A second generation of solar neutrino experiments were carried out in the 1990s to search for the more abundant low energy pp neutrinos. They were located in the former Soviet Union (SAGE) and in Italy (GALLEX). They used gallium as an absorber, with a total mass of 60 Mg and 30 Mg (60 and 30 metric tons) at the two sites, respectively. The detections take place through a reaction similar to (1) which converts the gallium nucleus $^{71}_{31}$Ga to a germanium nucleus $^{71}_{32}$Ge. This reaction has a threshold of only 0.23 MeV so it allows detection of the basic pp neutrinos.

The rates of the pp reaction are predicted with better accuracy than the ^{8}B reaction which is highly temperature dependent. The detections thus serve as a direct check on the most basic reaction believed to be operating in the sun.

Results from the two gallium experiments again indicate a deficiency of neutrinos; only about 1/2 of the expected flux is detected. This increased confidence greatly that the solar neutrino problem (deficiency) is real. A similar deficiency was also reported from a very different type of experiment, the Japanese Kamiokande which was sensitive only to the ^{8}B neutrinos.

Super-Kamiokande

Super-Kamiokande (SK) is a much enlarged version of the original Kamiokande experiment. It is located 1000 m underground in the Japanese Alps, in a mine of the Kamioka mining company. Beginning in 1996, it has monitored the energetic ^{8}B neutrinos from the sun. It is a huge (40 m diameter and 40 m tall $\rightarrow$ 50 000 m^3) tank of pure H$_2$O viewed by 11 200 large (diameter 0.5 m) photomultiplier tubes (PMT; Section 6.2) that line the sides, top and bottom of the tank.

Electron neutrinos from the sun scatter elastically off the electrons in the H$_2$O, with the most common reaction being

$$\nu_e + e^- \rightarrow e^- + \nu_e \tag{12.5}$$

The electron is given a significant portion of the solar neutrino energy, e.g., $\sim$10 MeV, and it emits visible light of a nature called *Cerenkov radiation* which is detected by the PMT. The neutrino energy threshold of SK is much higher than the chlorine experiment, about 5 MeV compared to 0.8 MeV, so it detects only the energetic ^{8}B neutrinos. Results from SK also indicate the rate of solar neutrinos is lower than the solar model predicts.

The SK experiment also yields rough arrival directions of the neutrinos. In the reaction (5), the electron preferentially recoils in a direction approximating that of the neutrino momentum. The pattern of light detected by the PMT lining the walls reveals this direction, roughly. The measured directions from the original Kamiokande confirmed that the neutrinos arrive on average from the solar direction with an uncertainty of $\sim 20°$. This was a welcome confirmation of the hypothesis that the sun is the source of the neutrinos being detected in these experiments.

Cerenkov radiation

The directional information from these instruments is obtained from the directionality of the Cerenkov radiation emitted by the recoil electrons. It is emitted in preferred directions relative to the track (or velocity vector) of the emitting particle. The radiation originates along the track due to the electromagnetic interaction of the electron and the material through which it passes. These disturbances can travel no faster than the speed of light in water (speed c divided by the index of refraction n). If the emitting particle travels faster than this, but of course less than the speed of light c in a vacuum, the wavelets emanating from the track set up a wavefront as

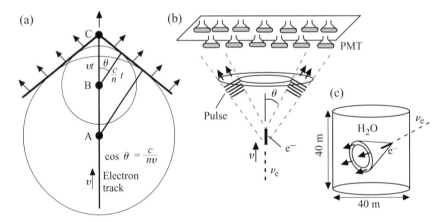

Figure 12.2. (a) Principle of directionality of Cerenkov radiation. The wavelets that originate from points along the electron track travel at the speed of light in the medium, c/n where n is the index of refraction. If the electron is moving faster than this, the wavelets combine to create an outward moving shock front at angle θ. The angle θ increases as the speed v of the electron increases relative to c/n. (b) Short burst of Cerenkov radiation from the short track of a recoil electron. The fast moving "smoke ring" propagates along a cone of directions to intersect the planar array of photomultiplier tubes (PMT) as a circle or ellipse. Higher energies produce bigger rings. (c) Sketch of Cerenkov cone in Super-Kamiokande which contains 50 000 tons of pure water and is lined with 11 200 PMT.

shown in Fig. 2a and explained in the caption. It is the electromagnetic equivalent to the shock wave formed by a supersonic jet airplane.

The Cerenkov light is thus emitted in a cone of directions about the direction of the electron momentum. Because the electron range in the water is relatively short, a few centimeters, the light illuminates (ideally) a circle on the side of the tank (Fig. 2b,c). This light is detected by the PMT lining the sides of the tank that it strikes. If the electron direction is not normal to a flat wall of PMT, the illuminated pattern will be distorted, to an ellipse if the wall is flat. The shape and size of the distorted circle, the total amount of light collected by each PMT, and the time of arrival of the light signal at each PMT (with nanosecond precision) together enable the astronomer to solve for the electron track direction, its energy, and its position within the tank.

The measured direction for a single event will have a large uncertainty because electrons scatter into different directions as they travel through the water. Furthermore, the electron recoil direction will only approximate the neutrino direction because the outgoing neutrino carries off some of the original momentum. However, the uncertainty in the average direction from many such events from a single source decreases as more events accumulate.

Neutrino oscillations and more

The puzzling deficiency of neutrinos from the sun could be resolved through the decay or the transformation of the neutrinos en route to the earth. Although the neutrino has long been considered to be a zero-mass particle (like the photon), particle theories can easily accommodate neutrinos with small masses. In this case, the wave function of a given flavor neutrino is actually a superposition of several mass eigenstates. This is a quantum mechanical phenomenon well known in particle physics.

In this mixing of states, the ν_e neutrinos could oscillate among the three flavors during their flight from the sun to the earth. A ν_e could be transformed to another flavor such as the ν_μ, and possibly back again. It is proposed that the transformations can be greatly enhanced by the interaction of the neutrinos with the electrons in the solar material; this is the *MSW effect* after the proposers, Mikheyev–Smirnov–Wolfenstein. This process could well lower the detected rate of ν_e at the earth to agree with the observations.

Recent (1998 and 2001) results from SK, the Sudbury Neutrino Observatory (SNO), and the Japanese KamLAND suggest strongly that such oscillations are indeed the explanation of the low detected rates. However, there remain apparent inconsistencies among the results of these and other experiments so more studies are required to find the exact nature of the oscillations. With that knowledge, precise checks of the solar models could be carried out.

Observations continue into the current millennium with ongoing and new experiments such as the aforementioned SNO in Canada with a kiloton of heavy water (D$_2$O), BOREXINO, ICARUS, and GNO in the Gran Sasso tunnel in Italy, AMANDA and ICECUBE which use the south polar icecap as a detector, and also ANTARES and NESTOR which use the water of the Mediterranean Sea. A reconstructed SK (after an accident destroyed many of the PMT) is restarting operations.

One major goal of neutrino astronomy is the detection of flashes of neutrinos from (extragalactic) gamma-ray bursts (GRB). These flashes of gamma rays are likely due to massive stellar collapses directly into a black hole (hypernova) or compact star mergers into a black hole. Like the flash of neutrinos from supernova SN1987A, they would give us a direct view into the very core of the collapse. Some of the aforementioned experiments are potentially sensitive enough to detect them.

We will find in Section 4 that hypernovae and compact star mergers could yield detectable gravitational waves. Both types of astronomy could thus probe the fundamentals of these cataclysmic events.

12.3 Cosmic ray observatories

The term *cosmic rays* sounds like a grand term to describe all radiation from the cosmos. In fact, it has taken on a specific and restricted meaning. It refers, for the most part, to high energy charged particles (mostly protons) that travel through the Galaxy, some of which arrive at the earth. It also refers to the progeny (descendants) of these particles that are produced in nuclear collisions in the earth's atmosphere. Many of these "secondary" cosmic rays are charged particles but neutral particles and photons are also present. One component of the "primary" cosmic rays is well known to radio astronomers. It is observed as the diffuse radio background which is synchrotron emission from high energy cosmic-ray electrons spiraling about magnetic field lines in the Galaxy.

Cosmic rays were so named by Millikan in the 1920s because his experiments confirmed earlier results by Hess, in 1912, that mysterious ionizing radiation originated high in the atmosphere, or beyond. The motivation for the Hess experiment was the excess and puzzling radiation observed with *electroscopes*, e.g., the leaf electroscope. These are devices, often seen in freshman or high school labs, that measure charge deposited on the leaves; the puzzling radiation causes them to discharge more rapidly than was expected.

Early studies of this radiation were carried out with detectors carried to the tops of mountains, taken even higher in balloons, or sunk deep into lakes. Indeed, the radiation studied in this way turned out to be the first signals detected by man from distant regions of the Galaxy, other than meteorites and visible light. Thus, the name "cosmic rays" is indeed appropriate.

Primary and secondary fluxes

Storage in the Galaxy

The general picture that emerged from subsequent research is that charged particles (protons with an admixture of heavier atomic nuclei), are accelerated to high energies by processes not yet well determined, but probably in part in supernovae and the shock waves of their aftermath. They then travel through the Galaxy, trapped (stored) therein by the galactic magnetic fields about which they spiral under the influence of the $v \times B$ force. Some cosmic rays are so energetic they could not have been trapped in the Galaxy. They may well have originated outside the Galaxy.

Collisions with the protons of the interstellar gas cause some of the heavier nuclei in the cosmic ray flux to break up (*spallation*). The element abundances of primary cosmic rays, as measured at energies $\lesssim 1$ GeV with satellites, differ substantially from the solar abundances (Table 10. 2). Most notably, cosmic rays have a substantial excess of the relatively rare light elements, Li, B, Be, which are products of the spallation reactions.

With knowledge of the gas density in interstellar space ($\sim 0.5 \times 10^6$ atoms m^{-3}) and the cross sections for the spallation interactions, the modified abundances yield the amount of material the typical ~ 1 GeV cosmic ray has traversed (~ 60 kg/m^2) which at speed c yields a typical age ($\sim 10^7$ yr). These cosmic rays have been in the Galaxy for a much longer time than it would take them to traverse its 10^5LY diameter. This is the basis of the above statement that they have been stored in the Galaxy.

Some of the cosmic rays enter the solar system and arrive at the earth's magnetosphere. If they are not sufficiently energetic, the earth's magnetic field ($v \times B$ forces) will cause them to be reflected back out to space. Otherwise they will penetrate the magnetosphere. This requires energies of up to 15 GeV depending on their initial direction of arrival and location relative to the earth's magnetosphere. Penetration at the earth's magnetic equator requires the most energy.

The cosmic rays that thus arrive at the top of the earth's atmosphere are called *primary cosmic rays*. At the higher energies they are a good sample of the cosmic rays trapped in the Galaxy. The primary flux at the top of the atmosphere at mid latitudes is about 1×10^4 particles s^{-1} m^{-2}, or one per square cm per second.

Nuclear component

The primary protons (and heavier nuclei) find the atmosphere to be quite opaque. High in the atmosphere, they undergo nuclear collisions with the nuclei of nitrogen and oxygen nuclei. Many lower energy particles, the *secondary cosmic rays*, are produced in these interactions. The average amount of atmospheric matter traversed (kg/m^2) by a primary proton before it strikes a nucleus is only 8% of the full atmospheric "depth" (matter thickness) of 10 300 kg/m^2. This is the number of kilograms in a column of 1 m^2 cross section; more on these units is in Section 10.4.)

Some of the secondary particles produced in the nuclear interactions were previously unknown to physicists and chemists because they exist for only a short time after their creation; microseconds or less. They decay to lighter particles which, in turn, may also decay to even lighter particles, one of which may be the well known and stable electron. Figure 3 illustrates some of the interactions and decays that take place in the atmosphere. The unstable particles are called *mesons* and they come in numerous varieties.

Specifically π mesons (pions) are preferentially produced in the primary (first) nuclear interaction. The pion is a nuclear active particle with mass 15% of the proton mass ($m_\pi c^2 = 140\,\mathrm{MeV}$). It can be charged (+ or –) or neutral. The neutral π^0 meson decays extremely rapidly (10^{-16} s) into two gamma rays. A charged pion decays to a muon and neutrino in $\sim 10^{-8}$ s, or it may suffer a nuclear collision with another nitrogen or oxygen nucleus, generating even more such particles.

The primary proton, if it has sufficient remaining energy, will similarly interact with other nuclei as will neutrons and protons ejected from the first interaction. Thus, if there is sufficient energy, a core of nuclear-active particles will propagate along the initial incident direction until its energy is depleted. Most do not reach sea level, but they can be detected at high altitudes in the cores of the extensive air showers discussed below.

Electromagnetic component

A gamma ray from a π^0 decay may interact with the electric field of an atmospheric nucleus and create a pair of electrons (e$^+$, e$^-$) in a process called *pair production*, previously encountered in Section 10.2. In turn an electron passing close to another nucleus will be accelerated by its electric field and thereby emit a gamma ray, or several of them, in the *bremsstrahlung process* (Section 11.3). This process repeats itself if the energy is sufficient.

Such electrons and gamma rays propagate downward as an *electromagnetic component* with negligible nuclear interactions. However, this *soft component* of the cosmic rays is relatively easily absorbed in the atmosphere and in matter, especially in materials of high atomic number, via the just described electromagnetic interactions. There is also energy loss due to the charged particles ionizing the atoms of the matter through which they pass.

Muon component

The muons from the π^+ and π^- decays are essentially heavy electrons with mass 207 times that of the electron ($m_\mu c^2 = 106\,\mathrm{MeV}$). A muon decays to an electron

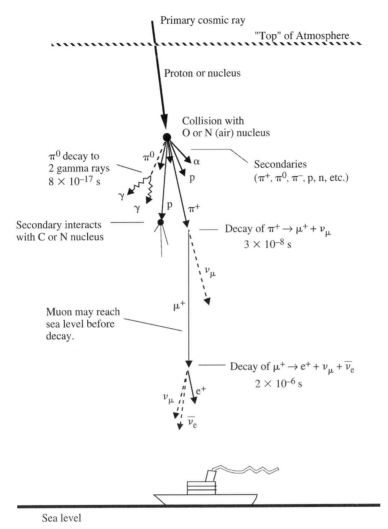

Figure 12.3. Sketch showing a high energy primary cosmic proton entering the atmosphere, its interaction with an atomic nucleus (oxygen or nitrogen), and the creation and decay of π and μ mesons (pions and muons), The scattering angles are shown greatly exaggerated. Large fluxes of muons reach sea level and penetrate deeply into the earth.

and two neutrinos with a mean life of 2.2 μs. This time gets extended by *relativistic time dilation* with the result that many muons reach sea level. The flux of muons at sea level is about one per cm^2 per minute. They are passing through you and ionizing your bodily atoms as you read this.

Muons are a *hard component* of cosmic rays at sea level. The muons do not interact with nuclei so they penetrate deep into the earth, losing energy only gradually via ionization losses of about 0.2 MeV per kg/m^2 of material traversed. Thus a muon would lose about 2 GeV in energy as it traverses the atmospheric depth of 10 300 kg/m^2 which is equivalent to about 10 m of water. Extremely energetic muons can penetrate hundreds of meters and more into the earth. The neutrinos produced in the π and μ decays will, as we learned in Section 2 above, mostly pass right on through the earth.

Cosmic ray astronomy

Cosmic rays are not the ideal tool for doing astronomy because the particles, being charged, are deviated from their initial directions of travel by the magnetic fields of the Galaxy and the earth. Thus cosmic rays from many different sources will be mixed together when they arrive at the observer. In fact, they arrive more or less uniformly from the entire sky; no known discrete sources of charged particle cosmic rays, other than the sun itself, have been found.

However, at the very highest energies $\gtrsim 10^{19}$ eV, there is hope that one may find directions of enhanced cosmic ray emission because the magnetic bending is less. The flux of such particles is quite low, about 1 per km^2 per year at energies $\geq 10^{19}$ eV, and decreasing rapidly as energy increases. This makes difficult the acquisition of good statistical precision (Section 6.5) in measures of their characteristics.

Nevertheless, cosmic rays are an important component of interstellar space, with an energy density comparable to those of the interstellar magnetic field, the microwave 3 K background, and starlight, ~ 1 MeV/m^3. They also provide direct chemical analysis of the highest energy phenomena in the Galaxy. The study of these real bits of matter from distant astrophysical objects has provided substantial insight into their origins and history through the study of the element abundances in the primary flux.

Extensive air showers

A primary cosmic ray of very high energy ($\gtrsim 1$ TeV; 10^{12} eV) creates an *extensive air shower* (EAS) in the earth's atmosphere. (SI prefixes are given in Table A3 of the Appendix.) The very high energy leads to the production of so many secondary particles that collectively they gained this name. The word "extensive" refers to the fact that the particles spread out laterally over as much as several hundred meters and up to several kilometers for the highest energy events (Fig. 4). The highest energy EAS detected have (so far) energies up to 300 EeV (3×10^{20} eV). Those above about ~ 1 EeV are sometimes referred to as ultra high energy cosmic rays (UHECR). The greater the energy of the primary particle, the greater is the number of particles in the resultant EAS.

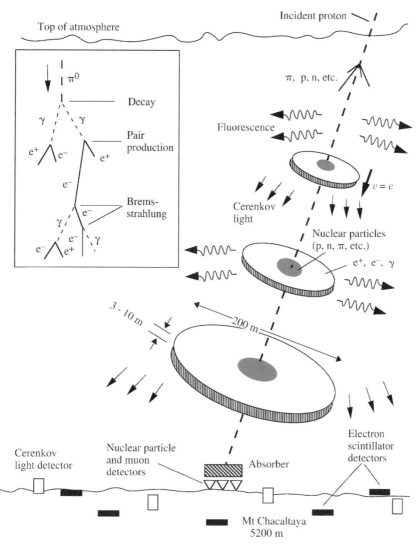

Figure 12.4. Extensive air shower showing the large pancake (disk) of electrons and gamma rays and the nuclear active particles (neutrons, protons, pions, etc.) in the core. The electromagnetic interactions in the disk are shown in the inset. Detectors on the ground can provide (*i*) a measure of the total energy by sampling the electrons and gamma rays at various points in the shower front and (*ii*) the arrival direction through precise measurements of the arrival times of the pancake because, in the figure, the right-hand detectors will be hit first. Fluorescence light from nitrogen is emitted isotropically and can be seen at large distances, whereas Cerenkov light is mostly directed downward. This sketch is based on the former Bolivian Air Shower Joint Experiment (BASJE) at 5200 m altitude designed to measure small showers ($\sim 10^{14} - 10^{15}$ eV) in a search for showers initiated by primary gamma rays.

Growth and decay

The nuclear components (protons, neutrons, etc.) near the center of the EAS are continuously undergoing new interactions with the atmospheric nuclei and thereby continuously feeding energy into the spreading electron/photon component. The mechanism for this is, as described just above, the production of π^0 mesons in nuclear interactions, their immediate decay to gamma rays ($\pi^0 \rightarrow 2\gamma$) and the subsequent successive generations of pair production of e^+, e^- and bremsstrahlung emission of gamma rays. See the inset box in Fig. 4.

As gamma rays beget electrons which beget gamma rays which beget electrons, etc., etc., the number of electrons and protons grows and spreads laterally because of electron scattering by Coulomb interactions with the atmospheric atoms. Most of these particles are highly relativistic because of the huge primary energy. Thus their velocities are all about the same ($v \approx c$), and the particles remain within 3 to 10 m of one another along the propagation direction. The result is the growing pancake of electrons and gamma rays shown in Fig. 4.

Eventually, of course, the initial energy of the primary is shared among so many particles that this multiplication must cease and the shower dies out. At maximum size, an EAS initiated by a primary particle of energy 1 PeV (10^{15} eV) will contain $\sim 10^6$ electrons, and a 100 EeV (10^{20} eV) shower will contain about 10^{11} electrons at its maximum size.

Cerenkov radiation and fluorescence

EAS may be detected in two other ways. The relativistic electrons passing through the atmosphere travel faster than the speed of light in air and thus emit visible *Cerenkov radiation* (Section 2). Most of the electrons travel in the downward direction, but with some scattering in their directions of travel. Since the speed of light in air is almost c, the electron can travel only marginally faster than this. Hence the Cerenkov light is directed mostly in the direction of the electron's instantaneous direction of travel; $\theta \approx 0$ in Fig. 2b.

The result is that a telescope impinged upon by this radiation will ideally record a $\sim 1°$ blurred spot in the sky (actually the atmosphere) which indicates the arrival direction of the primary. The image in a telescope displaced from the shower axis will be elongated because it views the EAS track from the side by means of Cerenkov photons from scattered electrons. Cerenkov light penetrates the atmosphere easily, being visible light. Thus relatively small EAS, with primary energies as low as 1 TeV, which dissipate their electromagnetic component high in the atmosphere, can be detected at ground level via their Cerenkov radiation. See more below.

The passage of EAS charged particles through the atmosphere results in the ionization of the nitrogen and oxygen atoms. Upon recombining and relaxing to a stable state, nitrogen emits faint visible and near-ultraviolet light (wavelengths 300–400 nm). This *fluorescent* radiation is emitted isotropically and hence can be detected with focusing mirrors at large distances from the shower core. Of course, clear dark nights are required for the detection of both fluorescence and Cerenkov light.

Detection of EAS

The growing pancake of electrons and gamma rays, $\gtrsim 100$ m in diameter, makes possible the study of the relatively rare high energy primaries. In effect, the atmosphere "develops" the primary proton to a large size so that it can be detected and located with relatively few detectors (e.g., large scintillation counters) spread at wide intervals over the ground, Fig. 4. Otherwise, it would be impossible to study these events given the low flux quoted above (~ 1 km^{-2} yr^{-1} at ≥ 10 EeV). If the primary proton were never developed by the atmosphere, it would be necessary to blanket an entire 1 km^2 surface with proton detectors in order to detect the ~ 1 event per year of energy at least 10 EeV!

Scintillation detectors might contain a large ($\gtrsim 1$ m^2) thin (~ 100 mm) piece of plastic in a light-tight container and doped with the appropriate additives so the ionization by EAS electrons traversing the plastic causes it to emit visible light. This light can be detected with a photomultiplier tube (PMT) which produces an electrical signal proportional to the amount of light impinging on its sensitive front window. We encountered other examples of scintillation detectors in Section 6.4.

Since most electrons traverse the plastic completely, and if one knows their direction of travel (usually the case), the magnitude of the pulse of light detected is a measure of the number of electrons traversing the detector. If one samples the EAS shower front at, say, five or more locations it is possible to estimate the total number of particles in the EAS and thus to estimate the total energy of the initiating primary particle.

The direction of travel of the EAS may be deduced, for the arrangement of Fig. 4, from the relative times of arrival of the EAS at the several detectors. Precise timing (nanoseconds) can yield the direction of tilt of the pancake. The normal to the pancake is the arrival direction of the primary proton. Accuracies of a few degrees are obtainable. (It is convenient to remember that light travels 0.30 m in 1 ns.)

Alternatively, or in addition, one can use water Cerenkov detectors to detect the electron fluxes. The EAS electrons emit Cerenkov light as they traverse the water, and the light is collected by one or more PMT. If the water is sufficiently deep that it absorbs most of the energy of the electrons and gamma rays at that position in the EAS, the pulse of light is a good measure of the integrated (over time) energy flux (*fluence*, J/m^2) at that position in the EAS.

Muons can be detected with detectors placed several meters below ground which shields them from both the nuclear and electron components. Muons are a good measure of the nuclear content since they are mostly the decay products of nuclear active particles.

It is through such measurements that we know of the existence of protons in the Galaxy with energies up to about 300 EeV. For comparison, man-made accelerators produce particles of energies of "only" ~1 TeV, a million times less. The primaries of EAS are the most energetic particles known to humans.

Fly's Eye

The detection of fluorescence from EAS has the advantage, as noted, that one can view an EAS from a great distance, 10 or more kilometers, if one has sufficiently large collecting mirrors to detect the weak near-ultraviolet signal.

The HiRes Fly's Eye (for "high resolution") experiment in Utah, USA, has two sets of mirrors on two mountain tops separated by 13 km. Each mirror is of diameter 3 m, has a field of view of diameter ~20°, and is focused onto an array of 256 hexagonal PMT, each of diameter 40 mm corresponding to 1° on the sky. Each PMT is like one pixel in your CCD camera but with much worse resolution. Another mirror is like another camera, viewing a different part of the sky. The mirrors at each site (numbering 42 and 21 respectively) view around the entire 360° azimuth of the sky. Together the two sets monitor the sky above a land area of ~3000 km².

As an EAS streaks through the atmosphere, it is imaged successively onto the PMTs at the foci of one or more of the mirrors. The total fluorescent light received from the entire track of the EAS is a measure of its total energy if the distance to it is known. The distance is obtained from the measured angular rate of the EAS motion across the field of view together with the knowledge that the EAS propagates at the speed of light. The precise arrival times of the fluorescent signal at each PMT are recorded for this purpose.

These times also allow one to solve for the tilt of the shower axis from the vertical. If the shower axis is tilted so that it is approaching the cluster of mirrors, the signal from the bottom of the EAS will arrive sooner relative to that from the top than would be the case for a vertical shower.

The analysis is actually somewhat more complicated. In practice, the entire set of PMT times and signal strengths is used in a multiparameter fit to determine the location, arrival direction, and total energy of the EAS. Observation of the same EAS by the second set of mirrors on the other mountain top serves to refine the shower parameters. The system is sensitive to EAS energies ≥ 0.1 EeV and collects ~300 events per year ≥ 10 EeV. One well documented event of 320 ± 90 EeV has been reported.

Auger project

The extremely ambitious international *Pierre Auger project* for the detection and study of the EAS initiated by the UHECR is now under construction in Argentina (Fig. 5a) at an altitude of about 1400 m where these EAS reach their maximum development. When completed in 2005, construction will begin on a sister site in Utah. Each site will have 1600 stations with water Cerenkov detectors 1.2 m deep and separated by ~1.5 km over an area of ~3000 km^2. The site will also include four groups of Fly's Eye detectors on the perimeter of the site. The combination of surface detectors and fluorescence in one array together with its huge collecting area makes the Auger project truly unique.

Many aspects of the UHECR EAS can be measured with Auger. The large area of the site will yield sufficient numbers of EAS to look for "hot spots" of arrival directions on the sky and to obtain the chemical composition of the primaries through the manner in which the EAS develop in the atmosphere. The two sites at mid latitudes will yield good coverage over the entire sky as the earth rotates (Fig. 5b). This is important in searching for the hot spots.

One major puzzle that Auger will address is that EAS are observed at 50 EeV and beyond where they should be sharply attenuated by collisions with the low-energy photons (~10^{-3} eV) of the cosmic microwave background (CMB; Section 11.3). In such an interaction, a π meson will be created, thus extracting considerable energy from the primary. This is known as the GZK effect, after Greisen, Zatsepin, and Kuzmin who first called attention to it. The maximum range for such particles in

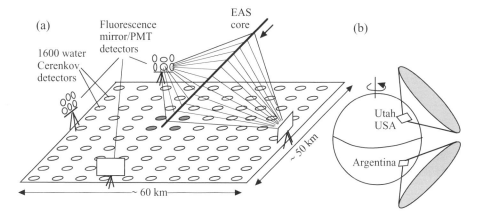

Figure 12.5. (a) This author's conception of one of the two facilities of the Auger observatory which will study ultra high energy cosmic rays (UHECR) through their development as extensive air showers. (b) Location of the sites in the northern and southern hemispheres at altitudes of ~1400 m. This gives all-sky coverage over a 24 hour period. The southern site is under construction and should be completed in 2005 at which time construction should commence on the northern site.

the CMB is about 150 MLY, a relatively local region to the Galaxy. (The Galaxy diameter is ∼0.1 MLY and the visible universe radius ∼15 000 MLY; Section 9.5).

One might thus conclude that the ≳100-EeV particles should be created within this 150-MLY distance. The deflection angles of these extremely high-energy particles would be small in the magnetic fields of the earth, Galaxy, and extragalactic space. Thus, one might expect to find a highly energetic astronomical object, such as a jet from an active galactic nucleus or a recent supernova within a few degrees of the arrival direction of such a primary particle. To date, such sources have not been found. Auger can address this problem with an all-sky map of EAS arrival directions together with accurate measures of primary energies.

Gamma-ray primaries – TeV and EeV astronomy

In contrast to charged particles, energetic gamma rays should travel along more or less straight lines in the Galaxy like other forms of electromagnetic radiation. For example, gamma rays are produced in collisions of the high energy cosmic ray protons with the gas in the plane of the Galaxy through the creation and decay of neutral pions, $\pi^0 \rightarrow 2\gamma$. These gamma rays have been detected at MeV energies. They arrive preferentially from the galactic plane.

Also, it is known that the Crab nebula and blazars (AGN with jet-like behavior) are sources of gamma rays with energies ∼1 TeV. Astronomers use large mirrors focused onto PMTs to observe the Cerenkov light from EAS as described above and select those initiated by gamma rays. This is called *TeV astronomy*.

At higher energies, >10 EeV, one would like to identify the rare EAS with gamma-ray primaries among the ∼10^4 times more abundant EAS initiated by charged particles. One expects such EAS to be lacking in nuclear and muon components.

The EAS array sketched in Fig. 4 is based on the Bolivian–Japanese–US "BASJE" array on Mt Chacaltaya in Bolivia as it operated in the 1960s. It was at high altitude (5200 m) so as to be sensitive to smaller, more frequent EAS of energies ∼100 TeV. The shielded detectors shown were intended to exclude the electromagnetic component and hence to detect muons in the EAS. It could also detect the nuclear component. If the EAS core impinges on top of the shielding (as shown), the absorber material has the appropriate thickness to fully develop the nuclear component (if there is one) into a shower of particles which give, large signals in the scintillators beneath it.

Showers selected as low in muons and with no nuclear component could well be initiated by gamma-ray primaries. If so, they should show an enhancement in the galactic plane of the (MW) Galaxy. Unfortunately the BASJE and other EAS experiments have failed to detect such an asymmetry. Again the large numbers of well documented EAS of high energy detected with the Auger project could perhaps also resolve this puzzle.

12.4 Gravitational-wave observatories

Gravitational waves ("G waves") are hypothesized to travel through space at the speed of light much as electromagnetic radiation does. They are a consequence of Einstein's general theory of relativity (GR), but they have never been detected. They arise, in the theory, from an accelerating mass, specifically from an oscillating quadrupole moment of the mass distribution of a material object. The strength of the wave (the *strain*) is derivable from the second time derivative of the quadrupole moment. This is analogous to an accelerating charge giving rise to a transverse propagating electric field.

The existence of G waves has been confirmed, in a sense, by the observed loss of orbital energy by a binary star system consisting of two neutron stars. One of the two stars is a radio pulsar which allows one to track its orbit precisely, through Doppler shifts of the pulses. The system, the *Hulse–Taylor* (H-T) pulsar, is found to lose energy at exactly the rate predicted by GR for such a system radiating G waves. Thus we have seen the source of the waves but not the waves themselves.

General usage is that "gravity wave" is an oscillation that occurs in fluids, whereas the cumbersome "gravitational wave" is used for changes in the gravitational field as described by Einstein. Too bad; history rules. Here, I choose to use "G waves".

Here we describe the H-T pulsar, estimate the signal strength expected from the last seconds of coalescence of such a system, describe two types of detectors, bars and interferometers, and finally outline ongoing efforts to obtain sufficient sensitivity to detect the expected signals.

Orbiting neutron stars

A *neutron star* is a possible end state of a star. Its nominal mass is $\sim 1.4\ M_\odot$ and its radius ~ 10 km. This is an extremely compact object; a mass comparable to the mass of the sun is contained in an object the size of Manhattan. Neutron stars were first discovered in 1967 as *radio pulsars*. They spin and send out a beam of radiation that sweeps across the earth once each rotation, as does the searchlight beam in a lighthouse. An astronomer thus detects a pulsing radio source. There are ~ 1500 pulsars now known, most in the Galaxy. They are lighthouses of radio emission.

Most radio pulsars are isolated neutron stars but some are in binary systems with a normal star and a very few are in binary orbits with another neutron star. The latter are called *binary pulsars*. The first discovered was the H-T pulsar. As the pulsar orbits its companion, the detected radio pulses are delayed or advanced depending on its location in the orbit – farther away or closer to us. In this way the orbit can be tracked with precision. Another view of this technique is that one tracks the line-of-sight velocity through the Doppler shift of the pulse rates.

Hulse–Taylor pulsar

The H-T binary pulsar was discovered in 1974; its coordinate name is PSR B1913 + 16 but it often called simply *the binary pulsar* because for many years it was the only one known. Its pulse (spin) period is $P = 59$ ms and its orbital period 7.75 h. It is distant about 16 000 LY. The Doppler variation of the pulsing frequency indicates a rather eccentric orbit with eccentricity 0.617. The companion is a second neutron star which does not pulse and therefore is not directly detected. In fact its gravitational effect on its pulsing companion is a perfectly valid detection.

The short orbital period of only ~8 h indicates a small orbit. This and the large measured Doppler velocities mean that the neutron stars are undergoing large accelerations. This acceleration causes them to radiate energy in the form of G waves according to Einstein's theory of general relativity. They lose energy and spiral ever so slowly closer together with increasing angular velocities, just as an artificial satellite spirals toward the earth as it loses energy to atmospheric friction.

The changes in the orbital parameters of the H-T pulsar have been tracked since the 1974 discovery (Fig. 6). They are exactly in accord with the predictions of GR. In particular, the time of periastron (the closest approach of the stars to one another in their mutual eccentric orbits about the center of mass) advanced 26 s from 1975 to 2000, relative to that expected for a constant orbital period. Furthermore, the plot shows that, during the 25-year period, the data points follow very closely the track predicted by GR.

The GR effects allow the inclination of the orbit relative to the line of sight and the two star masses to be individually determined; Newtonian physics yields only a combination of them. The masses of the pulsar and its companion are extremely accurately obtained, $1.442 \pm 0.003\ M_\odot$ and $1.386 \pm 0.003\ M_\odot$ respectively. Another similar pulsar in a binary, PSR B1534 + 12, has been monitored with similar results. A third, PSR 2127+11C resides in the globular cluster M15.

Energy loss rate

This orbital decay is detectable because the very massive neutron stars are whipping around each other in a compact orbit (indicated by the short period). The masses are being highly accelerated and hence they emit sufficient gravitational radiation to bring about a detectable advance in the orbit phase. This detection is equivalent to the discovery of gravitational radiation predicted by the general theory of relativity, even though such radiation has never been detected directly.

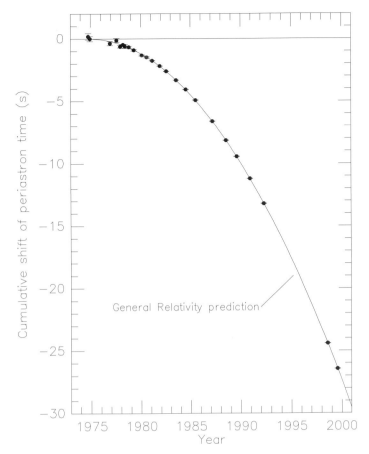

Figure 12.6. Decay of the orbit of the Hulse–Taylor binary pulsar, PSR 1913 + 16. By 2000, the phase of the orbit relative to the periastron had advanced by ~26 s from that expected for a constant orbital period at its 1975 value. This indicates the orbital speed has increased ever so slightly as is expected if total energy (kinetic + potential) is being lost; the two neutron stars are spiraling closer together. The decay of the orbital period follows precisely the track expected if gravitational radiation emitted by the rapidly accelerating neutron stars is carrying away the energy. [Provided by J. H. Taylor & J. M. Weisberg, 2000]

The rate of total (kinetic and potential) energy loss dE/dt due to gravitational radiation requires detailed calculations taking into account the eccentricity of the orbits and other effects. The eccentricity makes a large difference because the radiation energy loss is a strong function of the separation of the stars; it varies hugely around an eccentric orbit. The useful value one can calculate is the time average over one orbital period.

With these caveats about its applicability to the H-T pulsar, we write here the energy loss expression for two stars in *circular* binary orbits. The total mass is

$M = m_1 + m_2$, the reduced mass is $\mu = m_1 m_2/M$, and the separation between the stars is a. The expression is based on a weak field approximation to Einstein's field equations which is applicable if the gravitational fields are not too strong.

$$\frac{dE}{dt} = -\frac{32}{5}\frac{G^4}{c^5}\frac{M^3\,\mu^2}{a^5}$$

$$\underset{m_1 = m_2 \equiv m}{\longrightarrow} \quad -\frac{64}{5}\frac{G^4}{c^5}\left(\frac{m}{a}\right)^5$$

(W; energy loss to
gravitational radiation;   (12.6)
circular orbits)

This is the total power lost to the system by radiation. The radiation is not isotropic; it is maximum at the pole of the orbit, but is substantial in the orbital plane. The result here is thus an integration of the radiated energy flux density over all directions. Finally note the strong dependence on separation, $\propto a^{-5}$.

Gravitational waves

Gravitational waves may be loosely compared to electromagnetic waves. The latter originate with oscillating electric charges and are detected through their effect on other electric charges, e.g., conduction electrons in an antenna. Similarly, G waves originate in the oscillations of masses and are detected by their effect on other masses. In each case, the waves travel at the speed of light, $c = 3 \times 10^8$ m/s, and may be described with a frequency and wavelength, according to $c = \lambda\nu$.

Distortion of space

In GR, gravity is considered a distortion of the space-time fabric. Light rays are bent when they pass near a massive object (e.g., the sun). We might be tempted to say that gravity exerts a force on the photons. However, it is more appropriate to say that space is warped, and that light rays define "straight" lines, known as *geodesics*. When a G wave passes us, space is momentarily distorted.

A ring of test particles in empty space will respond to a passing wave as shown in Fig. 7 where the view is along the propagation direction. In the first polarization (left side), the ring will successively be flattened, circularized, and elongated as shown. In the other polarization, the oscillatory flattening will be rotated at 45° to the first.

The passing wave will produce a *strain h* in space defined as twice the fractional change in diameter (length) of the ring,

$$\frac{h}{2} \equiv \frac{\Delta\ell}{\ell}$$

(Strain definition)   (12.7)

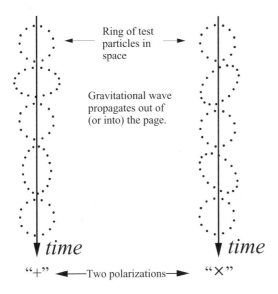

Figure 12.7. Ring of test masses in free space oscillating when a monochromatic (single frequency) gravitational wave is passing into or out of the page. Time runs downward (arrows). The particle oscillations are transverse to the propagation direction as in electromagnetic radiation. The two linear polarizations are shown. Two circular polarizations not shown are linear combinations of the linear polarizations.

where ℓ is the length across the ring and $\Delta \ell$ is the change in the length due to the wave. The strains associated with the two polarizations shown in Fig. 7 are denoted as h_+ and $h_\times$ respectively.

Quadrupole radiation

The gravitational radiation predicted by general relativity is *quadrupole* (or higher-moment) *radiation*. This describes the angular distribution of the emitted radiation. It arises from a mass distribution with an accelerating quadrupole moment. Dipole radiation is not possible because, in part, there is only one sign of gravitational mass; one can not assemble a dipole distribution of gravitational mass because one would need + and − masses.

A mass has a quadrupole moment if it looks like a discus or a football, as well as a host of other non-symmetrical shapes. A sphere has no quadrupole moment. The discus and football (without the pointed ends or sharp edges) approximate the shapes of the *oblate spheroid* and the *prolate spheroid* respectively. The degree of departure from a spherical shape is described by the *quadrupole moment tensor **Q***.

The tensor $\mathbf{Q}$ has nine components Q_{jk} where j and k each take on values 1, 2, and 3 which may correspond to the x, y, z axes of a Cartesian coordinate system. For a set of discrete masses, a representative component of $\mathbf{Q}$ is

$$Q_{jk} = \sum_{A} m_A \left(x_j x_k - \frac{1}{3} \delta_{jk} r^2 \right) \qquad \text{(kg m}^2\text{; component } jk \text{ of} \qquad (12.8)$$
$$\text{quadrupole tensor)}$$

where each term of the summation refers to one of the masses, r is the distance of the object from the chosen origin (often the center of mass), x_j and x_k are the j and k position components (e.g., the y and z components) of mass element m_A, and δ_{jk} is the *Kronecker delta symbol* which equals unity if $j = k$ and zero if $j \neq k$. Note that the units are those of the moment of inertia, kg m^2.

It is instructive to construct from (8) the 3×3 quadrupole tensor for several simple mass distributions such as a dumbbell or four masses in a square pattern. For a symmetric distribution, e.g., six masses symmetrically placed on the positive and negative legs of the three axes, the expression (8) yields zero for all nine components. Similarly a sphere would yield zero values. In contrast prolate and oblate distributions yield non-zero components. We give an example below.

It actually only takes five numbers to describe the "quadrupoleness" of a mass distribution. Quadrupoles can be described with spherical harmonics with index $\ell = 2$ and there are five of them ($2\ell + 1 = 5$). Our tensor is symmetric in the off-diagonal elements and hence three of the nine values are redundant. Also the trace of the matrix (sum of the diagonal elements) is constrained to a fixed value (zero for the normalization of (8)) which makes another component redundant. The five remaining independent values are those needed. (Do not confuse the "ℓ" in this paragraph with that for the "length" elsewhere in this chapter.)

A rapid oscillation of the shape of the mass, and therefore its quadrupole moment, e.g., from prolate to oblate to prolate, etc., will generate gravitational waves. It is the second time derivative of the tensor $\mathbf{Q}$ that yields (not derived here) the strain tensor $\mathbf{h}$ with components h_{jk} at a distance R (m) from the oscillating mass. Thus, without proof, and for SI units,

$$h_{jk} = \frac{2G}{c^4} \frac{\partial^2 Q_{jk}}{\partial t^2} \frac{1}{R} = 1.65 \times 10^{-44} \frac{\partial^2 Q_{jk}}{\partial t^2} \frac{1}{R} \qquad \text{(SI units;} \qquad (12.9)$$
$$\text{strain tensor)}$$

The derivative of a tensor is obtained simply by taking the derivative of each of its components. This tensor contains the information needed to extract the strength of the two polarizations h_+ and $h_\times$ observed at some arbitrary distance and direction. This "acceleration" of $\mathbf{Q}$ is the analog of the acceleration of electric dipoles which yield electromagnetic radiation.

Merger of neutron-star binary

The small value of the coefficient in (9) means that the values of h are very small unless very massive bodies are considered, that is astronomical bodies. Consider two neutron stars spiraling around one another in a binary system that is in its death throes (Fig. 8a). Let them be in circular orbits almost in contact with one another so their surfaces almost touch. (Tidal forces would begin to distort the stars a few orbits prior to this.) Application of Kepler's third law yields an orbital period of only ~ 1.0 ms.

One can refer to the orbits of the two stars as a "binary circular orbit" (in the singular) because momentum conservation demands that the orbits of the two stars about the center of mass have identical eccentricities. Otherwise the center of mass would not be stationary (or uniformly moving).

Variable quadrupole moment

The quadrupole moment tensor at time t for this case is easily constructed from (8) if we approximate the mass distribution as two point masses each of mass m in a circular binary orbit of angular frequency ω (rad/s) in the x, y plane. Each mass is at a distance r from the origin (Fig. 8a); the origin is taken to be the center of mass. The separation of the stars is $a = 2r$.

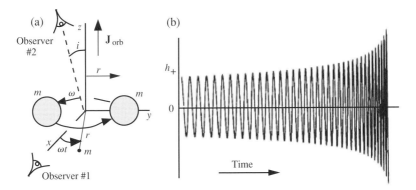

Figure 12.8. (a) Binary system of two neutron stars in a close circular orbit. Observer #1 is in the orbital plane (inclination $i = 90°$) and detects a gravitational wave with linear polarization. An observer normal to the plane ($i = 0°$) would detect circular polarization. Observer #2 is at arbitrary inclination i. (b) Calculation of the waveform of the strain h vs. time in the orbital plane in the last fractional second before the two stars merge into a black hole, for circular orbits. The rapid increase of frequency and amplitude is a "chirp" of gravitational radiation. The frequencies are in the audio range, ending at about 1 kHz. [(b) From K. Thorne, in *Proc. 1994 Snowmass Summer Study*, eds. E. Kolb and R. Peccei, World Scientific, 1995; gr-qc/9506086]

At the instant of time when one of the masses is at angle ωt from the x axis, the tensor will be, from (8),

$$Q_{jk} = 2mr^2 \begin{bmatrix} \cos^2 \omega t - \frac{1}{3} & \cos \omega t \sin \omega t & 0 \\ \cos \omega t \sin \omega t & \sin^2 \omega t - \frac{1}{3} & 0 \\ 0 & 0 & -\frac{1}{3} \end{bmatrix} \quad (12.10)$$

(Quadrupole tensor; circular binary orbit)

where we factored out the common term $2mr^2$. The components of the tensor contain the time variation of the quadrupole moment as the masses orbit at frequency ω. The strain tensor (9) requires that each term be differentiated with respect to time twice. The result is

➡ $$h_{jk} = \frac{2G}{c^4 R} 4\omega^2 mr^2 \begin{bmatrix} -\cos 2\omega t & -\sin 2\omega t & 0 \\ -\sin 2\omega t & \cos 2\omega t & 0 \\ 0 & 0 & 0 \end{bmatrix} \quad (12.11)$$

(Strain tensor for binary, circular orbit)

Strain

An additional tensor operation is required to extract the desired observable scalar quantities, the strains h_+ and $h_\times$ at different observer positions. For an observer in the equatorial plane, it turns out that $h_+ = h_{11}/2$ and $h_\times = 0$, where the h_{11} component is the upper left term in the tensor matrix. Thus, for our observer in the equatorial plane,

➡ $$h_+ = -\frac{2G}{c^4} \frac{\omega^2 (2mr^2)}{R} \cos 2\omega t \quad \text{(Strain for equatorial observer;} \quad (12.12)$$
$$\text{circular orbit, } m = m_1 = m_2)$$

$$h_\times = 0$$

Here, $\omega = 2\pi/P_{orb}$, where P_{orb} is the orbital period, and again m is the mass of each star, ω is the orbital angular velocity, r the radius from the center of mass (1/2 the separation), and R the distance to the observer.

The strain h_+ in (12) will oscillate at twice the orbital frequency. It is apparent in (12) that a large mass, a large orbital radius, and a high orbital frequency all increase the amplitude of the space distortion. At the pole of the orbit (inclination zero), h_+ and $h_\times$ have equal amplitudes, each twice that given for h_+ in (12). The result is circular polarization.

Compare (12) to (9) and note that the second time derivative of the tensor $\ddot{Q}_{jk}$ reduces, for this particular situation, to $-\omega^2 I_z \cos 2\omega t$ where $I_z = 2 mr^2$, the moment of inertia about the z axis (Fig. 8a). This is the same result we would have obtained (except for the minus sign) if we had naively used the second time derivative of the moment of inertia about the view (x) axis $\ddot{I}_x$ in (9) instead of $\ddot{Q}_{jk}$ and had also ignored the geometrical factor of 2 introduced in the conversion from h_{11} to h_+. The moment of inertia about the x axis oscillates as $I_x = 2m(r \sin \omega t)^2 = mr^2(1 - \cos 2\omega t)$. The x-axis observer sees the separation of the two masses oscillating between zero and $2r$ twice each orbit, i.e., with angular frequency 2ω. The second time derivative of I_x substituted into (9) yields the result (12), less the minus sign.

This shows us that mass motions relative to the observer's line of sight give rise to the observed radiation for this particular situation. One can use this result to estimate the magnitudes of the strain from a system with moment of inertia I to be $\sim (G/c^4)(\omega^2 I/R)$ if the mass distribution does not have a substantial spherical component.

If the mass distribution were purely spherical, such as a sphere with oscillating radius, there would be a large $\ddot{I}$, but $Q_{jk} = 0$ for a sphere, so also $\ddot{Q}_{jk} = 0$. (This notation indicates that each component of the tensor is zero.) Thus, in spite of the large $\ddot{I}_x$, there is no gravitational radiation. If in doubt or in need of a precise result, use the proper quadrupole tensor expressions.

For completeness, we quote here the expressions for h_+ and $h_\times$ for the more general case where the two masses m_1 and m_2 are not necessarily equal and the angle between the observer and the pole of the orbit (inclination i; Fig. 8a) need not be $90°$. The orbits are again circular and, as before, ω is the orbital angular velocity, $M = m_1 + m_2$ and $\mu = m_1 m_2/M$.

$$h_+ = \frac{2G}{c^4} \frac{\omega^{2/3} \mu (GM)^{2/3}(1 + \cos^2 i)}{R} \cos\left(2 \int \omega \, dt\right) \qquad \text{(Strain; circular}$$

$$h_\times = \frac{4G}{c^4} \frac{\omega^{2/3} \mu (GM)^{2/3} \cos i}{R} \cos\left(2 \int \omega \, dt\right) \qquad \begin{array}{l}\text{orbits at} \quad (12.13) \\ \text{inclination } i, \\ m_1 \neq m_2)\end{array}$$

We have inserted a phase shift of π radians (thus removing a minus sign) in each of these expressions (13) to be in accord with convention. The angular position of one of the stars, formerly indicated by ωt, should actually be the integral form shown because ω is varying with time due to the inward spiraling of the stars. Finally, these expression make use of Kepler's law, $GM = \omega^2 a^3$ to remove the star separation $a = r_1 + r_2$ from the final expression (13). The reader can confirm that these expressions reduce to those of (12) for the specified circumstances.

Detection in Virgo cluster

Substitute the values for our system, namely $m = 1.4 \, M_\odot, r = 1 \times 10^4$ m, $P_{\text{orb}} = 1 \times 10^{-3}$ s, $G/c^4 = 0.8 \times 10^{-44}$ (SI units), into (12) to obtain the result,

$$h_+ = \frac{360}{R} \cos 2\omega t \qquad \begin{array}{l}\text{(Strain for coalescing neutron stars;} \quad (12.14) \\ R \text{ in meters)}\end{array}$$

In our Galaxy, only a few cases of neutron star mergers are expected per 10 000 years. Planned experiments should be able to study several thousands of galaxies in the nearby Virgo cluster of galaxies. At a distance of $R \approx 60$ MLY $= 6 \times 10^{23}$ m to the far side of the cluster, the expected strain for an equatorial observer is

$$h_+ = 0.6 \times 10^{-21} \cos 2\omega t \tag{12.15}$$

This is somewhat higher than more careful estimates. Even so, it is an extremely small fractional displacement, but detectors are close to reaching these levels of sensitivity and beyond.

Final chirp

Many years from now, the neutron stars of the H-T pulsar will have spiraled in to be within several neutron star radii of one another, and the losses to GR will become huge. The final seconds of a neutron-star binary would thus be characterized by a rapid and accelerating decay of the orbit. The orbital frequency and amplitude would increase at rapidly increasing rates and then suddenly go silent, like the *chirp* of a bird or cricket (Fig. 8b). This would be the last signal from the binary. Detection of this chirp requires sensitivity (15) and is a prime objective of G-wave astronomers.

We remind the reader that gamma-ray bursts may be another ramification of such mergers. Also neutrino flashes (Section 2) should be forthcoming from them.

Detectors

The chirps of merging neutron stars are probably the most reliably predicted source of G waves. There are, however, other possible sources of G waves, namely supernovae and hypernovae, normal stellar binaries, gamma-ray bursts, and even the early universe. Searches with past and current detectors have failed to detect G waves, but this is not surprising in view of the predicted weakness of the signals. Here we describe briefly the two principal types of detectors.

Resonant bars

The first searches were carried out with large cylindrical bars (Fig. 9a) operated in isolated and sometimes cooled environments to shield them from vibrations and ambient disturbances. The natural resonant frequency of the bar is where it is most sensitive. The first such experiment was carried out in the 1960s. It made use of a 1400 kg aluminum bar which was resonant at 1660 Hz, roughly the frequency one might expect from closely orbiting neutron stars or in a supernova collapse. Cryogenically cooled bars have greater sensitivity.

Laser interferometers

Long baselines between test masses are desirable because the absolute displacements for a given strain are proportionally larger. A favored technique now is to place two masses several kilometers apart (in a long vacuum tank) and to measure their separations with a laser interferometer. The geometry is similar to the famous Michelson interferometer (Fig. 9b). Laser light from a common source is split and sent down the two arms of the interferometer. At the ends of the arms, mirrors send the light waves back to the common point where the two beams interfere. Interference fringes are the result. If one arm is lengthened momentarily by a passing G wave, say by $\lambda/10$, the round trip light path is increased by $\lambda/5$. The fringes would translate by 20% of a fringe cycle.

The US Laser Interferometer Gravitational-wave Observatory (LIGO) has arms of length 4 km. A strain of 10^{-21} over a distance of 4 km yields a displacement of

$$\Delta\ell = \frac{h}{2}\ell = \left(\frac{10^{-21}}{2}\right)(4 \times 10^3) = 2 \times 10^{-18} \text{ m} \qquad (12.16)$$

which is 1000 times less than the size of a proton, $\sim 10^{-15}$ m!

Note that the wavelength of the laser light used at LIGO (1064 nm $= 1\times 10^{-6}$ m) is huge by comparison; exotic interferometry techniques are clearly required. This would seem to require measurements of a fractional fringe shift of $\Delta\lambda/\lambda \approx 2(2 \times 10^{-18})/(1 \times 10^{-6}) \approx 4 \times 10^{-12}$. Multiple bounces ($\sim 100$) of the laser beam effectively lengthen the path leading to a fringe shift $\Delta\lambda/\lambda \approx 4 \times 10^{-10}$. Detection of such a small shift is essentially a signal-to-noise problem; one must measure the maxima of the fringe intensity with this precision. With sufficient

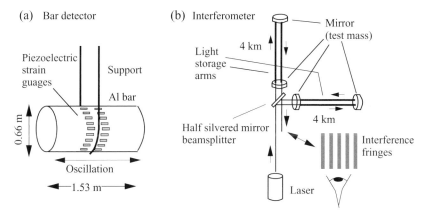

Figure 12.9. (a) Sketch of a bar detector as used in the first serious search for gravitational waves by Joseph Weber. (b) Sketch of a modern interferometer for gravitational-wave detection with the 4-km arm lengths of LIGO.

numbers of photons accumulated, it can in principle be done. Of course, the measurements require that noise sources exceeding these levels be suppressed.

Multiple antennas

A gravitational signal may come in the form of a single brief pulse. Unwanted events due to local phenomena can be discarded if two detectors are operated independently and at a large distance from one another. A genuine gravitational pulse would be detected almost simultaneously in both whereas a spurious local pulse would appear in only one. The neutrino flash from the supernova 1987A was highly credible because two detectors saw it at the same time, one in Japan and one in Ohio.

In fact, multiple observatories allow one to deduce the arrival direction of a pulse of gravitational radiation as the plane wave will arrive at the different stations at slightly different times. The principle is the same as that used to determine the arrival directions of EAS (Section 3). The time delay between two stations provides a line of possible source positions that is a circle (great or small) on the sky. A third detector can provide a crossing line of position (another circle) to obtain two localized celestial positions while a fourth can narrow it down to one position. Gamma-ray astronomers use this technique with satellites to find the celestial positions of gamma-ray bursts.

For this reason the LIGO project has two stations separated by 3000 km, one in the state of Louisiana and the other in Washington State in the USA. In addition, the VIRGO French–Italian observatory near Pisa with 3-km arms and the GEO-600 German–UK instrument can help verify signals and resolve celestial positions. Planned observatories of comparable sensitivities are the Australian AIGO and Japanese TAMA.

LIGO should be able to detect neutron-star mergers in the Virgo cluster when it reaches its design sensitivity. A planned major upgrade in a few years should make it 10 to 15 times even more sensitive. Other instruments will undergo similar upgrades. In the meantime, the several observatories continuously monitor the gravitational sky to ensure we do not miss the unexpected pulse of gravitational radiation from a nearby supernova or gamma-ray burst.

The LIGO and similar instruments are currently sensitive to G-wave frequencies in the range $\sim$100 to $\sim 10^4$ Hz, the lower limit being due to seismic ground motions. With improved isolation of the detectors and test masses, they should be able to approach $\sim$1 Hz, a hard limit due to mechanical filtering and local mass motions (human, atmospheric, ground vibrations). Such instruments can study rapid motions such as spiraling neutron stars, gamma-ray bursts, and supernovae events as described.

Low frequency antenna in space

Lower frequency studies (seconds to hours) will be carried out by the ambitious NASA/ESA Laser Interferometer Space Antenna (LISA) program to be launched in ~2010. It will consist of three satellites that will define the legs of an interferometer. The separations will be huge ~5×10^9 m, (3% of an astronomical unit) and their relative positions must be measurable to 10^{-12} m or to 10^{-5} fringes. Low frequency G waves are expected from motions of stars in longer-period binary orbits, the merging of two massive black holes (active galactic nuclei) into a supermassive black hole, massive objects circulating around and spiraling into the black holes of active galactic nuclei, and space-time ripples originating in the early universe.

Problems

12.2 Neutrino observatories

Problem 12.21. (See Section 10.4 for cross sections, etc.) (a) Derive a relationship for the creation rate r (s^{-1}) of ^{37}Ar nuclei in the Homestake tank (ignoring ^{37}Ar decays) as a function of (*i*) the solar flux of ^{8}B neutrinos $\mathscr{F}_\nu$ (neutrinos $s^{-1}m^{-2}$) at the earth in the Homestake energy range, (*ii*) the average (over all energies) cross section σ (m^2) for absorption of a neutrino by a single chlorine nucleus, (*iii*) the mass density ρ (kg/m^3) of the C_2Cl_4 fluid, (*iv*) the volume V of the tank, and (*v*) the mass m_p of the proton. Consider the probability of absorption to be so low that the beam is not significantly attenuated by the fluid. (b) What is the expected number of interactions per second, per day, per 35 d mean life of ^{37}Ar, and per three ^{37}Ar mean lives if $\mathscr{F}_\nu = 7 \times 10^{10}$ s^{-1} m^{-2}, $\rho = 1600$ kg/m^3, $\sigma = 3 \times 10^{-47}$ m^2, $V = 400$ m^3, and $m_p = 1.7 \times 10^{-27}$ kg? [Ans. (b) ~1.5 day^{-1}]

Problem 12.22. (a) Show analytically that competing creation and decay processes in the Homestake neutrino tank yield a net number $N(t)$ of $^{37}_{18}$Ar atoms that, starting from zero, initially increases rapidly and linearly and then increases more and more slowly toward a maximum equilibrium value. Proceed as follows. Define a creation rate r (atoms/s) and take the mean decay time (mean lifetime) to be τ. Write down the appropriate differential equation for a differential time interval dt and integrate it to obtain $N(t)$ in terms of r, τ and time t. Let $N = 0$ at $t = 0$, and use the fact that τ^{-1} is the probability per unit time for the decay of a single atom to occur. Make a sketch of $N(t)$. (b) What is $N(t)$ in the limits of short times ($t \ll \tau$) and long times ($t \gg \tau$)? Demonstrate that the latter is a measure of the creation rate r. Would a measurement at $t = \tau$ provide a value of r? (c) Discuss the arguments for sweeping the tank every ~100 days ($t \approx 3\tau$). Be quantitative. Consider for example the long term average rate of accumulated events. (d) What is the number N at $t = 3\tau$ (~100 d) if the creation rate r has the value you obtained in (or saw in the answer to) Problem 21? [Ans. (a) $N = r\tau[1 - \exp(-t/\tau)]$; (d) ~50]

12.3 Cosmic ray observatories

Problem 12.31. The principle of relativistic time dilation states that the mean life at rest, $\tau = 2.2$ μs, of a fast moving muon will be extended to τ' according to

$$\tau' = \gamma\tau \qquad \gamma = \frac{E}{mc^2} \qquad\qquad (12.17)$$

where E is the total energy (rest energy + kinetic energy), and $mc^2 = 106$ MeV is the muon rest energy. What is the kinetic energy E_k of a muon (in GeV) that will just make it to sea level at the end of its (extended) mean life? Assume the muon is created at a depth of 1000 kg/m^2 in the atmosphere, or about 18 km above sea level. You may approximate its speed of travel as about equal to the speed of light c and neglect any energy loss due to ionization during passage through the atmosphere. To what extent are these assumptions warranted? [Ans. ~3 GeV]

Problem 12.32. A relativistic muon passes vertically through you while you are standing up. (a) About how much energy is dissipated by it inside your body, in units of MeV? The energy deposition rate per kg/m^2 is about the same as in the atmosphere; see text. (b) Roughly, how much energy is deposited in your brain by cosmic ray muons in one second? (See text for muon flux.) How many molecules (mostly water) of your brain are likely to suffer an atomic ionization in that one second? (It takes 13 eV to ionize either an H or an O atom.) By what age might you expect to have all the molecules in your brain ionized, if recombination is neglected? [Ans. ~200 MeV; 10^{12} yr]

Problem 12.33. A fairly large EAS due to an incident gamma ray reaches its maximum size of 1×10^7 electrons, positrons, and gamma rays as it reaches sea level (atmospheric depth 10 300 kg/m^2). Consider all of the created particles to still be present at the maximum. What approximately is the mean interaction length in the atmosphere, in kg/m^2, for either the pair production or the bremsstrahlung process; assume they are equal. This is known as the "radiation length" in air. Hint: assume that each interaction doubles the number of particles. [Ans. ~ 400 kg/m^2, or ~1/25 of the atmosphere]

Problem 12.34. (a) What is the approximate time delay expected between the arrival time of the electrons at the detectors under the right edge of the EAS shown in Fig. 4 compared to those at the left edge? The arrival direction of the primary proton is tilted 15° from the vertical. (b) How accurate must the timing be to attain a precision of $\Delta\theta$ in the (left–right) arrival direction of an EAS of width D arriving from a zenith angle θ? Evaluate your expression for $\Delta\theta = 1°$ for arrival directions $\theta = 0°$, 15°, and 30°. (c) Do you think these accuracies are attainable? Discuss. [Ans. ~200 ns; 10–15 ns]

12.4 Gravitational-wave observatories

Problem 12.41. Construct from (8) the quadrupole tensor for several elementary distributions of discrete masses. That is, evaluate (8) for each of the 9 components $Q_{11}, Q_{12}, Q_{13}, Q_{21}, \ldots, Q_{33}$. It is convenient to present the 9 values in a matrix of 3 rows and 3 columns. The goal is to give a feel for the meaning of "quadrupole moment" and the tensor that describes it. Comment on what you learn; e.g., what does the Q_{33} term represent, what components are systematically zero and what kind of mass distribution would give non-zero values, and do the diagonal elements have a systematic relationship? The mass distributions consist of discrete masses, each of mass m placed on the 3 axes (x, y, z) as follows: (a) Two masses on z axis at $z = \pm r_0$ (symmetry along one axis). (b) Four masses, two as in (a) and two on y axis at $y = \pm r_0$ (symmetry along two axes). (c) Six masses, four as in (b) and two on x axis at $x = \pm r_0$ (symmetry along all three axes). (d) Same as (c) except the two on the z axis are closer to origin at $z = \pm r_0/2$ (oblate distribution). (e) Same as (c) except the two on the z axis are further out at $z = \pm 2r_0$ (prolate distribution). (f) Six masses, same as oblate distribution (d) except the pattern is rotated $45°$ about the z axis, so the four masses in the x, y plane are no longer on the axes and have both x and y components. [Ans. Q_{33} values are (in units of mr_0^2): 4/3; 2/3; 0; -1 ; 4; -1]

Problem 12.42. (a) What is the approximate orbital period of two neutron stars orbiting each other just barely in contact? Let each star have mass $m = 1.4 M_\odot$ and radius $R = 10$km. Consider all the mass of each star to be at a point at its center. Ignore effects of general relativity and tidal forces. Use selected laws by Newton or Kepler. (b) Construct from the general quadrupole tensor (8) the quadrupole tensor (10) as a function of time t for a binary system consisting of two neutron stars of equal masses m, each in a circular orbit in the xy plane with orbital angular frequency ω. The masses rotate about the z axis at radius r and are aligned along the x axis at $t = 0$ (see Fig. 8a). Again assume point masses. (c) Find the strain tensor (11) for a binary from the general expression (9) and your answer to (b). (d) Find the units of the tensor component h_{jk} from (9). Is your answer what you expect? [Ans. $\sim$1 ms; Eq. (10); Eq. (11); —-]

Problem 12.43. Find, from (13), the amplitudes and frequency ν (Hz) of each strain polarization (h_+ and $h_\times$) at the earth due to gravitational radiation from the Hulse–Taylor binary pulsar at its present period (7.75 h), *if* its orbit were circular. The masses of the two neutron stars are $\sim$1.4 $M_\odot$, and the earth is distant $\sim$16 000 LY in the direction $47°$ from the orbital pole. Do you think this could be detected with the currently operating detection systems, e.g., LIGO? [Ans. $\sim 10^{-4}$ Hz, $\sim 5 \times 10^{-23}$ for each polarization]

Problem 12.44. (a) Calculate from (6) the numerical value of dE/dt for a binary pulsar containing two neutron stars, each of mass $m = 1.4\ M_\odot$, in circular orbits of period the same as the H-T pulsar, $P = 7.75$ h. (Use selected laws of Newton or Kepler to obtain the needed parameter a.) Compare to the solar luminosity. (b) Find an expression for the characteristic decay time $\tau = E/(dE/dt)$ of the system in terms of the variables given in (6) where E, the total energy (kinetic plus potential), is taken from Newtonian mechanics. What is the physical significance of the characteristic time for such a binary system? Evaluate your expression for our binary and compare your answer to the age of the universe, ~15 Gyr. (c) From the rate of energy loss, estimate the advance of the time of periastron in 25 years. Make simplifying assumptions as needed, e.g., that the rate of energy loss does not change appreciably in 25 years. Hints: find a relation between ΔP and Δa and also one between ΔE and Δa to find the period change after 1/2 the 25 yr. (A fun aside: how much does the separation decrease in 12.5 yr?) Compare to the 26 s actually measured for the H-T pulsar. How can you explain the difference? (d) Recalculate the phase advance for a circular orbit that is at the closer periastron distance of the H-T pulsar, $a' = 0.383a$. Again compare to the H-T pulsar phase advance and comment. [Ans. ~ 10^{24}W; $\lesssim 10^{10}$ yr; ~− 2 s; ~−25 s]

Credits, further reading, and references

Sources for the material in this volume include the published resources listed here by chapter. These are excellent starting points for further information on the covered topics. Following this, we give other useful reference material. Figure credits are given in the figure captions. Personal credits are given in the acknowledgments.

Credits and further reading

1   Astronomy through the centuries
Various astronomy texts, e.g., M. Hoskin, *The Cambridge Concise History of Astronomy*, Cambridge University Press, 1999

2   Electromagnetic radiation
Various physics and astronomy texts
G. Bekefi & A. Barrett, *Electromagnetic Vibrations, Waves and Radiation*, MIT Press, 1977
Absorption: M. Zombeck, *Handbook of Space Astronomy & Astrophysics*, 2nd Ed., Cambridge University Press, 1990

3   Coordinate systems and charts
Various astronomy texts
Catalogs: Strasbourg Astronomical Data Center (CDS) on internet

4   Gravity, celestial motions, and time
Motions: Various astronomy texts
Eclipses: M. Littmann & K. Willcox, *Totality*, University Hawaii Press, 1991;
    P. Guillermier & S. Koutchmy, *Total Eclipses*, Springer/Praxis, 1999
Time: *Explanatory Supplement to the Astronomical Almanac*, ed. K. Seidelmann, University Science Books, 1992; US Naval Observatory website

5   Telescopes
Image formation: Various astronomy texts; physics optics texts
Techniques: C. R. Kitchin, *Astrophysical Techniques*, Institute of Physics, 1998; *X-ray Astronomy*, eds. R. Giacconi & H. Gursky, Reidel, 1974
Diffraction: G. Bekefi & A. Barrett, *ibid*.
Adaptive optics: J. Beckers, *Ann. Rev. Astron. Astrophys.*, eds. G. Burbidge, D. Layzer, & A. Sandage, **31**, 13 (1993)

Speckle interferometry: "Speckle Imaging: a boon for astronomical observations", S. K. Saha, *Proc. Conf. Young Astrophysicsts of Today's India*, 2001; also astroph/0003125

6   Detectors and statistics
Proportional counter: G. W. Fraser, *X-ray Detectors in Astronomy*, Cambridge University Press, 1989
CCDs: CCD physics: C-K. Kim in *Charge-coupled Devices and Systems*, eds. M. J. Howes & D. V. Morgan, John Wiley & Sons, 1979, pp. 1–80
EGRET: G. Kanbach *et al. Space Science Reviews* **49**, 69 (1988)
Statistics: R. Evans, *The Atomic Nucleus*, McGraw-Hill, 1955; L. Lyons, *Data Analysis for Physical Science Students*, Cambridge University Press, 1991; P. Bevington & D. K. Robinson, *Data Reduction and Error Analysis for the Physical Sciences*, 2nd Ed., McGraw-Hill Inc., 1992

7   Multiple telescope interferometry
E. Fomalont & M. Wright, in *Galactic and Extragalactic Radio Astronomy*, 1st Ed., eds. G. Verschuur and K. Kellermann, Springer-Verlag, 1974, p. 256
A. R. Thompson, J. Moran, & G. Swenson, Jr., *Interferometry and Synthesis in Radio Astronomy*, Kreiger, 1991

8   Point-like and extended sources
Various astrophysics texts, e.g.,
B. F. Burke & F. Graham-Smith, *An Introduction to Radio Astronomy*, 2nd Ed., Cambridge University Press, 2001
J. D. Kraus, *Radio astronomy*, 2nd Ed., Cygnus Quasar, 1986

9   Properties and distances of celestial objects
Temperatures: D. Clayton, *Principles of Stellar Evolution and Nucleosynthesis*, McGraw-Hill, 1968
Distances: various texts and literature, e.g., G. Jacoby *et al.*, *PASP* **104**, 599 (1992)

10   Absorption and scattering of photons
Processes: W. Tucker, *Radiation Processes in Astrophysics*, MIT Press, 1975; G. Rybicki & A. Lightman, *Radiative Processes in Astrophysics*, Wylie-Interscience, 1979
Cross sections: Physics radiation texts and M. Zombeck, *ibid.*
Extinction: B. Savage & J. Mathis, *Ann Rev. Astron. Astrophys.* **17**, 73 (1979)
Abundances: N. Grevasse & E. Anders, in *AIP Conf. Proc.* **183**, 1989, ed. C. J. Waddington

11   Spectra of electromagnetic radiation
Continuum spectra: G. Rybicki & A. Lightman, *ibid.*
Lines and radiative transfer: E. Bohm-Vitense, *Introduction to Stellar Astrophysics*, two volumes, Cambridge University Press, 1989
R. J. Rutten, *Radiative Transfer in Stellar Atmospheres*, Utrecht University lecture notes, 8th ed., on internet, 2003

12   Non-photonic observatories
J. N. Bahcall, *Neutrino Astrophysics*, Cambridge University Press, 1990
J. B. Hartle, *Gravity*, Addison-Wesley, 2002
W. C. Haxton, "The solar neutrino problem", *Ann. Rev. Astron. Astrophys.* **33**, 459 (1995)
M. Boratav, "Probing theories with cosmic rays", *Europhysics News* **33**, No. 5 (2002)
M. W. Friedlander, *A Thin Cosmic Rain: Particles from Outer Space*, Harvard University Press, 2000
T. K. Gaisser, *Cosmic Rays and Particle Physics*, Cambridge University Press, 1991

C. Ohanian & R. Ruffini, *Gravitation and Spacetime*, 2nd Ed., Norton, 1994
S. Hughes, "Listening to the Universe with gravitational wave astronomy", *Annals Phys.* **303**, 142 (2003); astro-ph 0210481

Research results

Review articles in *Annual Review of Astronomy and Astrophysics*; each volume contains a list of titles of the previous 12 volumes
Articles in amateur astronomy magazines such as *Sky and Telescope (see* "News Notes") and *Astronomy*
Occasional review articles in *American Scientist*, *Physics Today*, *Scientific American*
NASA's *Astrophysical Data System* (ADS) on the internet; search for review articles on a given topic or for a specific journal article

Encyclopedias

S. Maran, *The Astronomy and Astrophysics Encyclopedia*, Van Nostrand Reinhold, 1992
R. Meyers, Editor, *Encyclopedia of Astronomy and Astrophysics*, AIP, 1989
P. Murdin, *Encyclopedia of Astronomy and Astrophysics*, 4 volumes, Nature Publishing Group, 2001

General reference

A. Cox, ed., *Allen's Astrophysical Quantities*, 4th Ed., AIP, 2000
K. Lang, *Astrophysical Formulae*, 3rd Ed., Astronomy and Astrophysics Library, 1999
J. Mitton, *Cambridge Dictionary of Astronomy*, Cambridge University Press, 2001
I. Ridpath, *Norton's 2000.0 Star Atlas and Reference Handbook*, Longman, 1989
R. Zimmerman, *The Chronological Encyclopedia of Discoveries in Space*, Onyx, 2000

General astrophysics texts for science majors

B. W. Carroll & A. Dale Ostlie, *Introduction to Modern Astrophysics*, Addison Wesley, 1996
M. Harwit, *Astrophysical Concepts*, 3rd Ed., Springer, 1998
F. Shu, *An Introduction to Astronomy*, University Science Books, 1982
A. Unsold, *The New Cosmos: An Introduction to Astronomy and Astrophysics*, 5th Ed., Springer, 1999
M. Zeilik & S. Gregory, *Introductory Astronomy & Astrophysics*, 4th Ed., Saunders College Publishing, 1998

Appendix

Units, symbols, and values

Table A1. *Base units: Système Internationale (SI)[a]*

Quantity	Name	Symbol
Length	meter	m
Mass	kilogram	kg
Time	second	s
Electric current	ampere	A
Thermodynamic temperature	kelvin	K
Amount of substance	mole	mol
Luminous intensity	candela	cd

[a] National Institute of Standards and Technology (NIST) physics website. http://physics.nist.gov/cuu/Units/index.html

Table A2. *Some SI derived units[a]*

Quantity	Name	Symbol	Equivalent units
Plane angle	radian	rad	$m/m = 1$
Solid angle	steradian	sr	$m^2/m^2 = 1$
Frequency	hertz	Hz	$cycles/s = s^{-1}$
Force	newton	N	$kg\ m\ s^{-2}$
Pressure	pascal	Pa	$N\ m^{-2}$
Energy	joule	J	$N\ m$; $kg\ m^2\ s^{-2}$
Power	watt	W	$J\ s^{-1}$
Electric charge	coulomb	C	$A\ s$
Electric potential	volt	V	$J\ C^{-1}$
Resistance	ohm	Ω	$V\ A^{-1}$
Magnetic flux density	tesla	T	$N\ A^{-1}\ m^{-1}$

[a] NIST physics website, *ibid.*

Table A3. *SI prefixes*[a]

Log factor	Prefix	Symbol	Log factor	Prefix	Symbol
−3	milli	m	3	kilo	k
−6	micro	μ	6	mega	M
−9	nano	n	9	giga	G
−12	pico	p	12	tera	T
−15	femto	f	15	peta	P
−18	atto	a	18	exa	E
−21	zepto	z	21	zeta	Z
−24	yocto	y	24	yotta	Y

[a] NIST physics website, *ibid.*

Table A4. *Energy-related quantities*

Quantity	Symbol	Unit	Definition[a]
Specific intensity	$I(\nu, \theta, \phi, t)$	$\text{W m}^{-2}\,\text{Hz}^{-1}\,\text{sr}^{-1}$	
Spectral flux density	$S(\nu, t)$	$\text{W m}^{-2}\,\text{Hz}^{-1}$	$S = \int I\,d\Omega$
Flux density	$\mathscr{F}(t)$	W m^{-2}	$\mathscr{F} = \int I\,d\Omega\,d\nu$
Power	$\mathscr{P}(t)$	W	$\mathscr{P} = \int I\,d\Omega\,d\nu\,dA$
Energy	U	J	$U = \int I\,d\Omega\,d\nu\,dA\,dt$
Fluence	$\mathscr{E}$	J m^{-2}	$\mathscr{E} = \int \mathscr{F}\,dt$
Surface brightness	$B(\theta, \phi, \nu, t) = I$	$\text{W m}^{-2}\,\text{Hz}^{-1}\,\text{sr}^{-1}$	$B = I$
Volume emissivity	$j(\boldsymbol{r}, \nu, t)$	$\text{W m}^{-3}\,\text{Hz}^{-1}$	$I = \int j\,dr/(4\pi)$

[a] Definitions presume that area element traversed by flux is normal to the propagation direction as in Fig. 8.5.

Table A5. *Physical constants*[a]

Universal	
Light speed in vacuum	$c = 2.997\,924\,58 \times 10^8 \text{ m s}^{-1}$
Permeability of vacuum (exact)	$\mu_0 = 4\pi \times 10^{-7} \text{ N A}^{-2}$
Permittivity of vacuum	$\epsilon_0 = 1/\mu_0 c^2 = 8.854 \times 10^{-12} \text{ s}^4 \text{ A}^2 \text{ m}^{-3} \text{ kg}^{-1}$
Gravitation constant	$G = 6.673 \times 10^{-11} \text{ m}^3 \text{ kg}^{-1} \text{ s}^{-2}$
Planck constant	$h = 6.626\,069 \times 10^{-34} \text{ J s}$
Particle	
Electron charge	$e = 1.602\,176 \times 10^{-19} \text{ C}$
Electron mass	$m_e = 9.109\,382 \times 10^{-31} \text{ kg} \,(\approx 10^{-30} \text{ kg})$
in eV $(m_e c^2/e)$	$= 0.511 \text{ MeV}$
Proton mass	$m_p = 1.672\,622 \times 10^{-27} \text{ kg}$
	$= 938.272 \text{ MeV}$
Neutron mass	$m_n = 1.674\,927 \times 10^{-27} \text{ kg}$
	$= 939.565 \text{ MeV}$
Atomic	
Fine-structure constant	$\alpha = \mu_0 c e^2/2h = 7.297\,353 \times 10^{-3}$
	$= 1/137.036$
Bohr radius	$a_0 = 0.5292 \times 10^{-10} \text{ m}$
Physicochemical	
Avogadro constant	$N_A = 6.022 \times 10^{23} \text{ mol}^{-1}$
Atomic mass constant $m(^{12}C)/12$	$m_u = 1.660\,539 \times 10^{-27} \text{ kg}$
Boltzmann constant	$k = 1.380\,650 \times 10^{-23} \text{ J K}^{-1}$
Molar gas constant	$R = 8.3145 \text{ J mol}^{-1} \text{ K}^{-1}$
Stefan–Boltzmann constant	$\sigma = 5.670 \times 10^{-8} \text{ W m}^{-2} \text{ K}^{-4}$
Conversions	
Electron volt	$1.0 \text{ eV} = 1.602 \times 10^{-19} \text{ J}$
Standard atmos. pressure	$1.0 \text{ atm} = 101\,325 \text{ Pa (N m}^{-2})$
Standard gravit. acceleration	$1.0 \, g = 9.806\,65 \text{ m s}^{-2}$
Temperature and energy $kT\text{(eV)}$	$T(\text{K}) = 11\,605 \times kT(\text{eV})$
Photon frequency and energy $E(\text{eV})$	$\nu(\text{Hz}) = E(\text{eV}) \times 2.418\,0 \times 10^{14}$
Photon energy and wavelength	$\lambda(\text{nm}) \times E(\text{eV}) = 1\,239.842$

[a] Values are rounded to precision deemed useful for this text; most are more precisely known.

Sources: P. Mohr & B. Taylor, CODATA Recommended Values of Fundamental Physical Constants, 1998, *Rev. Modern Phys.*, **72**, No. 2 (2000). Also NIST website http://physics.nist.gov/cuu/Constants/index.html

Table A6. *General astronomical constants*[a]

Quantity	Value
Astronomical unit (semimajor axis of earth orbit)	$1.0 \text{ AU} = 1.496 \times 10^{11} \text{ m}$
Light (Julian) year	$1.0 \text{ LY} = 9.461 \times 10^{15} \text{ m} (\approx 10^{16} \text{ m})$
Parsec	$1.0 \text{ pc} = 3.086 \times 10^{16} \text{ m} = 3.262 \text{ LY}$
Solar mass	$1.0 \, M_\odot = 1.989 \times 10^{30} \text{ kg}$
Solar radius	$1.0 \, R_\odot = 6.955 \times 10^{8} \text{ m}$
Solar luminosity	$1.0 \, L_\odot = 3.845 \times 10^{26} \text{ W}$
Earth mass	$1.0 \, M_\oplus = 5.974 \times 10^{24} \text{ kg}$
Earth radius, mean	$1.0 \, R_\oplus = 6371.0 \text{ km}$
Earth radius, equatorial	$1.0 \, R_{\oplus,\text{eq.}} = 6378.1 \text{ km}$
Moon mass	$1.0 \, M_{\text{moon}} = 7.353 \times 10^{22} \text{ kg}$
Moon radius, mean	$1.0 \, R_{\text{moon}} = 1738.2 \text{ km}$
Moon orbit semimajor axis	$1.0 \, a_{\text{moon}} = 384\,400 \text{ km}$

[a] *Allen's Astrophysical Quantities*, 4th Ed., ed. A. N. Cox, AIP Press, 2000.

Table A7. *Constants involving time*[a]

Day	
Mean solar day	86 400.00 UT (mean solar) seconds $\approx 86\,400.002$ SI seconds
Sidereal day, relative to vernal equinox	86 164 .09 UT seconds[b] = 23 h 56 m 04.09 s UT
Year	
Julian year (exact)	365 25 d = 31 557 600 s (SI)
Julian century (exact)	36 525 d
Tropical year (equinox to equinox)	365 2422 d
Sidereal year (fixed star to fixed star)	365 2564 d
Anomalistic year (perihelion to perihelion)	365 2596 d
Julian date	
1900 Jan. 0.5	JD 2 415 020.0
2000 Jan. 0.5	JD 2 451 544.0
2100 Jan. 0.5	JD 2 488 069.0
Standard Julian epochs	
J1900.0	JD 2 415 020.0 = 1899 Dec. 31, 12 h TDB
J2000.0	JD 2 451 545.0 = 2000 Jan. 1, 12 h TDB
J2100.0	JD 2 488.070.0 = 2100 Jan. 1, 12 h TDB

[a] *Allen's Astrophysical Quantities, ibid.*
[b] UT time is essentially mean solar time. The sidereal day is the same value in SI seconds, within the given precision. See Eq. (4.13) and the following discussion.

Index

Boldface page numbers indicate that there is a section or subsection on this topic, beginning at the indicated page. A range of boldface page numbers indicates a major section or chapter on this topic. An italic page number indicates a figure or a table.